John Hasse
15785 Portis
Northville, MI
48167

(313) 420-2054

HIGH PERFORMANCE

Johns Hopkins Studies in the History of Technology

Merritt Roe Smith, Series Editor

The instant before the big show: Lions Drag Strip, Long Beach, California, 1970. The photoelectric beam shining on Jim Dunn's spokes from the left indicates that he is staged. Not without reason are the starting lights called a Christmas tree. (Photograph by Mert Miller, courtesy National Hot Rod Assn.)

HIGH PERFORMANCE

The Culture and

Technology of

Drag Racing

1950 – 1990

ROBERT C. POST

The Johns Hopkins University Press   Baltimore and London

The Johns Hopkins University Press
2715 North Charles Street
Baltimore, Maryland 21218-4319
The Johns Hopkins Press Ltd., London

Library of Congress Cataloging-in-Publication Data will be found
at the end of this book.

A catalog record for this book is available from the British Library.

CONTENTS

PREFACE

Performance: A musical, dramatic, or other entertainment . . . . The manner in which or the efficiency with which something reacts or fulfills its intended purpose.
RANDOM HOUSE DICTIONARY

Talk about the purpose of the thing . . . is a statement of opinion and can never be anything else.
DAVID PYE, THE NATURE AND AESTHETICS OF DESIGN

his is a book about a human activity whose actors view it from many perspectives: as a sport, a business, recreation; as a means of identity, confirmation of mastery—a matter of "satisfaction, meaning, and self," says one of the sociologists who has addressed drag racing.[1] More specifically, it is about a device central to that activity, a species of automobile, and about opinions regarding the purpose of that device, which are as diverse as the ways of viewing the activity itself. Since the activity involves an idiomatic technology, some practitioners will tell you that the purpose has to do with progress, or with efficiency. Most historians of technology are wary of "progress talk" already. Suffice it to say that, like purpose, talk about efficiency can never be more than a statement of opinion either.

According to the opinion that initially prevailed among drag racers, the most efficient machine was whatever could be driven from point A to point B, a quarter of a mile away, as fast as possible, or quicker than anything else. As the quirky but shrewd Mr. Pye puts it, however, "We can never select the one result we want to the exclusion of all others."[2] Thus, some people held to opinions that efficiency also entailed safety considerations, cost considerations, and other constraints on the designer's freedom of choice, even commercial considerations. Today, to a racer who counts on his or her sponsors to help pay expenses or who has perhaps attracted an angel with deep pockets, pleasing those sponsors or that angel takes precedence over mere technological matters. To an impresario who promotes drag racing events the important thing is the size of the audience, and this has not invariably meant spotlighting machines that are fastest. Nothing currently gets attention like "big numbers," yet neither drag racing's audience nor its sponsors have always thought of purpose and efficiency in just those terms.

All of drag racing is saturated in the language of mechanical technology, naturally, but scarcely more so than with the imagery of the theater. The race cars come out in pairs, then they stage and prepare to perform. The very word *performance* has a delicious ambiguity, and I have taken that ambiguity as one of my themes. Some conceive of performance in the context of engineering; for others the crucial referent is entertainment. The show can be as scrubby as a small-time carnival, or it can be the stuff of high drama; call it "legitimate" theater, perhaps, but it is still theater.

In the beginning that part was different. The essence was a straightforward technical challenge—no roar of the crowd, no angels, agents, or roadies. The rewards were profoundly internalized. In the 1990s, though drag racing has hundreds of minor venues at which this has not changed much, success in the big time is usually contingent upon striking an optimum balance between the two definitions of *performance* enumerated in the epigraph. And yet, even now, "success" has a significant financial dimension for only a slim minority of racers. For most of them the rewards are *still* profoundly internalized.

I have framed what follows partly as a complement to the concept of technological enthusiasm, a concept initially elaborated a generation ago. In 1966 Brooke Hindle suggested that technology was perhaps "not so much a tool or a means as it was an experience—a satisfying emotional experience." A few years later Eugene Ferguson offered a compelling set of observations about what motivated "enthusiastic technologists." Meantime, Thomas Hughes had begun to enunciate his idea of "technological momentum," which took full wing in his 1989 masterwork, *American Genesis: A Century of Invention and Technological Enthusiasm, 1870–1970*. John B. Rae indicated that a shared trait among early automotive entrepreneurs was an enthusiasm so powerful that "they seem to have preferred to go broke manufacturing automobiles than to get rich doing something else." In a similar vein, a historian of the aircraft industry has recently shown how "manufacturers stayed in business for irrational, noneconomic reasons, basking in the glamour and prestige of a high-technology industry while consistently losing money." Such concepts as this, of course, go back at least to Joseph Schumpeter, who likened the entrepreneur's "will to conquer" to sport, where the "financial result is a secondary consideration."[3]

The constellation of sporting activities involving conveyances provides a rich source of evidence. Clearly, the quest for "a satisfying emotional experience" often overwhelms the odds for optimizing economic returns. In that regard, much the same book as I have written here could be written about the culture and technology of a dozen different varieties of "motorsports," or, indeed, about the Soap Box Derby, in which there are no motors, no application of any kind of power, and the technological challenge consists of finding subtle ways to minimize inertia and aerodynamic drag. Just like drag racing, the derby is replete with "legends, superstitions, tactics, psychological warfare, computer-designed cars, big-time sponsors, and luck."[4] For someone interested in technological enthusiasm this activity is surely evocative. But I have chosen to address drag racing partly because of my own experience. I know a good bit of what follows because I saw it with my own eyes. I grew up in southern California, where it all began. As a teenager, I cobbled together drag racing machinery in my parents' driveway. To their horror, had they known, I regularly participated in unlawful street races. I was not there when the first drags conceived as a commercial enterprise were staged, but I was present at one of the initial events, and to this day I remember it vividly.

To begin with, there was the noise. I had heard unmuffled exhaust before but never engines that were "full house," far too souped-up to be fit for the street. Then there was the smell—smoking rubber, of course, but more distinctive were the fumes emitted by engines running on "fuel." Gasoline was *a* fuel, to be sure, but it was not *fuel*. While fuel was a subject of profound mystery, what I learned soon enough was that the most pungent and eye-smarting exhaust was produced by a compound called nitromethane, "nitro" for short, CH_3NO_2. Nitro was used as a commercial solvent. It was also used in model airplane engines—and in rockets. People in traditional realms of auto racing used it gingerly because its destructive tendencies were so hard to control. Drag racers needed only to make a quarter-mile sprint, however, and they found they could blend nitro with methyl alcohol in ratios of 50 percent and more. A "big load" could enhance horsepower by half again. But nitro was apt to preignite and detonate even in a quarter mile. At the drags "fuelers" were often hitched up, waiting to be towed home for engine repairs, long before the elimination rounds began in the afternoon.

And then there was the speed. I had driven some pretty strong machinery in street trim, prewar Ford coupes; the best could turn 100 miles per hour in the quarter mile, with elapsed times around 14 seconds, and, believe me, that kind of acceleration got your attention. On the drag strip I saw spartan fuel-burning roadsters and "rail jobs" clocking 120s in less than 12 seconds, and to me that was absolutely fascinating.

Now, I have declared something fundamental to my point of view. I started out as an enthusiast myself, and to some extent I still am one, even though I regret the way notions of purpose have changed since the early days. After commercialism became a driving force, a symbiosis with the tobacco industry became drag racing's crucial tie to corporate America.[5] That to me is reprehensible. Yet I cannot say that drag racing has been any more reprehensible in this regard than other sporting activities: If the nabobs of tennis (tennis!) did not take the high ground, how are we to condemn the men who controlled drag racing?

Moreover, even though drag racing has been thoroughly commodified, it seems clear that people would still push the technological envelope even in the absence of market forces. Indeed, they do push it. As drag racing was in a headlong rush to embrace big-time backers, a hardy band of enthusiasts kept returning to El Mirage, a lonely dry lake in the Mojave Desert, and to Utah's even lonelier Bonneville Salt Flats, in order to conduct speed trials, without any financial incentive at all: "They scrimp, save and sacrifice in an effort to gain their own little piece of fame," wrote a reporter for a Utah newspaper in 1989.[6] Far from being anything pecuniary, their motivations are essentially "romantic and emotional," in a phrase of Eugene Ferguson's which I will quote again.

I do not take much stock in the Frankfurt School's bleak strictures about technology, about domination for the sake of domination, but I am quite aware of what is held to be politically correct. Anyone who does not much like automobiles, who agrees with Ellen Goodman that "the car is to the environment what the cigarette is to the body,"[7] is surely not going to think fondly of drag racing. Quite apart from booze 'n' butts, the hucksterism carried to wretched excess, many drag racing devotees are decidedly deficient in refined sensibilities. So be it. I can only reiterate my interest in the explanatory power of technological enthusiasm; if we are to start with "the intuitively compelling idea that technology may be *the* truly distinctive idea of modernity," then it seems to me that we may be able to learn a great deal from analyzing the behavior of people who find something inherently compelling about technology.[8]

But I also need to reiterate that I, too, am an enthusiast. Forty years after first seeing a rail job perform on a rural airstrip, I remain as delighted by dragsters as I am by steam engines and square-riggers. Moreover, I admire attitudes that prevail among drag racers—the intensity, the sly humor, the gritty resilience, the open-mindedness—in technological realms, naturally, but, more than one might anticipate, in cultural realms as well. Among other things this has served to impel women into mainstream roles to a degree far beyond what prevails in most similar activities.

Yet it will be obvious that much of what follows has a sharp edge. People who expected expansive praise may well be annoyed. Before assuming the stance of critical historian, then, I would like readers to get some idea of the poetics this activity can inspire: "Drag racing is the product of as-yet unsung heroes of everyday life. It represents a group of anonymous master mechanics whose level of craft, and of technological and mechanical genius, springs from native intelligence and [a] kind of enterprising, exploratory spirit. . . . The machinery is stunning."[9] These are the words of a woman whose participation in drag racing coincided with her pursuit of a doctorate in art history. In the 1970s Virginia Anne Bonito sought "to learn the mysteries and to find the magic combinations of fuel, clutch, and tire." Then she worked with the crew of a fueler. She continues to follow drag racing, in part, because she remains enchanted by the very idea of such an activity, just as I do.

During the early days I watched intrepid pioneers such as Harold Nicholson, Art Chrisman, Holly Hedrich, Joaquin Arnett, Calvin Rice, and Emery Cook push speeds up to 130, then 140, 150, and 160. For a time I tried to emulate what they were doing but finally had to admit that I had (in Bonito's words) neither the intelligence nor the spirit. So I stood aside and kept in touch with drag racing's evolving theater of machines as a spectator. If that were *all* I had done, I might not be very well prepared for the task at hand. The literature abounds in accounts rich in anecdote but shy on analysis. Along the way, however, I spent time pondering the truth of the maxim that "history is not merely what happened; it is what happened in the context of what might have happened." I came to appreciate why "technical designs cannot be meaningfully interpreted in abstraction from the human fabric of their contexts."[10] Hence, my stance is at once engaged and detached. I trust this yields a useful mode of inquiry.

In the documentation I have sought to indicate what I have learned by reading and analyzing what others have written. I have also learned a great deal from historical actors such as Virginia Bonito. I was fortunate to get to know the two men who costarred in the drama of drag racing (not always amicably) for most of my forty-year period, Don Garlits and Wally Parks. To those two, and to Pat Garlits and Barbara Parks, I owe thanks for many courtesies; Don and Pat did me the singular honor of critically reading my typescript. Special thanks also to Greg Capitano, who watches over Don's Museum of Drag Racing in Ocala, Florida. At Wally Parks's National Hot Rod Association (NHRA) in Glendora, California, special thanks are due to Neil Britt, Leslie Lovett, and Teresa Long. And a thank-you, too, to several former denizens of the NHRA—John Raffa, Tammy Ferrell, Joe Sherk, Steve Earwood, and Dave Densmore.

It was a pleasure to conduct formal interviews with several of the key players—Parks and Garlits, Ed Iskenderian, Don Montgomery, Marvin Rifchin, and the late Keith Black, among others. I am grateful to everyone who participated in the Drag Racing Oral History Project at the National Museum of American History, especially Don Jensen of Kihea, Hawaii, who provided me with a rich store of his own recollections as well as recorded dialogues with his old cohort from northern California.

Among the community of racers, past and present, I appreciate the generous assistance of Dale Armstrong, Joaquin Arnett, Dave Bishop, Fuzzy and Jana Carter, Art Chrisman, Kathy Donovan, Roy Fjastad, Ray Godman, Red Greth, Ernie Hall, Don Hampton, Chet Herbert, Harry Hibler, Eddie and Ercie Hill, Jerry Jardine, Wayne King, Bruce Larson, Harry Lehman, Tom Madigan, Bill Martin, Marty McDonough, Lanny and Tony Miglizzi, Shirley Muldowney, Mike Nagem, Bob Neal, Gary Newton, Don Nicholson, Pete Ogden, Setto Postoian, Bill Shultz, Jim Skorupski, and Henry Velasco. One memorable Saturday evening, on the San Diego Freeway, Jim Lytle unwittingly gave me a push. On a Saturday at the Texas Motorplex many years later Billy Meyer helped a crew from the Smithsonian when we were tired, and Darrell Gwynn helped when we were thirsty. On several occasions Sonny Messner was a congenial guide to the southern California specialty shops. Be it the 1950s or the 1990s, I could invariably count on C. J. Hart for sage advice.

While I was starting this book, I was also involved in a pair of related exhibitions for the Smithsonian. The first, a retrospective on the career of Garlits, was installed largely through the heroics of John Stine, a man who has never let me down. The second, called "A Material World," is still ensconced at the first-floor transept of the National Museum of American History, featuring a unit on dragster technology which was made into something special by Jeff Howard and Karen Loveland: Jeff mounted a Swamp Rat, and Karen produced a lovely movie.

Dave Wallace, Jr., who could very well have written this book better than I, read the entire typescript and saved me from countless errors. He also suggested that I get a critique from Ed Sarkisian in Rhode Island. Ed, Dave told me, was a man of vast knowledge and keen insight, and he was right on. Thanks, too, for help from Ed's wife, Marge, and from Dave's wife, Connie Strawbridge, and his friend Pete LaBarbera. Dave's brother, Sky, with help from Bret Kepner, Phil Elliott, and Chris Martin, searched out the data in the appendixes. For assistance with photographs, in addition to several people already mentioned, I owe thanks to Jane Barrett, Carolyn Hall-Salvestrin, Ruben Lovato, Jeff Tinsley, Bill Turney, and Fred Wohlfarth. For high performance in an editorial role, I've not seen anyone to equal Elizabeth Gratch.

Colleagues in the historical profession to whom I must express appreciation include Joe Corn, Steve Cutcliffe, Eugene Ferguson, Jim Flink, Paul Forman, Robert Friedel, Stanley Goldberg, Bob Gordon, Joe Guilmartin, Brooke Hindle, Ed Layton, Otto Mayr, Art Molella, Carroll Pursell, David Shayt, Bruce Sinclair, Roe Smith, Jeffrey Stine, Walter Vincenti, Rudi Volti, and Jack White. I will never forget the kindness of the late John B. Rae. Ted Park, Connie Wilson, Joan Mentzer, Alan Blum, Jim Kelly, and Jim Norris had words of wisdom at timely junctures. I learned much about the Australian scene from Steve Munro of Tullamarine, Victoria. It was always fun to reminisce with friends from street racing days: Keith Thompson, C. V.

Hansen, Frank Carey, and especially my one-time partners, Peter Massett and Vaughn Bowen, Jr. Sandy Williams shared my curiosity about what makes racers tick. Don Haworth was one of several old schoolmates who kept me liberally supplied with clippings from the *Los Angeles Times* and other southern California newspapers, and I also owe it to Don that I attended the premiere U.S. Fuel and Gas Championship in 1959. That was just before I met my wife, Dian, and, had I known her then, she would have gone too, of course. My ultimate thank-you, then, is to Dian, for sharing my enthusiasm for all these years.

INTRODUCTION

A NEW THEATER OF MACHINES

The life of an American is, indeed, only a constant racing.

FRANZ ANTON RITTER
VON GERSTNER, 1839

rag racing is an activity with a history so brief that people still around were there at the start. They can recall how it began as a hobby among young men infatuated with speed and power—"hot rodders," they were called. They have seen it become a compelling spectacle with a complex web of commercial relationships. It remains a hobby for some but is a very serious matter for others. Millions of people attend drag races every year. Billions of dollars change hands because of drag racing. Yet no historian has ever addressed the subject, and responses toward the very idea of doing so tend to be much the same as one Europeanist encountered when he began writing about horse racing: "How odd . . . how unprofessional and what a misuse of his time and ability."[1] And, unlike horse racing, drag racing is still more or less peripheral to mainstream popular culture.

Oh, plenty of people are aware of it because *Sports Illustrated* and *USA Today* both take it seriously, because it gets repeated exposure on television, and because Hollywood made movies like *American Graffiti* as well as a feature about one of its stars, Shirley Muldowney. Most visitors to the National Museum of American History pause to contemplate "Big Daddy"

Don Garlits's sinuous "Swamp Rat XXX," once able to move off "almost as fast as an aircraft carrier's catapult can fling a fighter into the wild blue yonder."[2] Dragsters are the fastest race cars on earth, bar none. They are the most powerful and the most dramatic. Still, there are metropolitan dailies that never mention Don Garlits or Shirley Muldowney, even though they may fawn over Rick Mears and Dale Earnhardt, stars of the Indianapolis 500 and the southern stock-car circuit. The *Los Angeles Times* pays considerable attention to drag racing, but, at the other extreme, the *Washington Post* never prints one word of routine coverage. My daily paper, the *Baltimore Sun,* runs a few inches every so often, about the same as would be accorded a duckpin tournament in Hagerstown. Actually, it makes no difference to me whether I can read about drag racing with my morning coffee, but I am puzzled by indifference to an activity that, truly, echoes a dominant theme in American social history.

Think back to the last century. The transatlantic packet, the stern-wheel steamboat, the express locomotive, and all such storied contrivances from the American past were aspects of a "technology of haste" which was linked, in Daniel Boorstin's words, to "rewards that others might grab if you were not there before them."[3] The question "How fast will it go?" was one that gained currency as Americans became preoccupied with ways of "saving" time (or, in the case of clipper ships, "clipping" it). "Getting there first" brought a tangible payoff.

The same insularity that made time valuable enough to try to save also fostered the Ingenious Yankee—isolated by distance and thrown back on his own resources—who had to become adept at tinkering. Inevitably, he began to "improve" the design of things, and with a vehicle that usually meant making it faster. Soon he developed a profound affection for the speedy machine, be it an iceboat or a prancing 4-4-2 locomotive on the Central. This affection grew even stronger in the new age of internal combustion among such descendants of the Ingenious Yankee as the aerial barnstormer and the auto racer. In the latter twentieth century, as "getting there" fast made less difference in most forms of commercial transport, to enthusiasts simply "going" fast meant more and more.

This is not to say that speed, by itself, had not made quite a bit of difference even to those whose primary pursuit was profit. With commercial conveyances that were designed for speed, this characteristic often owed something to the sheer exhilaration of making time; Captain Laughlin McKay once drove his brother Donald's *Sovereign of the Seas* on a South Pacific reach of 1,478 miles in four days. Sometimes speed was actually the predominant motivation, and sometimes it predominated at the *expense* of commercial advantage, not to say rationality, if that term be equated with economic self-interest. Prime but by no means unique examples are the clipper ships and oceangoing side-wheelers of the 1840s and 1850s, which were a profitable proposition only under such a rigorous set of external constraints that freewheeling entrepreneurs were left beached when that context shifted even slightly.

In certain realms of transport speed has remained crucial to commercial advantage—the strategy of firms such as Federal Express testifies to that—but there are inevitable trade-offs. Generally, speed no longer plays a determining role. However people or goods move about the United States on the eve of the twenty-first century, they move slower than they might. Jetliners are throttled back, and no appeal to international gamesmanship proved sufficient to bring the Mach-3 SST to life, no last-ditch plea that it was "essential to America's continued leadership in the field of commercial aviation."[4] Speedy passenger trains are pretty much a foreign phenomenon. The pace of a river tow is excruciatingly slow, even on the very waters where the *Natchez* and the *Robert E. Lee* once raced all-out with safety valves tied down. In *Looking for a Ship* John McPhee tells how the captain of the container-laden *Stella Lykes* ran her slowly "to keep her alive." On the road there are 55 MPH signs in places where the posted speed was once 65 or 70. Truckers habitually flout the law, of course, but virtually no form of transport maintains the schedules it could if it were not constrained by practical considerations.

Yet one still hears people eagerly ask the question "How fast will it go?" Speed matters even as the initial rationale has largely been lost. And, when it comes to vehicles designed *just* for speed, Americans are usually on the leading edge. To be sure, an affection for speed as an end in itself is not a unique national characteristic and never has been. The fastest airplanes were once built by Germans; the fastest bobsleds are still. Australians have always been hot rodders at heart, and in 1983 they also proved that American predominance in 12-meter sailboat racing was not forever. The land speed record has been a British passion ever since Malcolm Campbell first bettered 300 miles per hour on the Bonneville Salt Flats in 1935.

Grand Prix, in some regards the most sophisticated of all forms of racing, has attained only a tenuous acceptance in the United States. But that is precisely the point. Formula 1 cars are engineered for a broad range of functions, not just speed; cornering necessitates slowing down quickly and slowing down a lot. Americans tend to favor their speed contests neat—in the case of autos, racing around superelevated ovals that are essentially one continuous straightaway, or, even more peculiarly, they actually *do* go in a straight line. The contestants end up not where they started but in a different place altogether, just like a clipper ship or express locomotive.

With an Indianapolis 500 that draws one of the largest crowds of any sporting event in the world, Americans are clearly not lacking in enthusiasm for "closed-course" competition, where the starting line is also the finish line. But along with the various ovals, loops, and road circuits, large and small, there are hundreds of point-to-point courses, where the finish line, point B, is somewhere different from the start, point A. Occasionally, B is a long way away, as in the coast-to-coast competition that persists in sub-rosa affairs like the Cannonball Run. Yet, for many enthusiasts, a test of "endurance" has nothing like the immediate appeal of flat-out speed. The course can be perfectly straight and drastically abbreviated.

Even before Bonneville became a prime venue for chasing international speed records, southern California hot rodders were racing across the dry lakes of the Mojave Desert, notably at a place called Muroc, where the first organized event was held in 1931. The next year the top time at Muroc was 118 miles per hour; before things were suspended "for the duration" an enthusiast named Bob Rufi had topped 140. That was nothing special compared to Sir Malcolm's "Bluebird," but people like Rufi, it turned out, were just getting warmed up.

During the war the government appropriated Muroc as part of the site of a test facility for the air force and the National Advisory Committee for Aeronautics. (It was later named Edwards Air Force Base in honor of a test pilot who died in a Northrup Flying Wing.) So, while Chuck Yeager was pursuing Mach 1 high above the old race course, the hot rodders moved to El Mirage, near Adelanto, a crossroads in the high desert country of San Bernardino County. Events were held under the auspices of various loose-knit sanctioning bodies, foremost being the Southern California Timing Association (SCTA), which dated from 1937. Note that the activity was "timing" rather than racing. By nature dry lakes were dusty, and the driver of a car trailing in a race simply could not see where he was going. Dire consequences were predictable. Hence, the SCTA had restructured the activity. The starter sent drivers off singly, and speeds were timed by means of a pair of photoelectric cells after an acceleration run of about a mile.[5]

In 1949 the SCTA made arrangements to hold time trials at Bonneville in August, the event to be called "Speed Week" and repeated annually. The deal was negotiated by SCTA's executive secretary, Wally Parks, in concert with Robert Petersen and Lee Ryan, general manager of Petersen's magazine publishing firm (Petersen's monthly *Hot Rod* was emerging as a phenomenal journalistic success).[6] Hot rodders began making a trek to Bonneville each summer, and within a decade the fastest among them were starting to crowd those Englishmen like Campbell.[7]

Some hot rodders relished the idea of racing only "the clock." But the clock is, after all, an abstraction, and most of them got more satisfaction out of actually competing wheel to wheel with someone else. Oval tracks were an option, but precluded the thrills of flat-out speed. For that there were abandoned airfields and paved flood-control channels, but public streets were obviously handiest. Street racing entailed plenty of danger, needless to say, but one positive lesson the racers learned was that they could test their mettle and mechanical prowess quite nicely in a short sprint. So it was that drag racing was defined at the midpoint of the twentieth century. Efficiency would be defined in two ways. One was getting from A to B quickly, and eventually the primary challenge would become one of minimizing *elapsed time* (ET) (as was likewise the case with clipper ships and express trains). That was what won races. But sheer speed counted for a lot, too. The standing mile record, held by a German, had stood for decades at about 200 miles per hour. By the mid-1960s so-called dragsters were clocking 200 miles per hour in a quarter-mile, 440 yards, which became the conventional distance. A quarter-century later they were flirting with 300, and a close race was awesome.[8]

Though perfectly useless in any "rational" sense, a dragster is by any measure a mechanical marvel. John Kouwenhoven saw in dragsters the persistence of "vernacular design," one of the great creative forces in American life.[9] To be sure, the demands on a *driver* are exceedingly specialized: Dragsters race only in pairs, and a "lap" entails executing a straightaway burst lasting only a few seconds, with no cornering, and braking required only to get stopped safely before running out of room and hitting something. A big part of drag racing's appeal lies in the very brevity of the encounter, the drama inherent in a sprint; at a track meet, after all, people go out for refreshments during the 8,000-meter run but stand and cheer wildly during the dashes.

Yet there is much to the appeal of drag racing which is subtle. The mental and mechanical skills of a racer who can maintain a winning edge are all but unfathomable to an outsider. Beyond that there is something else. Virtually every form of racing is pursued within an explicit context of self-imposed limitations: It is intrinsically more difficult to go fast on wheels than it is to go fast through the air, while it is especially challenging to maintain adhesion while gathering speed. Over the years various self-proclaimed authorities announced that the "laws of physics" stood in the way of any further progress in the acceleration of dragsters. They were always wrong, and racers, most of whom were religiously pragmatic, delighted in such testimony to the shortcomings of abstraction. Still, to power a rapidly accelerating fueler (the 1990 version pulling five Gs—i.e., five times the force of gravity—during the initial "launch") through a pair of rubber tires was to operate under a profound technological handicap. Drag racers relished the thought that what dragsters did on asphalt or concrete was, on paper, simply not possible.

Additional handicaps could be imposed if people so chose. To permit competition by all sorts of vehicles drag racing has a multiplicity of "classes," each with discrete limitations regarding such matters as the location and displacement of the engine and the body configuration. Hence, there are dozens of class winners at a major event, and, strung out well below the ultimate marks, drag racing has hundreds of official records, both speed records (mile-per-hour clockings at the finish line) and elapsed-time records from starting line to finish; top speed and elapsed time are *independent variables,* each clocked separately. Every record is regarded respectfully, for there is no denying the keen ingenuity displayed by all those vehicles that are designed to accelerate quickly and go fast but cannot accelerate as quickly or go as fast as the quickest and fastest vehicles do.

Indeed, part of the folklore is that "unlimited" records are simply a matter of spending a lot of money while records attained under stringent technical restrictions are more precious. There may be a bit of truth to this. Some exceedingly elegant machines have resulted from contending with stringent handicaps deliberately written into the rule book. Here one begins to approach the ultimate in realms in which conveyances are not permitted to have mechanical power at all; even in an age of 300 MPH dragsters, an Olympic bobsled or a "human-powered vehicle," a pedal-powered HPV, are wondrous devices. I have already paid regards to the technology of the Soap Box Derby.

For an aficionado the performance of an HPV (60 MPH) or a Soap Box racer (30 MPH), or any drag racing machine in one of the limited classes, lacks for nothing in the realm of drama or the display of mechanical ingenuity. More books about drag racing surely ought to follow this one, and some other author might well concentrate on the diverse ranks of racers who are *purely* hobbyists, some of whom frequent the same tracks as the 300 MPH machinery, all of whom partake of the same intense enthusiasm. But I have elected to focus on the pacesetters, on those machines that come most readily to mind when one thinks of "high performance," in any of its definitions. Usually, if not inevitably, the key has been nitromethane. The use of nitro turned drag racing into a money-making proposition for some racers; for others it removed profits from the realm of rational hope. But, without doubt, it is what made the turnstiles click fastest. The fury produced by an engine on a big load of nitro is a sensation one cannot begin to convey in words, although certain photographs in this book are suggestive of its impact on bystanders. An academic acquaintance who went to a drag race only once, in 1964, still recalls that prime facet of his experience: "I had never before heard *artificial* thunder."[10] To many fans the thunder of nitro may be almost as addictive as the actual competition is to racers.

I mentioned earlier that drag racing is fairly peripheral to mainstream popular culture. But this is not to say that it lacks loyal legions of followers—far from it. One day in February 1992 I found myself standing at the guardrail beside a drag strip. The location was the Los Angeles County Fairgrounds in Pomona, and I was looking across the track at a wall of fifty thousand people, on their feet for the first round of fuel racing at the opening event of what the National Hot Rod Association calls the Winston Championship Series. As each pairing charged away from the starting line and through the timing lights, I knew that some of these people would feel a tinge of disappointment if the electronic scoreboards recorded numbers slower than 290s, or at least 280s. It occurred to me that I might just be standing on the exact same spot where I stood one day forty years before.

There might not have been one hundred people in the stands and only occasionally a run in excess of 100 miles per hour. *Progress* is a word that has fallen into disrepute among historians of technology, ironically perhaps, though certainly not without reason. In posing this comparison across time, I am not necessarily talking about progress, but I am most certainly saying that a lot has changed and that this change is powerful testimony to the kind of momentum an arcane technological pursuit is capable of attaining. To move along I now want to take a more sustained look at the way it was at the beginning.

HIGH PERFORMANCE

WARMING UP

I had no idea it would get as big as it did, but I didn't see any reason for it not to always be a good moneymaker.

C. J. HART, 1981

n a crisp Sunday morning in 1949 a group of hot rodders converged on a stretch of two-lane road north of Santa Barbara. The road ran westerly towards the ocean from California's Coast Highway, Highway 101. Ordinarily, it provided access to a landing field at Goleta, but on this April weekend a half-mile had been closed off with portable fencing. Although the site was well known among local street racers, this was a special occasion—a match race between two out-of-town celebrities, both of them dry lakes veterans, Tom Cobbs and Fran Hernandez. Cobbs had been winning races all around Los Angeles in his Ford roadster, a 1929 Model-A body channeled over a '34 frame. The engine was a '34 V-8 with a Roots blower from a GMC diesel truck or bus fitted on top as a supercharger. Cobbs had challenged Hernandez, who raced a fenderless but otherwise stock-bodied '32 Ford three-window coupe with a new Mercury V-8 that had been over-bored and stroked to $3\frac{3}{8} \times 4\frac{1}{8}$, 296 cubic inches compared to Cobbs's 249. But there was no blower on top, just three Stromberg carburetors on a special manifold.

There were marked contrasts between the two racers themselves as well as their hot rods. Cobbs was called "a clever engineering sort who could

At Goleta in 1949 Tom Medley, later a mainstay of the Hot Rod *magazine staff, leans out to snap two fenderless coupes speeding toward the finish line. (Courtesy Don Montgomery)*

afford, as heir to tobacco fortunes, to experiment and to test on Stu Hilborn's dynamometer." Hernandez, who managed Vic Edelbrock's place on West Jefferson Boulevard in Los Angeles, was "a scrappy master of machine shops." Cobbs hung out in the beach town of Santa Monica with Hilborn, who manufactured fuel injectors for dirt-track racers, and Jack Engle, who was one of the first southern Californians to go into business regrinding Detroit camshafts, changing lobe profiles to alter valve timing. Hernandez's buddies were Bobby Meeks, who worked for Edelbrock, too, Ed Iskenderian, a onetime apricot pitter from Fresno who had a cam grinding shop just down the street from the Edelbrock Equipment Company, and Lou Baney, who rebuilt engines in a shop on South Normandie.[1] Nominally, Cobbs's roadster was in "legal" trim and could be driven on the streets, but Hernandez's coupe lacked such niceties as headlights and mufflers, so he had towed it in with a pickup.

Other hot rodders—nearly all of them young men around twenty, with just a few girlfriends in evidence—showed up to participate, to drag it out with one another, but the Hernandez-Cobbs match was the feature. Everyone crowded up close for a good view, either at the starting line or near the finish, where there was a hump and the roadway narrowed to cross a culvert. The course that had been marked off allowed racers three-tenths of a mile to accelerate and sufficient room to stop before coming to a sharp turn beyond the culvert. Hernandez's coupe was balky about starting, so it had to be hand-pushed and fired on compression. When it finally kicked over, the exhaust fumes immediately betrayed the presence of something other than gasoline. Cobbs may have been surprised, but Hernandez already had a reputation as one of the select few who were expert in setting up Stromberg carbs for nitro.

Side by side, a few feet apart, Cobbs and Hernandez edged towards a white line across the blacktop, where the starter stood holding a flag on a wooden stick pointed towards the ground. Then, just as all four front tires touched the line, the starter yanked his flag skyward. Open headers roared and Hernandez jumped out in front while the roadster spun its tires, filling the air with clouds of white smoke. Although Cobbs finally regained traction and was closing the gap toward the end, Hernandez's deuce crossed the culvert a length ahead. He quickly gathered his things, while his friends bolted a tow bar to the frame of his coupe and hitched it to the pickup. Then he was gone.

Word of the outcome quickly got around, and hot rodders rehashed it long afterwards, a diversion known as bench racing. Cobbs had changed to lower rear end gears, thinking (mistakenly) that this would give him an advantage out of the chute—could he have won with "lakers gears" like Hernandez had? Did that "Jimmie" blower really produce 10 pounds of boost, as some people said? What kind of load was Hernandez running, anyway? The collective memory later coalesced as a tale titled "The Day Drag Racing Began," which was reprinted time and again.[2] While eyewitnesses could attest to its essential accuracy, it had all the makings of a classic legend. The details need not be taken literally.

Clandestine drag racing had been going on for some time, of course, but what was unique about this particular event is that officials of the Santa Barbara Acceleration Association had sought, successfully, to have the California Highway Patrol confer approval: The races at Goleta were not against the law. That nicety aside, one might denominate "the day" as almost any day (or, more likely, any night) in the late 1940s and the place as being Arrow Highway in the San Gabriel Valley, Riverside Drive in the San Fernando Valley, Culver Boulevard in West Los Angeles, Baker Street between Harbor and Bristol in Orange County, or any one of many other spots—in Texas or Florida, even in Michigan, Illinois, or Ohio, as well as in California —where hot rodders raced unlawfully. Drag races that were not only legal but also in keeping with commercial conventions were not established until a year after "the day" at Goleta, at an air strip in Santa Ana, California.

Soon enough similar events were being held in Caddo Mills, Texas, Zephyrhills, Florida, and elsewhere, but Santa Ana (which for some reason the racers always pronounced Santee Ana) is the place where drag racing had its inception as a commercial enterprise—or, to be more precise, where the *staging* of races had its commercial inception, since the making and selling of hot rod parts was already established in the hands of people like Vic Edelbrock and Ed Iskenderian.

The paved bed of the Los Angeles River remained a favorite spot for clandestine drags long after efforts had been initiated to provide sites for racing legally. Here, in March 1955, police disperse racers near the Fourth Street Bridge. (Hearst Collection, courtesy Department of Special Collections, University of Southern California Library)

Although the reader will find me yielding to temptation in the narrative that follows, seeking to denominate "firsts" is not a terribly profitable exercise. What is important to establish here is that some time close to the midpoint of the twentieth century an activity was invented—or, again for precision, an activity invented previously was endowed with formal sanction. Given that activity's thoroughgoing transformation in the ensuing forty years, it is worthwhile to take a look at the Santa Ana Drags. The instigator was one Cloyce Roller Hart—"C.J." Hart, also known as "Pappy." Hart was thirty-nine in 1950, having left Findlay, Ohio, for California with his wife, Peggy, and their two young children after some colorful innings as a moonshiner and roustabout, among other things. Hart always cautioned that he "didn't invent drag racing." He had done a lot of it himself around Findlay, even as a teenager, and reckoned that dragging on public thoroughfares must have been going on "ever since there was cars!"[3]

But street racing took off by orders of magnitude in the postwar years. There were thousands of unmarried males, many of them ex-GIs, with plenty of spare dollars, enhanced mechanical skills, an assertive bent, and a love of speed. As the sociologist Bert Moorhouse writes, after the war "the 'hot rod' became significant in the lives of a large number of (mainly young)

At Santa Ana, California, in 1950, Howard Johansen stands beside his stripped-down Model-T roadster. The engine is a Ford V-8 with three carburetors and finned aluminum heads. In coveralls behind driver Ed Osepian is Nick Arias, Jr., who, like Johansen, would become a prominent manufacturer of this kind of equipment. (Courtesy National Hot Rod Assn.)

C. J. Hart is seen here with the timing equipment for the Santa Ana Drags. Hart's wife, Peggy, drove their Model-T roadster in competition, hence the helmet. On this particular Sunday afternoon, in the summer of 1953, the top mark was Art Chrisman's 136.98 while a couple of others had turned 126. (Courtesy Petersen Publishing Co.)

Dick Kraft poses enthusiastically with his "Bug" in 1950, arguably the machine from which the term rail job *derived. Minimum weight was the purpose, not safety and certainly not showy appearance. (Courtesy Don Montgomery)*

Americans."[4] Many of these young men also had a dubious sense of social responsibility, and the press relentlessly flaunted an image of the hot rodder as "a deliberate and premeditated lawbreaker."[5] No doubt the "hot rod menace" was overdrawn in the newspapers and in a genre of fiction purveyed by Henry Gregor Felsen and his followers. Nevertheless, street racing had clearly become a problem of major proportions by 1950, particularly in California. People were getting killed in hot rods, and by hot rods.

"Throw 'em all in jail" was one response. But among the ranks of citizens who tried to address the problem constructively the idea of "giving the kids a place to race safely" began to seem like better tactics. Indeed, this became the primary rationale for the establishment of drag strips nationwide in the 1950s, although it was probably not what was foremost in C. J. Hart's mind. Hart had been an inveterate street racer himself, and a good deal of what was really exciting about this activity was that it *was* illegal.[6] What nobody enjoyed, however, were the busts, an ever greater likelihood as the police infiltrated drive-ins and other hot rodders' hangouts to gain advance information and as "engaging in a speed contest" became the costliest of all citations. Besides, Hart thought, there could be money to be made in staging legal drag races.

Aircraft landing strips and taxiways, smooth and uncrowned, had always seemed like ideal places for drags, and in Santa Ana, where Hart operated a used car lot on Bolsa Avenue, there was a sleepy little airport behind Newport Bay. In concert with two partners, one of whom promoted motorcycle races and was in a position to provide insurance, Hart approached the airport manager and struck a deal to rent an unused runway every Sunday "for 10 percent off the top." All told the initial investment was less than a thousand dollars. Hart recalled that "about the only thing [he and his partners] had to put in there was an ambulance, and . . . some kind of concession for the food, and insurance, and trophies."[7] The first day for the Santa Ana Drags was Sunday, June 19, 1950.

Five days later the North Koreans came across the 38th parallel, and what ensued in the Far East was bound to put a crimp in any recreational activity attractive to draft-age males. Still, within a few months enthusiasts who usually spent their Sundays at El Mirage were showing up regularly at Santa Ana. There they could race one another as well as get timed. To supplant stopwatches Hart had designed a timing system with a pair of photoelectric cells that activated a clocking device set up in an old hearse parked at the finish line. Although it would gradually become clear that quick start-to-finish elapsed times were what won drag races, C.J.'s device recorded only top speed at the quarter-mile mark, or perhaps a little beyond—1,320 feet had not yet been established as conventional. The fastest run in 1950 was 120 miles per hour, clocked by a gutted '34 Ford

The Nicholson brothers' "?"—a coupe from which the top had been removed—is seen here in 1951 at the Paradise Mesa strip, near San Diego. The paired tires were an attempt to enhance traction. At right is the old lakes machine that became Art Chrisman's first dragster (see page 20). (Courtesy Don Montgomery)

with hand-formed bodywork in front, four rear tires, and a big question mark painted on the side. The "?" was campaigned by Harold and Don Nicholson, brothers from Pasadena in their early twenties. The boys from "Nick's Speed Shop" were already legendary around local drive-ins like Larry and Carl's for their street-racing exploits.

Or, rather, 120 was the fastest run by an *auto,* for Al Keys had topped 121 aboard two different fuel-burning Harley-Davidson motorcycles, Joe LeBlanc's "Beauty" and "The Beast" owned by Chet Herbert, who ground cams for automobile engines which utilized roller tappets, like Harleys. Weight did not matter a great deal at the dry lakes, where there was more than a mile to get up to speed, but already it was becoming obvious that weight was a key constraint on a drag strip. A hopped-up Harley 74 on fuel could put out a lot of power, while the overall weight was far less than any automobile, no matter how ruthlessly it was gutted, no matter how many nonessential parts and pieces were removed. Not that hot rodders had not begun trying, almost at once in fact. A lakes veteran named Dick Kraft had stripped a roadster all the way down to bare frame rails, calling it "The Bug," and a lot of other people would be thinking along the same lines, that less was more.[8]

Certainly, creating a "rail job" to gain an edge was much cheaper than investing in a full-house engine with such expensive refinements as a stroked crankshaft (i.e., reground with a longer stroke). At that time it seems unlikely that anyone had ever spent as much as a thousand dollars on a vehicle intended solely for the drag strip. The ease with which one could jump right into the thick of things contributed to an initial surge of participation. Several other California strips had opened by 1951, including one near Monterey Bay, at the Municipal Airport in Salinas, and another at New Jerusalem, near Tracy. On Armistice Day, 1951, a strip opened north of Stockton which was known as Kingdon and was managed by Bob Cress, a Stockton policeman. Yet another of the first strips was located in Saugus, just beyond the northern rim of the San Fernando Valley. It was run by Lou Baney (née Aloysius James Benedetto), a thirty-year-old confrère of the Edelbrock-Iskenderian circle.

Baney was an operator, a man whose organizational skills and entrepreneurial flair would make him a major presence in drag racing for many years. He had headed the Russetta Timing Association, a rival to SCTA. He had raced track roadsters on local ovals such as Carrell Speedway against men like Jack McGrath and Manuel Ayulo, who went on to the big-time on the "champ car" circuit that included the Indy 500. He had even tried to make a deal with the manager of the Santa Ana airport for staging drag races there but lost out when his partner got cold feet. After Hart's venture caught on Baney talked with Lou Senter, in whose shop he did engine work, about launching a rival strip. With five hundred dollars borrowed from Senter's brother and another five hundred dollars from Baney's dad, the two of them incorporated as Sports Events, Inc., and scheduled Sunday races at a place called Six-S Ranch Air Park. "The strip was short and a little too narrow," Baney recalled, "but it was better than nothing."[9]

Senter was a silent partner; Baney ran the show. Because he would have had to pay taxes on a dollar admission, he made it a penny less and "didn't have to get involved with the government." He advertised by posting bills around the San Fernando Valley, and Saugus sometimes drew fifteen hundred people, but the average Sunday turnout was only in three figures. C.J. did better than that, even though people in the San Fernando and San Gabriel valleys thought of Santa Ana as a long way to go. Orange County itself had a total population of only 220,000. Where a vast urban village of two million would sprawl by the 1980s, there was then mostly mile after mile of orange groves.[10]

Needless to say, the drags were decidedly small-time. Santa Ana and Saugus usually had about two hundred cars in competition on any given Sunday, the majority nominally stock but with a few dozen that had been modified, most of which looked pretty rough-hewn. Each strip had a system of classes, with stockers the slowest and fuel-burning roadsters the fastest. There were time trials before noon and elimination rounds after the lunch break, with trophies for class winners. At Santa Ana there was no prize money per se, although a winner who preferred cash could sell his trophy back to C.J. for its wholesale cost, $7. Baney offered a $25 war bond for the racer who could best the entire field, or $18.75 cash, which is what a bond cost. Occasionally, he would offer a $100 bond, or $75 cash. The racers always took the cash. "I don't think we ever actually gave away a War Bond," Baney remarked.[11]

Later, in the mid-1950s, Baney campaigned a Cadillac-powered dragster that ran speeds in the 150s and won some big events. Then he got deeply involved in the automobile business with Bob Yeakel, a classmate from Manual Arts High School, and he was away from the racing scene for several years. When he returned in the 1960s, he recalled, "the cars were going a whole lot faster," and a lot of other things had changed as well: "I almost didn't recognize it."[12] Drag racing had become a spectator sport; there were thousands of people in the grandstands. And often as not there would be a thousand dollars posted for the day's "top eliminator." As I suggested in the introduction, however, the shock would have been mild compared to what it would have been for someone who was absent for an extended period and then attended an event that was part of NHRA's Winston Championship Series.

Forty years from its beginnings at Santa Ana and Saugus, Salinas and New Jerusalem, the drags would be transformed into a lavish spectacle with full-blown television coverage, plush suites for sponsors, and multimillion-dollar gates at a nationwide series of four- and five-day events. It would have produced a stunning theater of machines, quite as ingenious in sum as any technology ever contrived in isolation from government patronage. Drag racing would have superstars, prophets, and a genius or two, with a few old-timers still on the scene as sages—Pappy Hart, Bob Cress, and Lou Baney among them. Fran Hernandez would have gone on to a big-time career at the Ford Motor Company. Vic Edelbrock, Jr., and Ed Iskenderian's sons would be running large-scale facilities for producing parts for the so-called automotive aftermarket.

Wally Parks first sat down to recount drag racing's history in the 1960s, and it has been related several times since, mostly as a tale of ever better performances (both technologically and theatrically), ever bigger rewards, larger gates, higher levels of "professionalization," a tale of incredible growth. To read *National Dragster* week after week, year after year, is to bathe in a saga of unilinear progress. This is a compelling saga but probably not a great deal different from that of many other activities that became something much more than anyone had anticipated at the outset, including many sporting activities. People like Hart, and like Parks, who founded the National Hot Rod Association in 1951 as a "semi-social car club" and then built it into a position of drag racing dominance that approached absolute, may be forgiven a tendency to muse about never having dreamed "it would get as big as it did." Yet a mere chronicle of growth is less meaningful, and much less interesting, than an analysis of differing concepts of purpose.

We know that there is no intrinsic logic to technological change, that it lacks imperatives of its own, although certainly not its own power. We know that technology is affected by "the politics of design." This provides an extremely valuable conceptual tool, but assigning agency to contests over power is an exercise that readily slips over into economic determinism. As a matter of perspective, we need to keep in mind an observation of Eugene Ferguson's from which I have already quoted a phrase: "To plumb the murky depths of human motivation with measuring rods precisely calibrated in economic terms is to miss the strong romantic and emotional strain in the narrative."[13] Along with everything else, drag racing has always had its romantic and emotional strain. Ferguson has also written that, "if we fail to note the importance of enthusiasm that is evoked by technology, we will have missed a central motivating influence in technological development."[14] Enthusiasm has been one of drag racing's primary engines of change.

*I*n 1950 the situation was this. A competitive activity involving a technological device, the motor vehicle, had been formulated and set into a rudimentary institutional context. All evidence indicated that the device was certain to change. So was the context, and change there was likewise certain to change the technology. Surely nobody would have thought about inventing, or at least about "organizing," something like drag racing if America had not been the kind of place it was in the postwar years and if young American men, a lot of them anyway, had not felt the way they did about cars. After noting that the invention of the automobile preceded any general perception of its necessity, George Basalla writes that "the artifacts that constitute the made world are not a series of narrow solutions to problems generated in satisfying basic needs but are material manifestations of the various ways men and women throughout time have chosen to define and pursue existence."[15] The technology and the institutional structure of drag racing both underwent a complex elaboration between 1950 and 1990, and my aim in this book has been to address that elaboration in terms of the way different people thought about matters of purpose, that is, how they chose "to define and pursue existence."

Initially, many people saw purpose solely in terms of devising machinery for getting from point A to point B fastest: Dick Kraft's Bug and Chet Herbert's Beast exemplify responses to the challenge as so defined. Others wanted that kind of performance *and* something else. Maybe they wanted something showy, not crude like the Bug and the Beast, something admirable for its polish as well as its speed. Some of them wanted a car in which they could still cruise the streets or one that at least looked like it could be driven on the streets, the idea being to preserve what marketing people would later term "product identity." Or perhaps they wanted a car that did not appear to be anything special at all, a "sleeper" to surprise the unwary adversary.

Some people cared more about flaunting novelty than about deception, but, retaining a "lakes" mind-set, they might also care more about the speeds they could clock than about winning races. It is important to keep in mind that speed and elapsed time are independent variables and that a machine capable of "big numbers" at point B was not necessarily quickest from start to finish. Speed was certainly a measure of "performance," and the best bragging rights long tended to inhere in the answer one could give to the question "How fast will it go?" Some people wanted to be left free to pursue what technology they would, irrespective of safety factors; others believed that safety should always be the overriding consideration in all matters of design and construction. With respect to the machinery itself, then, concepts of purpose were a matter of opinion.

Regarding the institutional structure of drag racing, power brokers like Wally Parks saw an opportunity to establish their authority to make rules and confer or withhold sanction according to their own precepts. Promoters like Hart thought in money-making terms, and so they sought to encourage the widest possible participation and to draw spectators as well as participants. After a while some competitors also began to consider the possibility that drag racing could be a way of making money. If people knew the secrets of superior "performance"—be it technological, theatrical, or some combination of the two—should they not be able to command a tangible reward? Don Garlits and Shirley Muldowney, among many others, perceived drag racing as a means of reinventing themselves in a novel context, one in which they were defined as professionals. But most people remained content to pursue challenges they conceptualized apart from any likelihood of profit. A lot of them simply liked proving the value of pragmatism. There is no doubt about drag racing being socially constructed, yet different cultures were "embedded" in different technologies, different technologies in different cultures.

José Ortega y Gasset has defined technology as "the production of the superfluous." Historically, there have been many technologies that were "merely flights of the creative spirit, materialized fantasies, projections from the realm of ideas into the real world." A splendid example is the mechanical clock, whose initial appeal at the beginning of the 1550s had nothing to do with utility or practicality; a clock, rather, "was a wonder of inventiveness, a triumph of craftmanship, an example of the particular beauty of machinery."[16]

Likewise, in the early 1950s another "materialized fantasy" was being elaborated. Drag racing served no practical purpose, and it produced profit only for those artisans and merchants who supplied hardware and for the impresarios who provided a stage upon which to perform. In 1990 it still served no practical purpose, but it was deeply embedded in commercial affairs, and for some people it yielded empowerment, political reward. For others the rewards still lay in realms we may call, after Ferguson, "romantic and emotional." At the beginning this was true for almost everyone. By going back to 1950, then, one is presented with an opportunity to examine the emergence of a new technology in a context almost entirely devoid of any motivations outside those that can best be termed artistic and existential. My initial focus will be on people who thought about purpose mostly in terms of getting from A to B most efficiently, as they conceived the term. What is important to remember here is that building a better drag racing machine was good for personal satisfaction and for status within a marginalized subculture, but it paid few dividends otherwise.

2

STAGING

We started with a clean slate. All that was needed was for the car to start and with luck stop afterwards, but that wasn't mandatory and many didn't.

DON JENSEN, 1989

orty years after drag racing began in 1950 its machinery took many forms. The most distinctive, the top-fuel dragster, had a long, slender frame of thin-wall tubing, with abbreviated body paneling and two inverted airfoils, one hugging the ground in front, the other high in the air thirty feet aft. An aluminum V-8 engine with a Roots supercharger and fuel injector was positioned ahead of the rear axle, the driver in a protective cage ahead of the engine. Next there were the funny cars, or fuel coupes, as some people preferred to call them. These had essentially the same power plant and the same sort of chassis, but they were less than half as long, they had enclosed bodies, and, most important, the relationship of the engine and driver was transposed from the dragster configuration. While they had cleaner aerodynamics than top-fuelers, they were also heavier: In 1990 the National Hot Rod Association stipulated a 2,175-pound minimum weight (including the driver) compared to 1,925 pounds. Hence, fuel funny cars were not quite as quick from start to finish; they took about a quarter of a second longer to cover the quarter mile, clocking 5.20 elapsed times compared to 4.90s. Their speeds were close, though usually a bit slower, 270s compared to 280s.

Different versions of both types of machine ran under a set of constraints aimed at keeping costs down by restricting competitors either to nitromethane or a supercharger. They could not use both. Ordinarily, the choice was to keep the blower and use methanol for fuel. In these so-called alcohol classes funny car and dragster marks were almost the same, the best of both types clocking 230s in around six seconds.

Then there were the pro stockers, which, according to the NHRA rules, could have no supercharger and nothing in the tank except gasoline. The rule book further stipulated that pro-stock engines had to be equipped with carburetors (no injectors) and displace no more than 500 cubic inches (the latter provision also applied to both fuel classes). Although they could have a frame fabricated from thin-wall tubing, they had to retain something close to the original configuration of a late-model Detroit auto and had to weigh at least 2,350 pounds. Thus handicapped, their best marks were around 190 in a little more than seven seconds.

Top fuel, funny car, and pro-stock constituted the "professional" tier; the two alcohol classes were the top-level "sportsmen." Arrayed around these faster ranks there were all those classes designed (as the NHRA rule book put it) "to accommodate the wide range of vehicles that are suitable and available for competition." This system enabled participation at almost any level of mechanical skill and financial outlay right on down to automobiles that were required to be, nominally anyway, as they had been delivered from the maker. The slowest record holders were in the 17-second, 80-MPH range.

Though close to a formula-*libre* realm compared to the other classes, the two types of fuelers were subject to plenty of restrictions under NHRA rules. For example, a dragster could not have a wheelbase shorter than 180 inches, and a funny car's could not be longer than 125, while its overall appearance had to mimic (abstractly, at least) that of a production auto no more than five years old. Equally germane were matters of convention, the constraints imposed by what historians of technology call normal design practice.

There was nothing in the rules that said a funny car could not be set up with the driver in front of the engine or a top-fueler with the driver behind, but none had been built that way since the early 1970s. There was nothing that said a dragster could not have full streamlining or gain the down-force essential for stability and traction at high speed by means of ground-effects devices rather than airfoils; those tactics were tried from time to time and sometimes appeared promising, but they remained outside normal design. There were enthusiasts who prized novelty as an end in itself, and there were periods when the quest for novelty seemed to be a widely shared aim. There were always anomalies. Nevertheless, design tended to normalize around "combinations" (i.e., systems) that were perceived as ideal for the state of the art. Racers were aware of all sorts of possible alternatives but tended to be conservative, out of financial necessity if not by instinct.

At the beginning the situation was different. The situation begged for conceptual novelty. Yet it took several years to break free of inherited frames of reference, to devise a form, any form, more closely congruent with the purpose of getting quickly from A to B. And it took several years more for one such form, denominated the slingshot dragster, to become a paradigm. The shift from slingshots to the mid-engine configuration might appear to have taken place quite suddenly, in a period of a little more than a year in 1971–72. Actually, there had been mid-engine dragsters even before slingshots. When the funny car hit the scene in the mid-1960s it was not a novel configuration either; rather, it perpetuated the tradition of full-bodied hot rods, deuce coupes and the like, which had been a drag strip staple from the start.

In another sense, however, the funny car was truly revolutionary. Funny cars borrowed dragster technology and were just as loud and fierce, but they *looked* something like everyday cars. That combination of attributes made them exceedingly popular with spectators and, later, with commercial sponsors as well. Initially, there was scant expectation of better performances in a technical sense (i.e., of covering the ground quicker) but tremendous promise of better performance in a theatrical sense, hence boosting box office returns. The emergence of the funny car marked the moment when commercial considerations began driving drag racing as directly as technological enthusiasm, although designers would subsequently rework these simulated automobiles into a form well suited to high performance in a technical sense as well. Accomplishing that was a matter of incremental change, an approach that became all the more common as escalating costs rendered it riskier to attempt any sharp departure from normal design.

By 1990 the cost of making an all-out quarter-mile run in a fuel funny car or dragster was well into four figures, even if there was no unusual attrition at all. And, with radical departures, chances of attrition were greatly magnified. With 4,000-horsepower engines even the tried-and-true was right on the ragged edge of havoc. Racers devised wonderful euphemisms for breakage—"tossing" a rod, "tweaking" a supercharger, "smoking" a piston, "ventilating" this or that—but their benign sonority did not mitigate the violence and costliness of any major malfunction. Some competitors focused on programs that were purely experimental, mostly people who relished an image of nonconformity. Some of the most comfortably financed racers were able to try unusual setups in private testing sessions, "renting" track time at midweek. The others simply could not afford a lot of experimentation.

Finances were always a constraint, of course, even at the start. Then there were no corporate minions willing to buy a piece of the action and few individuals who would play the role of angel just to share in the excitement. But limited budgets were not as daunting as the basic technical dilemma: Curious aberrations aside (Colonel John Paul Stapp's rocket-sled adventures in New Mexico would soon be famous), nobody had ever thought very much about designing a vehicle having acceleration as its *primary* purpose. Hot rodders and dry lakes veterans, even the keenest minds among them, had only a hazy sense of the full range of interrelated factors affecting optimum (technological) performance. What had happened was that the activity had been socially constructed before anyone understood how to design a machine well suited to that activity technologically.

To be sure, there were purposes that were never technological. For the community at large these were in the realm of social control: Organized drags, at least in theory, would diminish the incidence of street racing and thereby serve to mitigate a perceived menace. For those who participated directly, drag racing served to channel enthusiasm and vent competitive impulses in a lawful setting while at the same time fostering a shared sense of self-esteem among individuals who did not necessarily hold high station in the community at large.

Drag racing also had commercial purposes. While some strips were begun as nonprofit operations, others were managed by entrepreneurs who expected to make money from their efforts. And for manufacturers of "speed equipment" and their distributors, speed shops, drag racing offered a significant potential for expanding markets. Manufacturers and distributors had been promising improved performance on the dry lakes, at Bonneville, on dirt tracks—and on the street. But Santa Ana had been going for only a short time when Chet Herbert began running ads in *Hot Rod* telling how a certain racer had picked up 10 miles per hour after installing a Herbert roller-tappet camshaft; likewise, Earl Evans of Whittier, California, told how the aluminum cylinder heads and intake manifolds he designed, cast, and machined excelled on the drag strips as well as at the lakes, and ads placed by the Clark Header Company in nearby Downey featured Joe Mailliard's '34 Ford coupe running its exhaust equipment at Santa Ana.

By 1990 drag racing's upper crust looked to corporate sponsors to help them cover expenses, and some competitors were paid for endorsements. In the early 1950s there was scant hope of any monetary compensation at all. Oh, there were the war bonds, and certain racers were getting parts gratis in return for the commercial use of their names. Sometimes there was a small guarantee for some popular performer; Bob Cress, the manager at Kingdon, recalled inducing southern Californians like Don Montgomery to come up north with the promise of lodgings and dinner.[1] For most drag racers, however, all expenses were out-of-pocket. The initial efforts to devise vehicles to excel at acceleration may be seen as a classic exemplification of what Paul Goodman once called "pure technology."

Finding out what to do with that "clean slate" was not easy. As designers began to discern answers to simple questions, more troublesome problems emerged. "We can never select the one result we want to the exclusion of all others," says David Pye, an observation so keen it bears repeating. For example, to improve traction one could "lock" the differential gears—that is, render them nonfunctional by welding them together—but one then had to confront a critical trade-off in terms of directional stability. This sort of dilemma, setups that solved one problem while introducing another, was scarcely unique; it had been part and parcel of the inventive process for every new form of vehicle right on down to supersonic aircraft, which were in their developmental phase at almost the same time—but, of course, with entirely different conventions for funding.[2]

Compared to what could be seen flying high above Muroc, the first vehicles intended exclusively for drag racing looked plenty crude. Even *Hot Rod* had to concede that they gave "a first impression of hasty construction."[3] Partly this was due to budgetary constraints. Partly it was because, as Don Montgomery explains, "the most active street rodders were quickly adapting to racing at the strip, and their rapid modifications were making their cars ex–street rods. New cars were often built in one to two weeks. . . . Nice paint jobs and chrome were luxuries, especially when you were making major performance changes weekly."[4] In five years, from 1950 to 1954, speeds of the fastest machines improved about 25 miles per hour, to about 145. Speeds would improve more than that during a five-year period in the 1980s, from the 250s to the 280s, when diminishing returns should have been a major constraint. At first, however, almost everyone merely addressed the obvious, and the returns diminished rather quickly.

When *Hot Rod*'s editor, Wally Parks, began taking serious note of drag racing at the beginning of 1951 one of his first articles described an event staged by the Northern California Timing Association at Salinas. At least one other racer had been following Tom Cobbs's lead in marrying a GMC truck blower to a Ford V-8, and, with carburetor kits sold by Edelbrock, several were using nitro. Harold and Don Nicholson's fuel-burning roadster, up from Pasadena, was "clearly way beyond the rest of the field," but it suffered irreparable engine woes and was gone before the final eliminations. Dick Fullmer, a local, clocked 119.92 with a 315-cubic-inch flathead engine on "a roller skate." Having had weight "chopped by every means imaginable," it was said to have come across the scales at only 975 pounds.[5] One clear-cut route to enhancing performance had again been underscored, the same route taken by Dick Kraft. But it was not anything very profound: Who in the world didn't know that inertia is more readily overcome with something light than with something heavy?

In 1990 what was striking about the demeanor of most of the best top-fuel drivers, drag racing's elite, was their air of implacable calm. These people were not daredevil kids; their median age was, amazingly, close to fifty. In contrast, early drag racers were nearly all on the youthful side and not much attuned to the virtues of calm—nor moderation: With most things they tended towards the view that "a lot is good but too much is just enough." For example, in shedding weight, one could go far beyond taking off the headlights, bumpers, fenders, and windshield; one could remove the radiator, the front brakes, the firewall, the entire body. With a hole saw and torch one could make Swiss cheese out of just about every part of the chassis. One could even take an engine block and cut holes throughout the water jackets.

"Weight paring procedures have now attained the status of a fine art," *Hot Rod* reported a little later on.[6] Actually, the art was not fine at all. Roger Huntington, an automotive engineer who wrote frequently for popular consumption, had to warn zealous lighteners to cut holes only in the web of frame rails and I-beam axles, not the flanges.[7] For several years some of the fastest dragsters were conspicuously devoid of roll bars, an add-on that was not essential to a purist view of purpose. Radical techniques of minimizing poundage entailed a crucial trade-off. Yet racers would probably have remained amenable to the risk had not NHRA, "dedicated to safety," interceded. In 1990, even with the availability of a symphony of lightweight alloys and polymers such as nobody dreamed of using in the 1950s, the rules required that a top-fueler weigh a lot more than it would need to weigh if there were no margin of safety, maybe twice as much. It was all a matter of one's purposes and priorities. There was no doubt that stripping weight to the bare essentials could help (up to a point) in getting from A to B more efficiently, if one thought of efficiency in precisely those terms.

Naturally, teasing out more power could serve the same end. The hot rodder's standard tricks included increasing compression and displacement, changing valve timing, multiplying carburetion, juicing up ignition voltage, polishing intake ports, and smoothing the flow of exhaust gases. The quickest route to more "soup" involved fuel, but one had to be privy to the proper secrets. As early as the 1920s, a furnace oil had been patented containing nitrobenzene for "higher heating efficiency." In the 1930s Standard Oil patented fuels with nitrocarbons designated as "igniters" for increasing thermodynamic efficiency. The prewar German and Italian Grand Prix teams, Auto Union and Mercedes Benz, Alfa Romeo and Maserati, blended fuels containing small proportions of nitrobenzol. When patented (hence, public knowledge), the nature of fuels was disclosed with reference to things like "complex, symmetrically branch-chained esters and ketones," language beyond the ken of nearly all hot rodders, who often lumped everything other than gasoline and alcohol under the rubric "rocket fuel." With their time-tested trial-and-error tactics, however, they began delving into chemistry, and they became particularly enamored of nitromethane.

Nitromethane was a monopropellant that would ignite even in the absence of any other oxygen. There were several critical considerations when using it in an internal-combustion engine, foremost being an extremely rich mixture—2:1 or even 1:1 air-fuel ratios instead of the 12:1 or 15:1 ratios typical of gasoline mixtures. Among hot rodders evidently the first to understand this were the men at Vic Edelbrock's, who were running a nitro blend in a Ford-powered dirt-track machine at Gilmore Stadium as early as 1947. Nitromethane provided an edge the day that drag racing began two years later. Other early initiates included Holly Hedrich and Joaquin Arnett, both from the San Diego area. For those few who were conversant with the engineering literature formal papers on nitromethane as a racing fuel were starting to appear by 1950.[8]

Nitromethane actually has a very low thermal efficiency, only about 10 percent, pound for pound much less than gasoline, but when burned in huge quantities it could make 300 HP where there had only been 200. As with paring weight, however, there was a major trade-off. Even though less prone to preignition and detonation than the heavier nitros such as nitrobenzol and nitropropane, it could do plenty of damage if the circumstances were wrong, damage ranging from burned pistons to bent connecting rods, broken crankshafts, and total ruination. In drag racing, for which engines needed to be run only briefly, fuel efficiency (as the term is usually understood) was not a concern. Still, nitromethane entailed a direct cost entirely apart from its destructive tendencies. A gallon per run, at five to ten dollars per gallon, soon became the common rate of consumption (in the late 1980s ten or fifteen gallons at about thirty dollars per gallon was the norm). Even under optimum circumstances—no breakage—nitro was monstrously expensive. Yet nitro was also dubbed "the poor man's supercharger," and by 1951 Santa Ana was reporting speeds of 130 MPH by racers using "pop" in engines that did not appear to have entailed a heavy investment overall.

To be sure, reported speeds were often questionable. Aside from whether or not the timers were accurate and the distance was really 1,320 feet, skeptics pointed out that at Santa Ana, as at most strips, competitors were not flagged off from a dead stop. Rather, the procedure called for rolling starts, ten to thirty feet, a carryover from street-racing practice. The idea was, first, to relieve strain on drive-trains: Ford gearboxes and rear axles were notoriously fragile. Perhaps more important, rolling starts helped minimize "fishtailing" or "skating," a commonplace occurrence with the faster cars, with their differential spider gears deliberately rendered nonfunctional so that one tire would not tend to spin while the others stood still.

At first most competitors paid scant attention to such niceties as directional stability, "handling," as it was called. Getting badly crossed up, occasionally at least, was considered inevitable. Yet there was clear-cut evidence about how this situation might be alleviated. At many strips the victorious machine on any given Sunday would often be a motorcycle, not anything with four wheels. The Beast had a long reign at Santa Ana. At Kingdon three of the nine competitive classes were for motorcycles, and three of four meets at the beginning of 1952 were won by a "flying Harley Davidson stroker."[9] Given the import of weight, the suggestion seemed warranted that the best configuration for maximum acceleration might well be minimalist, only two wheels, not four (or six or three, ideas both tried periodically). But the superior acceleration of motorcycles was only partly due to a favorable power-to-weight ratio; perhaps even more important was an inherent facility for going straight.

A car would tend to veer from its initial trajectory as soon as one rear tire lost adhesion, and only an adept driver could correct without skating. The greater the distance between the rear wheels, the greater the veer; the less the angle of front-end caster (i.e., the closer the kingpins were to vertical), the more the danger of overcorrecting. With a motorcycle there was only one driving wheel, and the forks provided a high angle of caster. Ultimately, dragster designers would perceive what was to be gained by radically narrowing the rear tread and, in front, maximizing caster. By traditional precepts neither setup looked "right." Unless people could break free of

Art Chrisman is seen getting under way at Santa Ana; the tower marks the finish line. This photo dates from 1955, after Chrisman had switched from flathead Ford/Mercury engines to a hemispherical-head Chrysler Firepower. (Courtesy National Hot Rod Assn.)

convention, however, it would remain an open question whether the four-wheeler's directional instability would not render the cycle a better basic design for getting quickly from A to B.

*A*t Paradise Mesa, a strip southeast of San Diego which opened in 1951 on a site once notorious for "chaotic illegal sprint bashes," competition was about equally divided between two- and four-wheeled vehicles.[10] At the end of a day's racing it was often a cycle that emerged on top. Still, there were four-wheeled machines competing which looked a little like something purposefully designed, as opposed to having merely been "stripped to the bare essentials." These machines were classified as "modifieds," a traditional dry lakes designation for a one-seater built on a narrowed frame. One Paradise Mesa regular was a young man from a family of hot rodders, Art Chrisman, twenty-two years old in 1951. In an effort to stay ahead of the cycles Chrisman took a prewar modified, lengthened it 18 inches, relocated the engine a foot aft, and moved the cockpit right up against the cross-member. He did not change the rear tread width or the caster (the front and rear axles remained stock Ford), but he switched the fore-and-aft weight distribution to 30/70, just about the reverse of a conventional automobile. Arguably, this was the first dragster, and, amazingly, it was still around in running condition in 1990, a prime icon of the "nostalgia" movement that had flourished for a decade or more.[11]

Partly the nostalgia was for places like Paradise Mesa, lost, like most of the early southern California strips, to land developers before the end of the 1950s. (Santa Ana closed when the airport operation grew too busy to permit this diversion.) It is worth mentioning, however, that the promoter at Paradise Mesa, a San Diego car dealer named Fred Davies, was way ahead of his time in terms of what people would later call "presentation." The accuracy of times was beyond dispute, the equipment having been devised by J. Otto Crocker, a San Diego watchmaker who was enthusiastic about precision timing equipment; Crocker had been the chief timer at the Bonneville National Speed Trials since their inception in 1949. The starts were truly standing starts, and the trap for timing top speed ran from 64 feet inside the 1,320-foot mark to 64 feet outside (many strips placed the first light *at* the quarter-mile mark). Nobody ever questioned times announced at Paradise Mesa (by a seasoned sportscaster named Stan Bryan). And in a day when everything at any drag strip had a jerry-built look Davies dreamed of "permanent timing stands, safety barriers, underground wiring circuits for field communications."[12]

Such amenities were actually a long way off anywhere. Still, what Davies did manage to do, week in and week out, was stage exceptionally good shows, and it was at Paradise Mesa that many of the first drag racers emerged who became household names to hot rodders nationwide—Chrisman, Hedrich, and in particular a club known as the "Bean Bandits," whose members followed the lead of their honcho Joaquin Arnett in wearing unpolished cotton whites and straw hats. Although there is a suggestion that they invited condescension—"clown princes," they were called—these young men would emerge as drag racing's first folk heroes, the first participants to perceive the nuances of the word *performance*.[13]

Some of Arnett's early efforts were crude, but his machinery was undergoing continual transformation, and his originality as a designer and his craftsmanship were both quite evident in a car he debuted in 1953. Eventually, every single component of a dragster would be produced by specialty manufacturers, but in the 1950s parts were ordinarily salvaged from junkyards. Hence, what was most impressive about Arnett's work was the degree to which he built from scratch. No narrowed Chevy frame (like Chrisman had, like Garlits would use for "Swamp Rat I" in 1955), no reconfigured Ford body (like the vast majority of racers used)—rather, the chassis had been welded up from steel tubing, and the body was hand formed. The whole rig was extraordinarily tiny and light. In addition, Arnett was adept at extracting horsepower from a Ford flathead, even without enlarging displacement a great deal, one of Chrisman's tactics. He ran a 50/50 blend of nitro and methanol: "It's easier to mix that way," he explained. "A gallon of this and a gallon of that."[14]

This machine represented an important step in terms of fabrication techniques and maximizing the power-to-weight ratio, but there was another design factor that was even more important, namely, getting hold of the track, or traction. This was a more complex challenge, and neither Arnett nor anybody else had as yet met it particularly well. Partly it was a matter of redistributing static weight, so the placement of the engine closer to the rear axle helped. Partly it was a dynamic matter. One way to address this was to *raise,* not lower, the center of gravity.

Traction also depended heavily on factors that were essentially beyond the control of anybody like Arnett. With tires a "slick" tread worked best—something somewhat counterintuitive, like raising the center of gravity—by putting the maximum amount of rubber in contact with the pavement. But no retreader produced tires that racers regarded as much better than "bald" street tires. Following the Nicholson brothers, other racers tried mounting four rear tires, using rims welded together in pairs—notably, a Kingdon regular named Ralph Lynde, from Los Gatos, and Al "Romeo" Palamides, from Oakland. Naturally, traction was also affected by the condition of the race courses themselves. There were a number of variables, the two most critical being how they were paved and how recently. Concrete offered better "bite" than asphalt if the surface was free of oil but could be much worse if it was not. Old pavement of any kind was usually slippery.

The best indicator of bite was the sort of elapsed times that could be clocked from start to finish, but strips could not always be compared because many of them lacked elapsed time clocks. In California many partici-

Joaquin Arnett's "Bean Bandit" is seen here at Paradise Mesa on July 26, 1953. Race queen Bonnie Wallace presents Arnett with a trophy for top time of the day as his crew shares a laugh. (Hearst Collection, courtesy Department of Special Collections, University of Southern California Library)

J. Otto Crocker is seen here at El Mirage with the clocking device also used at Bonneville. Seated next to Crocker is Jim Lindsley, still an SCTA stalwart in 1990. (Photograph by Walt Woron, courtesy Museum of Drag Racing)

pants were not yet concerned with the reality that the key technical parameter was elapsed time, not top speed. Speed was what the lakes and Bonneville were all about, and old frames of reference were not easily cast aside. To repeat, a lot of pride inhered in big numbers, and, of course, these did provide a clear-cut measure of performance, even at the drags.

In parts of the country in which there was no tradition of top-speed timing drag racers put more stock in elapsed times, and the day would come when competitors from back East would begin challenging the Californians. But it would have been no contest at first. When legalized drags first came to Denver in July 1951 the best speed was barely over 90 miles an hour. That was at high altitude, where an engine had trouble breathing, but clockings were similarly unimpressive at Caddo Mills, Texas, and Mound City, Missouri, likewise opened in the summer of 1951. The next summer, when racing began at Akron, Ohio, and Half Day, Illinois, top speeds were 91 and 105, respectively.

Soon drags were being staged in Salt Lake City, Topeka, Colorado Springs, even in British Columbia and New England. But virtually nobody was yet building machines strictly for drag racing, whereas more and more of these were showing up in California. At Kingdon, Al Dal Porto, a Linden farmer, had turned 118. Ralph Lynde ran 119 at Fresno's Hammer Airfield. Otto Ryssman ran 119 at Paradise Mesa, Art Chrisman almost 122. At Santa Ana, Paul Leon topped 130. Some of these machines had been lightened "by extreme methods," and in the summer of 1952 several similar rail jobs turned up at a meet in Kern County sponsored by the "Smokers" of Bakersfield, later to establish a lasting place in drag racing annals by luring people from back East to come out and test their mettle.[15]

Famoso, site of the Smokers' annual U.S. Fuel and Gas Championship, became one of the most famous strips in the country, nearly as famous as another that opened that same summer in Pomona. After having staged races previously in nearby Fontana, the Pomona "Choppers"—led by a Pomona police sergeant named Bud Coons and with the good offices of Chief Ralph Parker and the city government—leased the parking lot of the expansive Los Angeles County Fairgrounds. At the first event there was a showdown involving Arnett and Bob Rounthwaite, from Glendale. Compared to Arnett's creation, Rounthwaite's was no beauty, a concession made explicitly in dubbing it "Thingie." It was, however, more functional for getting from point A to point B. It had no body paneling, not even a firewall, but more important the frame (tubing, like Arnett's) was considerably longer and the GMC engine was mounted high in front to elevate the center of gravity and enhance dynamic weight transfer.

Most dragsters used only two gears, leaving the starting line in second, with a shift to high partway down the strip, when the revs peaked. With its low frontal area Arnett's machine really stormed once in high, but its short wheelbase and weight distribution made it squirrelly getting out of the

A major break from convention, "Thingie" is seen here at Saugus, with designer Bob Rounthwaite at right. At left is Tom McLaughlin, like Rounthwaite a leading competitor at El Mirage in the late 1940s and early 1950s. This machine was later sold to Jake Smith, who switched from a GMC engine to a Mercury and added some skimpy but colorful bodywork (see page 28). (Courtesy Don Montgomery)

chute. Even though Arnett turned almost 130, he was outrun by Thingie, which was not as fast but powered away much quicker and straighter.[16] Rounthwaite's stark extremes had provided the most cogent evidence yet that the purpose of getting quickly from A to B required fundamentally novel design. He did not point the only conceivable direction, but he clearly suggested what was to be gained from breaking altogether free of convention.

By 1953 a matter of nomenclature was almost settled. In January, when *Hot Rod* ran a picture of Chrisman's machine on display at a Los Angeles car show, it was called a "glittering example of a 'dragster.'"[17] The next month the Bean Bandits scored a first, a photo on the cover of *Hot Rod*—now flirting with a half-million circulation—and the word *dragster* was no longer in quotes. *Dragster* it would be, though *rail job* would remain part of the idiom long after stock frame rails had been entirely superseded by lightweight tubing.

Arnett, Rounthwaite, and Chrisman reigned when Paradise Mesa staged its second annual Drag Festival. In preliminaries Arnett turned 127.40, Chrisman and Rounthwaite both turned 125.59. Arnett defeated Chrisman during early elimination rounds, but in the final he was once again "soundly trounced by 'Thingie' as soon as starter Paul Wallace dropped the flag."[18] Clearly, Thingie worked better for its intended purpose than the other two, however "right" they may have looked to the casual observer. By one means or another lakes racers usually sought a low profile. Rounthwaite, a bodyman, had designed one of the sleekest chopped coupes ever seen at El Mirage, but he could also envision what might be "right" in an entirely different context.

Also at the same winter event was a dragster with the cockpit located ahead of the engine. Its performance was only middling, though it clearly seemed capable of improvement. As for the basic concept, however, nobody could have then said whether or not it was "right." A few years later there would be evidence that it might be, but in the 1960s consensus would hold that it was altogether wrong. In 1990 racers regarded the mid-engine configuration as right in a top-fueler and wrong in a funny car, although some of the savviest innovators had suggested that they would like to test both conventions. As it was, the two radically disparate designs remained nearly equal in their facility for getting from A to B.

"Plenty of people still believe that 'purely functional' designs are possible," writes David Pye, "and believe that they themselves produce them, what is more! But none of them has divulged what an analysis of a function looks like and what logical steps lead from it to the design."[19] All you will get is talk about purpose, a matter of opinion. Some drag racers set a lot of store in the speeds they could clock, and speeds were definitely worth attention, especially since the concept of elapsed time was not so readily comprehensible anyway. While Arnett could not beat Rounthwaite, in 1953 his machine began pushing 140 miles per hour—a time still noteworthy at El Mirage, with better than a mile to get up speed—and that was what garnered him his feature in *Hot Rod*. Nevertheless, Arnett was eager to win races, too, and soon he had a "Mark II" version under construction, something a little more like Thingie.

Over the years drag racing technology always entailed a seesaw between power and traction; most of the time racers were able to make more power than they could effectively put "to the ground." On the mythic day that drag racing began Hernandez may have won the race because of his superior setup, or Cobbs may have lost it because he spun his tires. Although one had a supercharged engine burning pump gas and the other was naturally aspirated on a 3:1 blend of methanol and nitro, both probably had about the same horsepower.

Both had flathead V-8s of the sort debuted in 1932 and produced for twenty-one years for Fords and fourteen (1939–53) for Mercurys. A few flatheads were factory rated at 60 horsepower, most at 85, the last ones at 100. With basic skills and a few hundred dollars, any backyard mechanic could double their power by changing certain components. At greater expense blocks could be over-bored and crankshafts stroked, but there was a practical limit on displacement; to go much beyond 300 cubic inches was to risk catastrophic failure. While the valve-in-block port design did not permit optimum breathing, someone with the money to spend could buy a set of "Ardun" overhead valve heads, the brainchild of Zora Arkus-Duntov, who later designed engines for Chevrolet.[20] Some people knew how to increase a flathead's horsepower three or four times, but that took a knowledge of such esoterica as fuel and supercharging. There is evidence that an extraordinary supercharged flathead built by Don Yates put out 435 HP as installed in Chrisman's dragster in 1955, but that was decidedly on the ragged edge: There were pathways torched through the aluminum cylinder heads where they mated with the cast-iron block. Reliability had been traded away totally.[21]

In the early 1950s, besides Ford/Mercury flatheads, one would occasionally see something else in a drag racing machine, perhaps a flathead Cadillac V-8, much bigger but also much heavier, and sometimes in-line engines as well, even pre-1932 four-cylinder Fords, especially with overhead valve conversions. Both Rounthwaite and Don Montgomery ran six-cylinder overhead valve GMCs, and Montgomery also tried a Buick straight-8, as did Jarvis Earl, who had a mount similar to Thingie.[22] Any such engine was superior to a flathead in terms of volumetric efficiency. Cadillac and Oldsmobile had come out with wedge-head overhead valve V-8s in the late 1940s, Chrysler with its hemispherical-head ("hemi") Firepower in 1951, and later Buicks, Pontiacs, and Chevrolets, even Studebakers and Lincolns, had V-8s. Not all of these engines stood up well to the rigors of nitro, but every one had a higher factory horsepower rating and more potential than anything previously available. Overhead valve V-8s were turning up frequently at Bonneville by 1953 and 1954, but except for Ardun conversions they remained rare at the drags. At this stage dragsters simply did not need a lot more power, and the tried-and-true flathead remained the predominant power plant, even as every Detroit manufacturer had retooled for overhead valves.

A dragster is a technological system, quite a complex one, actually. System entails harmony. Racers had made headway in certain realms, but not much with respect to traction and directional stability. Harmonizing that part of the system was going to entail a basic reconceptualization. Several possibilities would emerge in the next few years as designers felt their way along. The quest would not take place, however, in a context that was purely technological. Just as much of the initial impetus had come from nontechnical realms, the technology was not headed anywhere in isolation.

First of all, during the same period that the people who built dragsters were seeking to normalize design the National Hot Rod Association would

Jake Smith's machine, according to a February 1953 item in a Los Angeles paper, had one purpose: "To get from here to there before anything else." But the paint job suggests that Smith regarded part of the purpose as entertainment. (Hearst Collection, courtesy Department of Special Collections, University of Southern California Library)

begin to assert authority to confer sanction. Although NHRA had not been founded with the idea of fostering organized drag racing, by 1953 this activity was ripe for picking by an organization with a leader of genuine vision. "Sanctioning" included providing for insurance and making rules. It also entailed publicity and promotion, and on that score NHRA's initial success was altogether phenomenal.

The first NHRA event was staged at Pomona on an April weekend in 1953, exactly four years after Goleta. On Saturday the turnout was two or three thousand, twice the normal Sunday gate at local strips; then, the next day a reported fifteen thousand spectators "watched spellbound as the cars stormed down the asphalt quarter mile." In the final they saw Arnett in his new dragster finally get Thingie. His elapsed time was also tops for dragsters, though the quickest ET of all was turned by Lloyd Krant's Harley, 10.93, despite being 10 miles per hour slower.[23]

At the second event staged under NHRA auspices, at Paradise Mesa in

Starter Paul Wallace gives the green flag to a pair of roadsters during the running of the first event sanctioned by the National Hot Rod Association, in April 1953. Though the crowd appears shy of the figures reported by NHRA and the press, it was sizable for the time. (Hearst Collection, courtesy Department of Special Collections, University of Southern California Library)

July, Krant again had the quickest ET, and he also won top eliminator (a term that was settling into the language just like *dragster*), even though his cycle was more than 15 miles per hour off the pace set by the fastest dragster. Could motorcycle technology dominate drag racing? There had been evidence of it right from the beginning. And, if that were the case, did matters need to be rethought in light of *commercial* considerations? Motorcycles had their own enthusiasts, naturally, but their numbers were not anywhere as large as the ranks of hot rodders. Plus a lot of hot rodders found motorcycles irritating.

As *Hot Rod* put it, the Pomona meet had shown what could be done "in all parts of the country for eager and willing hot rodders."[24] It showed something else much more clearly. NHRA had provided the hot rodders with a setting, nothing more, but it had provided itself, and a lot of others, with a clear confirmation of drag racing's potential for drawing large crowds to watch the performers. Those performers had their own purposes and interests, but "the best interest of drag racing and its future prosperity" could be construed in all sorts of terms. One could consider it in terms of image. One could consider it in terms of power and profit, shades of Marx and Veblen. Any such purpose obviously required that it be "handled properly."[25]

Technological enthusiasts would focus on the dilemmas posed by shortcomings inherent in the machines they had been devising. And they would make a lot of headway. They would invent and refine a configuration that they could take from speeds in the 140s in 1954 to speeds in the 230s in 1970, elapsed times in the 11s to elapsed times in the low 6s. Long before that, however, it would be quite apparent that "the single most powerful force in the sport of drag racing" was not a mechanical craftsman like Chrisman or Arnett, not an impresario like Hart or Baney, not a manufacturer like Iskenderian or Edelbrock. It was Wally Parks, founder and president of the NHRA.[26]

In the next chapter we will play out the last act of the purely technological quest for machinery for getting from A to B and in the one after that turn to the beginning of drag racing's transformation by men who associated the activity with a somewhat different set of purposes.

3

GATHERING SPEED

DRAGSTER (D) (Fuel type optional.)
. . . cars of the dragster variety are
specially constructed for all-out com-
petition.

NHRA DRAG RULES, 1956

Do we have to go fast for drag racing to
be successful?

DRAG NEWS, 1957

convenient model for analyzing the growth of design knowledge has been devised by Walter Vincenti, the premier historian of aeronautical engineering. Drawing on Donald Campbell's hypothesis about blind variation and selective retention, Vincenti posits that technologies have an early stage, at which "the knowledge sought is that of a workable general configuration," and a subsequent stage, at which "the configuration is settled and the object of design is a particular instance of it."[1] With aircraft this settled configuration was attained in the early teens. Designers then sought particular instances of it until the 1950s, when airplanes underwent a paradigm shift, to a second "normal" technology. Automobiles likewise had an experimental phase, then a first normal technology, then a second, And so did dragsters, a "particular instance" of automotive technology, the first paradigm emerging in the middle 1950s, the shift taking place in the early 1970s. Interestingly, however, the second normal technology—or second normal configuration, anyway—initially appeared in an almost complete form even before the first one and had a shadowy coexistence the whole time.[2]

As we have seen with Arnett and Rounthwaite, racers soon started fabricating their own frames. Some favored aluminum, even though welding a nonferrous metal was tricky and components might have to be bolted together. Others preferred to weld steel tubing, and by the mid-1950s a few were gaining familiarity with an alloy steel called chrome moly, which was about one-third stronger than mild steel and could be used with walls as thin as .050 inches. All such chassis, whether aluminum or steel, were not only lighter than stamped automobile frames; they also permitted more flexibility in arranging mechanical components.

As improvements in chassis design led to better traction, most racers would turn to overhead valve engines from flatheads, which had been stretched to the limit. The combination of "big inches" and a "big load" was literally explosive; with only slight modification a Chrysler or Cadillac could easily put out more power than any flathead. But a driver had other concerns. Directional stability remained so problematic that spinouts were not uncommon, and beyond that there was what the racers nonchalantly called a "flip," when a driver's life depended on the design and integrity of the structure surrounding him.

Pressed by the National Hot Rod Association, there had been some attempt to mandate basic protective apparatus such as roll bars, and, to try to ward off catastrophe, devices such as safety hubs to prevent the loss of a wheel should an axle break and scattershields for containment of disintegrating clutch and flywheel assemblies, a mishap that was becoming more and more frequent. Shrapnel could wreak havoc indirectly, by severing hydraulic lines and rendering brakes inoperative, or quite directly. In July 1955 the clutch in Otto Ryssman's dragster exploded at Santa Ana; Ryssman himself was only slightly injured, but a spectator was killed.[3] Ryssman was in and out of court for seven years.

That sort of thing was sobering to everyone, yet enthusiastic competitors tended to treat technical inspections as a game in which the object was to outsmart the inspector. They hated any mandate that meant adding more weight; hence, the lore is rich in tales of protective devices that may have appeared effective but would actually have failed to serve their purpose—roll bars fashioned out of electrical conduit, for instance. Still, most racers did have a sense that a correct knowledge of materials and skillful fabrication were essential to even minimal assurances of safety. And some of them perceived the need for an address to something more fundamental than power and weight: control and directional stability.

Part of the problem was adhesion between tires and pavement, or, rather, the deficiency thereof. Alex Xydias, proprietor of the So-Cal Speed Shop in Burbank, began marketing special "asphalt slicks" in 1953, recaps with seven inches of tread and purportedly yielding "four times the traction of a regular tire." Slick recaps were later improved by Bill Kretch's Inglewood Tire Company and by Bruce Alexander of Oakland, among others. But not until late 1957 did Marvin and Harry Rifchin's M&H Tire Company of

In a scene dating from late 1955 Glen Pengry (foreground) drives from behind the rear axle, while Tony Waters's cockpit is directly on top of the axle. One can get a sense of the different weight transfer characteristics. (Photograph by Richard King, courtesy National Hot Rod Assn.)

Watertown, Massachusetts, begin producing "pure" drag tires, M&H Dragmasters, molded from special soft compounds.[4] Tire technology would remain a perennial bottleneck, though never more so than in the years before 1957. Poor traction and attendant handling problems did, however, serve to curb any tendency toward complacency about questions of basic configuration.

Taking radical measures to lighten a vehicle without redistributing static weight could actually hinder acceleration. So could increasing power. The idea of twin engines had its enthusiasts right from the beginning, but putting two engines in front of a cockpit that was itself in front of the rear axle did not substantially alter static weight distribution, and it could exacerbate traction problems. Manuel Coehlo and Ken Droesbeke teamed up on a twin-engine machine with four-wheel drive, and others followed suit. While four-wheel drive obviated the need for concentrating weight on the rear tires, the trade-offs (increased overall weight and mechanical complexity) provided a perfect instance of Pye's observation about not being able to select the result we want to the exclusion of all others. Since the days when hot rodders had begun removing the fenders from Ford roadsters simplicity had been a widely shared ideal.

To get the engine close to the rear end the transmission could be coupled directly to the pinion shaft. Then the cockpit had to be put someplace unusual. It could be in front of the engine, and machines configured that way began turning up in the early 1950s. Or it could be on top of the rear cross-member, and that soon became a commonplace setup too. There was yet another possibility, but it entailed a conceptual leap. The cockpit could be cantilevered *behind* the rear axle. With the driver buckled in and the axle as a fulcrum, such a configuration could yield a static weight distribution of 90 percent on the rear tires, 100 percent under acceleration.

This configuration is usually dated to early 1954 and attributed to an El Monte hot rodder named Marion Lee "Mickey" Thompson, then twenty-five and employed as a pressman for the *Times.* But I trust that readers will not tire of hearing that one can find precursors to almost any such "first." In this instance Florida beaches are one good place to look, as far back as the turn of the century, in fact. In 1902 Ransom E. Olds had unveiled "The Pirate," designed specifically for speed runs on Ormond Beach, with the driver's location out behind the rear axle.[5] Fifty-one years later, in 1953, C. C. "Bill" Martin of Palatka took four-inch chrome moly tubing and fabricated a machine for drag racing on the sands of Jacksonville Beach. Traction on sand was obviously problematic, so Martin configured his dragster with

As Santa Ana's C. J. Hart watches, Mickey Thompson has moved about 8 feet, and the front tires are just barely in contact with the pavement. Compare this photo dating from October 1954 with that on page 41, taken later; Thompson has elected to have the injector stacks out in the open rather than fully faired in. (Hearst Collection, courtesy Department of Special Collections, University of Southern California Library)

Probably the earliest dragster configured with the cockpit behind the axle, and perhaps the earliest with twin engines as well, this machine dates from 1951. Carlos Ramirez, who shared with Joaquin Arnett driving duties for the Bean Bandits, is in the cockpit; Arnett is third from left; at left is Harold Miller, who crewed for various San Diego area racers until the 1970s. (Courtesy Mike Nagem and Ruben Lovato)

the cockpit all the way aft. The roll bar was ahead, the whole setup almost identical to his fellow Floridian Don Garlits's first dragster and a number that were built in California.[6]

Martin gave up drag racing in the latter 1950s to pursue a degree in mechanical engineering, then went into a career in boat design, whereas Thompson spent the rest of his life in search of automotive performance. Following his death, accounts of his accomplishments invariably noted that he was the first man to travel 400 miles per hour on land, at Bonneville in 1960. They also credited him with "revolutionizing" the design of dragsters, a revolution for which he had always been willing to take credit.[7] Nevertheless, Bill Martin was racing his machine on the sands of Jax Beach in the winter of 1953–54, and Thompson must have already seen California cars with the driver's seat behind the rear axle, perhaps one built for Arthur "Red" Jones by Gene LeBlanc, almost certainly another one built by Joaquin Arnett.[8] Whether or not Thompson emulated that specific idea, however, the important thing is that it was only one among several that he combined systematically.

When Thompson debuted his new dragster at Pomona it got everyone's attention long before it clocked low ET of the event, 11 seconds flat, and not just because it looked so crude compared to the Pomona Valley Timing

Association's own showpiece, which was also on hand. A craftsman named Ed Vogel had formed a set of gorgeous body panels for that dragster, whereas Mickey's fairing looked like an afterthought. If performance were taken to be a matter of getting quickly from A to B, however, it was no contest. Thompson's machine had a 97.5-inch wheelbase, but the tubular steel structure extended well beyond the rear axle. Out behind the driver sat "like a rock in a slingshot," as another driver, Art Chrisman's sometime partner Leroy Neumeyer, put it with evident admiration.[9] The metaphor was apt, for everyone agreed that this machine got off the line quicker than anything they had ever seen. There was hardly any weight at all on the front tires, but Thompson had taken other measures to maintain steering control under power. By tilting the front axle back 13 degrees, he established a kingpin inclination sufficient to make the wheels resist deflection (the same principle that governs the action of casters on furniture) and also reduced the danger of oversteering. More important, he shortened the rear axles and their housings so the tread was only thirty-six inches wide, which left the wheels "just as close together as the width of the driver's body would allow."[10]

Dragsters normally had locked (inoperative) differentials. This setup was regarded as necessary to keep one wheel from spinning while the other stood still, yet, as Thompson pointed out, it was "practically impossible for both rear wheels to get the same degree of traction at any point along the course from the time the car starts out of the chute 'til it crosses the finish line." With traction uneven, dragsters "were just desperately hard to keep going in a straight line." The most critical point during a run was "after the initial wheel spin period when the engine torque on the wheels is at its greatest in relation to the car's speed." Further:

> The normal reaction to this uneven push on the rear of the frame is for the chassis to pivot on the rear wheel with the least traction. Normally the front wheels would resist this pivoting tendency and hold the car on a straight course, but with the front wheels barely touching the pavement, they cannot exert sufficient resistance to prevent the pivoting action.

What Thompson had done was introduce a "blind variation," the sort of change that, in Donald Campbell's phrase, went "beyond the limits of foresight or prescience." In his mind's eye Thompson pondered "a chassis conventional in all other respects but with a single drive wheel on the center line of the frame. Such a chassis would not have any pivoting characteristics." By radically narrowing the rear tread, he concluded, "we have approximated the desirable handling features of a single drive wheel and retained the maximum traction of two wheels."[11]

More clearly than anyone else, Thompson had conceived of a dragster as a complex technological system. At first, however, not many others understood how the elements of his system complemented one another, and so they were only selectively emulated. The high angle of caster caught on quickly (eventually dragsters were set up with forty-five degrees and even more); the cockpit location caught on slowly—a year later there were a handful of other slingshots. Most racers shied away from the "weird" narrowed axle. It simply didn't look right.

A man of many parts, in 1955 Thompson took over Lions Associated Drag Strip on 223d Street in the Los Angeles harbor area—LADS, or Lions, a new drag racing venue that subsequently became the most famous on earth. In the eight years he managed Lions Thompson saw hundreds of dragsters make passes, and he saw speeds push 200 miles per hour as elapsed times dipped towards the seven-second zone. He also saw quite a few crashes and the demise of drivers who never dreamed it could happen to them.

It first happened early in 1956, to a young man named Dave Gendian, who, like Thompson, was from El Monte and had made quite a name as a street racer. Gendian had graduated from Ford coupes to fuel-burning road-sters and finally to a Chrysler-powered dragster built by a friend of his from Pasadena. It was a slingshot but had a standard-width rear axle, which was one reason dragsters had been so "hard to keep going in a straight line," as Thompson had explained. On top of that, like many early slingshots, Gendian's had the roll bar located ahead of the seat rather than behind, and there was no shoulder harness. And in a desperate attempt to improve traction a new pair of wheels and tires had been mounted with a hasty jury-rig. Thompson checked the setup as Gendian's friends prepared to push him up to the starting line for the first round of eliminations on Sunday morning, January 15. He pronounced it OK. But he would also have been the first to admit that he himself had driven machinery he knew to be unsafe, his only insurance being crossed fingers. Gendian lost a wheel and flipped. While the key welds held together, the roll bar did not protect his life.[12]

Even though the structure of Thompson's own slingshot did not look at all beefy—it had a space frame of small-diameter tubing rather than two-, three-, or four-inch members, as others had been using—it was much better designed and would probably have saved him in a crash. Aside from a fluke such as the loss of a wheel his machine would also have been less likely to crash because it handled so well. The slingshot owed its inception partly to safety considerations, to the better traction and directional stability it af-forded.[13] Ironically, it would later become obvious that drivers were left in one of the most dangerous positions imaginable. As better tires induced engine builders to wring out three and four times as much horsepower from Chryslers and other overhead valve engines as they could get from a Ford flathead, everything that was in front of the driver became potentially haz-ardous: A burned piston could result in the lubricant spraying out through the crankcase breathers, giving him an "oil bath" and leaving him unable to see where he was going; a backfire could blow the supercharger apart and shower him with flaming fuel; flywheels, clutches, and gearboxes (all of which were nestled between a driver's feet and legs) could disintegrate and demolish cast-iron housings with all the fury of a grenade. Dozens of drivers died in slingshots during the 1960s, yet the design prevailed until Don Garlits got maimed by an exploding gearbox, at Lions Drag Strip, and, while recuperating in a Long Beach hospital, focused his considerable brainpower on a different configuration, one that would put all potentially volatile components behind him, literally.

Yet this configuration had been familiar all along. Designers of dry lakes machines had often positioned the driver ahead of the engine, and mid-engine dragsters had shown up almost from the beginning. The first that could clearly run with the best was built by Bruce Terry and George "Ollie" Morris in 1954 and debuted by Morris at almost exactly the same time as Mickey Thompson finished his slingshot. With flathead power Morris won Santa Ana's fourth anniversary meet in July 1954 with a speed of 140 miles per hour, and later he turned 145, as fast as anyone had gone at that time.[14] Even as *Hot Rod* was predicting that Thompson's slingshot was "surely a preview of things to come," a very differently configured machine had emerged as a competitive threat.

By 1955, when E. D. "Dean" Brown began publishing a semiweekly tabloid called *Drag News,* mid-engine dragsters were showing up in considerable numbers—in California, of course, but even more so in the Midwest. Two of the most interesting had been built in Ohio. One was designed by Art and Walt Arfons, half-brothers from Akron, who eschewed automobile power plants altogether in favor of a 1,700 cubic-inch Allison airplane engine, such as had powered the P-51 Mustang of World War II and had been used by Gold Cup hydroplane racers since the late 1940s. Rated at nearly 1,500 HP, an Allison had double or triple the power of other engines then installed in dragsters, and in the next few years the Arfons brothers would turn speeds up into the 170s running only aviation gasoline. At 3,000 pounds or more an Allison-engined dragster substantially outweighed most others and had trouble getting off the line quickly enough to win many races. Yet the evolving line of "Green Monsters" attracted a lot of attention, and the Arfons brothers attracted even more attention when they turned from piston engines to jet power. For a variety of reasons aircraft engines never became part of the technological mainstream. Just as Don Garlits put Tampa on the drag racing map, however, Art and Walt Arfons did the same for Akron.

Ironically, at the same time that the Arfons brothers were working on their first dragster not far away Joe Scarpelli and Duane DePuy were building the "Cleveland Clipper," a machine that appears to have been squarely in the mainstream of *second* normal design. Scarpelli and DePuy patterned

With partner Holly Hedrich looking on, Red Henslee poses in his modified road-ster following a victory at Perryville, near Phoenix, in 1955. This photo was taken by none other than NHRA's Wally Parks. (Courtesy Petersen Publishing Co.)

Red Greth and Lyle Fisher (in cockpit) pose with their "Speed Sport" roadster, in its second incarnation in the early 1960s. Contrasts with the Henslee machine include a narrowed rear axle, supercharger, parachute for stopping, and roll bar for insurance. (Photograph by Bill Turney)

their dragster after Ferdinand Porsche's C-Type Auto Union, a Grand Prix machine of 1934, which had the cockpit in front of the engine, and they used a Chrysler Firepower, a hemi, one of the very first ever installed in a dragster. The Clipper ran close to 120 miles per hour on straight methanol, no nitro, and appeared to have a lot more potential, especially since the configuration presented substantially less frontal area, thus minimizing aerodynamic drag. Other mid-engine cars would soon prove capable of excellent top speeds, sometimes better than the best slingshots.

The Clipper did seem to have inherent problems, however. First was the old bugaboo of weight distribution. With a short wheelbase that put less than 70 percent of its weight on the rear tires it did not come off the line quickly, and often not very straight. And the configuration itself compounded that problem: With very little of the car's structure ahead of him along which to line up his sights the driver, Scarpelli, found that the rear end could "swing through a large transverse arc" before he realized what was happening and attempted to correct.[15]

Scarpelli soon switched to a slingshot, but for most of the 1950s there was usually one or another mid-engine machine that was in the hunt, as the expression went. Often these had highly modified Model-T roadster bodies. One of the fastest was campaigned for several years, with several part-

Lloyd Scott brings the "Bustle Bomb" off the line at Great Bend, Kansas, at the NHRA's first National Championships in the fall of 1955. The makeshift deflector over the carburetors is deceptive; overall, this was a carefully engineered machine. Support from two southern California manufacturers is evident. (Photograph by Eric Rickman, courtesy Petersen Publishing Co.)

Don Jensen's twin-engine machine of 1956, which combined the cockpit location of a modified roadster with a rear engine like the Bustle Bomb. In this shot Jensen is seated on the edge of cockpit. On his right are members of his Hayward, California, club, George Wulf (who owned the rear engine), Donn Blount, and Bryan Burnue. (Courtesy Don Jensen)

Mickey Thompson and Ed Losinski charge off the starting line at Pomona in 1955. Losinski's dragster mimics an Indy 500 machine; Thompson's is a harbinger of a new design paradigm. (Photograph by Gary Herbert, Hearst Collection, courtesy Department of Special Collections, University of Southern California Library)

ners, by Robert "Red" Henslee of Phoenix. Even faster was the one called "Speed Sport" run by friends of Henslee's from Tucson, Gary Greth (also nicknamed "Red") and Lyle Fisher.[16] Each held the all-time speed record briefly, the Phoenix machine at 157 miles an hour in 1956, Speed Sport at 169 a year later. Holly Hedrich, a onetime partner of Henslee's, ran another such machine well into the 1960s, occasionally turning in the 180s, as did Fisher and Greth in Speed Sport II.

For protecting the driver from flying parts, the mid-engine design had marked virtues. The roadster form also had aerodynamic advantages, for the engine and driver could be tucked away entirely, with only the driver's head in the airstream.[17] But, to borrow Nathan Rosenberg's expression, there was a technological imbalance. Indeed, the problem was quite literally one of balance. Unless every bit of weight possible was concentrated on the rear end, even a 300-HP flathead could spin the tires most of the way through the quarter-mile. As one chassis fabricator put it a few years later, tires were "the missing link" that racers "couldn't build up in our own shops."[18] With spinning tires a drag racing machine tended to lose direction; a driver seated at the rear could see this and correct, but up front very few could sense the problem quickly. Henslee tried setting up his clutch to slip and thereby minimize wheelspin, and that made some difference. Given all the right conditions, everything would work fine. But consistency was the perpetual hobgoblin, and mid-engine machines invariably had a reputation for "turning right or turning left" without warning. The final key to Garlits's success in the 1970s lay in the realization that a mid-engine setup necessitated a slow steering ratio to prevent overcorrecting. This is something that Scarpelli and DePuy understood even in 1954. But no mid-engine machine with a short wheelbase had optimum weight distribution, whereas a slingshot was marvelously effective in transferring virtually all the weight aft under acceleration.

There was, however, an even more direct way of getting weight on the rear end: putting an *engine* out behind. Transposing the slingshot configuration—engine behind the axle, driver in front—was improbable, because the front wheels would not have stayed down at all. But what if a dragster had *two* engines, one of them in front and one behind the rear axle? This was what Lloyd Scott did when he built the dragster dubbed the "Bustle Bomb." One of Scott's confrères was Larry Shinoda, a Santa Ana pioneer who had a keen instinct for chassis setups; early in 1955 a stock-bodied Ford roadster he had prepared for Scott turned 139 miles an hour. That was faster than all but the fastest dragsters: The inaugural issue of *Drag News* headlined a Santa Ana speed record of 147 by Bob Alsenz of Paramount, California, in an Ardun-Merc dragster out of Lakewood Auto Parts, and in March Ed Losinski equaled that time with a Chrysler engine in his dragster.

Neither of those machines was a slingshot, nor was the Bustle Bomb, yet

Scott's unorthodox creation embodied a more effective combination than any other dragster of its time. In front was an Oldsmobile engine; behind the rear axle was a Cadillac. Scott shifted a gearbox on the Olds, while the Cad powered right on through the entire run.[19] After shakedowns at Santa Ana, turning in the 140s, Scott trailered to Lawrenceville, Illinois, to the "World Series" championship of NHRA's rival, the Automobile Timing Association of America (ATAA). In winning top eliminator, Scott became the first dragster driver to top 150 miles per hour. Then at the end of September NHRA staged its inaugural National Championship Drags in Kansas, at Great Bend Municipal Airport. There, on a high-traction concrete surface, the Bustle Bomb ran even faster but showed its true forte in the realm of elapsed time. Previously, very few dragsters had excelled at both; the fastest machines tended to be sluggish getting off the line. Art Chrisman, who had the fastest single-engine entry, clocking 145.16 on opening day, turned a 10.98 elapsed time. Scott's was fully a half-second quicker, 10.48, and not long afterwards, back in California, the Bustle Bomb got an ET more than a second quicker than that, 9.44.

Scott campaigned the Bustle Bomb for only a short time, but later a young hot rodder from Hayward, California, designed and built a variation on the theme. Don Jensen had seen Scott run at Great Bend and liked his setup, although he elected to put his cockpit in front of both engines (his were both Cadillacs) rather than in between. He too clocked 150s, on pump gasoline, but his machine met its demise in a crash at Kingdon, and the combination of a mid- and rear engine was rarely tried again, and never by anybody as skilled as Scott or Jensen. For a time, anyway, there was nothing more efficient for getting quickly from point A to point B. But the efficiency came at the cost of complexity, and most racers perceived that there were lots of simpler things to be tried.

Tires were still deficient except on the best "biting" strips like Great Bend's concrete. They were improving, however: "Huge, wide slicks, run at relatively low pressure," wrote Roger Huntington, "are giving traction far beyond the wildest dreams of tire engineers."[20] As has almost always been the case, when available traction permitted, engine builders could find more horsepower. Many of them were finding it in Chryslers. Mickey Thompson turned 151 at the new San Fernando strip, as did Losinski on opening day at Thompson's Lions strip. Late in 1955 Red Henslee turned 148 at Perryville Air Strip near Phoenix, at the finals of the NHRA Nationals, which had been rained out in Great Bend. Thompson was driving his slingshot, of course, but Losinski's machine resembled a champ car, an Indy 500 racer, and Henslee was in his mid-engine modified roadster. Although Calvin Rice won top eliminator at Perryville in a slingshot, even after a year and a half that design had still not been widely emulated; slingshots were actually outnumbered by mid-engine machines. In what was probably the first direct confrontation between these two archetypes

Henslee outran Carlos Ramirez, driving a new Bean Bandits' dragster. It must have given even Mickey Thompson pause.

Technological choices are rarely clear-cut, and it took time for some competitors to concede the efficiency of slingshots. A handful never did concede it, steadfastly maintaining that the place for the cockpit was out in front. Yet, if one were to have conducted a poll in the 1960s, ninety-nine out of one hundred dragster racers would have stated that this was a setup that "did not work." And they would have been able to cite a number of reasons why. Such machines did not come off the line with consistency, and they were harder to keep under control consistently. Although it was unarguable that the cockpit of a slingshot was not a good place to be when broken parts started to let fly, slingshots would remain normal design for an entire drag racing generation, mid-engines mostly the domain of dreamers.

Whichever configuration they may have favored, enthusiasts kept pushing the records. Early in 1956 Leroy Neumeyer turned 150 at Lions, and a week later Bill Replogle, driving for Ernie Hashim of Bakersfield, clocked almost 154 at Famoso.[21] There had now been four 150-MPH dragsters, three with Chrysler engines, and the fastest also had a blower. Although times were occasionally publicized which seemed questionable, these were all credible, and there was no reason for any general skepticism. That was all about to change, however.

First, with the appearance of some truly stunning machines like Losinski's, and with 150s coming up more and more, drag racing was gaining appeal as a spectator sport, and the "promoter" was becoming an ever more prominent player. The NHRA now sanctioned drag strips from coast to coast, seventeen in California alone. Several of these had inaugurated Saturday night races. There were regular turnouts of several thousand paying customers at Lions. For a promoter it certainly did not hurt the gate to be able to advertise that somebody who had turned such-and-such a fabulous time would be on hand the coming weekend.

More important, manufacturers of engine components had begun seeking to capitalize on the exploits of racers using their products. This new commercialism quite naturally tended toward excess. Ad writers would seize on timing flukes, which were not an uncommon occurrence at even the best-managed strips. In sum, while the ads in *Hot Rod* and *Drag News* constitute an important historical source, they need to be treated with caution. Requiring utmost caution is the series run for years and years by Ed Iskenderian, who has been dubbed, not unfairly, "hot rodding's cam-grinding P. T. Barnum."[22] The first recipient of large doses of Isky's hype was an old-timer (he had run at Muroc before the war) named Jim Nelson, "Jazzy" Nelson, who lived in Venice, southern California's "beat" equivalent of San Francisco's North Beach. Nelson had designed a remarkable drag machine built on a narrowed Ford frame and clothed in a Fiat "To-

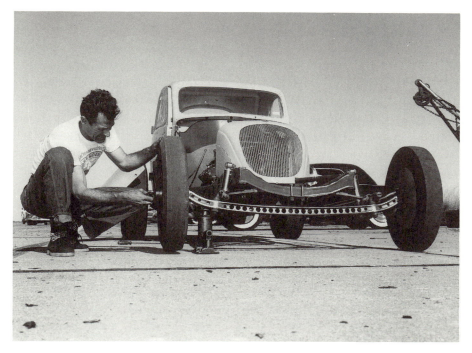

At Great Bend, Kansas, in 1955, Jim Nelson checks a wheel bearing on his Fiat-bodied coupe. The axle has been drilled to save weight. (Photograph by Bob D'Olivo, courtesy Petersen Publishing Co.)

Ed Iskenderian (right) poses with his foreman, Norris Baronian, and his work force in 1957. Isky's was one of the larger specialty operations of this period; he had just moved to this Inglewood location with four times the floor space of his previous building in Culver City. (Courtesy Iskenderian Racing Cams)

polino" body. He still used a flathead Ford engine.[23] Although his coupe rarely ran faster than 130, it was exceptionally quick, having turned 10.90 at Great Bend, when Chrisman's Chrysler-powered dragster could only muster a 10.98. By 1956 Nelson's was one of the few flathead machines that stood a chance in a race against a deep-breathing Chrysler or Cadillac.

Anyone who saw Nelson's Fiat in person or saw one of the numerous magazine features could not help noticing the advertising for Edelbrock cylinder heads and especially Iskenderian cams. Nelson ran a 404 model, whose timing would have been too extreme for a street machine, but Isky made less radical grinds that were suitable for everyday driving. And his business was starting to boom. Early in 1956 Jazzy Jim Nelson was given a clocking for a 9.10 at San Fernando. He had never run anywhere near that quick anywhere else, nor had anyone else with a flathead. Even Bill Replogle's best with a blown Chrysler was only 9.12.

Lots of odd things could happen with timing systems: photoelectric cells could be set up (deliberately or inadvertently) the wrong distance from one another; an errant piece of debris could accidentally trigger something; power surges could throw everything off; recorders could simply not register, or they could give false readings. Sometimes cars started way behind the "starting" line. At most strips, when a car clocked a time much better than it had ever run before, standard procedure was to wait to see if it was repeatable. But at San Fernando Jazzy's time went out over the public-address system, he was given a written slip confirming it, and Isky picked this up and ran with it: "The World's Fastest Accelerating Car," he proclaimed in the March 1956 issue of *Hot Rod,* "and powered by an Iskenderian '404' Cam." Isky wanted people to understand that this "fastest accelerating" designation included cars with any kind of engine, and, implicitly, that it included the Chrysler-powered Henslee machine, which his cam-grinding competitor Howard Johansen had advertised as the "World's Fastest Accelerating Roadster."

Actually, before 1956 manufacturers had rarely seen a great deal of value in this sort of thing: Isky and Johansen, along with Chet Herbert and a few others, were exceptional. Afterwards, commercial considerations became paramount, as manufacturers began not only giving free parts to favored racers but sometimes paying them as well. Did Jazzy Nelson truly turn 9.10 in his little flathead Mouse? Almost certainly not, but Iskenderian had not *invented* anything out of whole cloth. To someone with Isky's foresight and keen intelligence the commercial vistas must have been quite dazzling.

Meantime, the combination that Mickey Thompson had introduced was gaining adherents, but there was still nothing like a design consensus. Men like Don Collins of Portland, Oregon, still boasted of having built dragsters with "sprint car logic." Or, like another Oregonian, Ernie Hall, they took

Ernie Hall's 1957 dragster was a work of art, particularly in its use of 2¼-inch seamless steel tubing for the frame and .040 aluminum for the body. But the configuration was clearly imitative of a car intended for oval-track racing; the cockpit was in front of the axle and the tread was the same in the rear as the front, 58 inches. (Photograph by Peter Sukalac, courtesy Carolyn Hall-Salvestrin)

pains to endow their dragsters with the configuration and aesthetics of champ cars.[24] Indeed, some of the best constructed and best detailed (and most costly) dragsters of the mid-1950s mimicked the lines of champ cars. Some designers still maintained that extremes in weight distribution aft, narrow rear tread, and front-end caster were not essential. But in 1958 when a former Oklahoman named Gordon Alfred "Scotty" Fenn set up shop in Gardena, California, to manufacture chassis commercially—he called his firm Chassis Research—they were slingshots with narrowed rear ends and lots of caster. In fact, they had a different degree of caster for each front wheel, since the rotation of the engine and driveline tended to lift the right wheel before the left.[25]

One of Fenn's first machines, built before he moved to California, was campaigned by Lou Baney, who was now working as a service coordinator for the Yeakel Cadillac dealership in Los Angeles. Fenn and Baney won the NHRA's 1956 West Coast Regional Championship with a slingshot design Fenn called the "TE-440" (i.e., top eliminator, 440 yards). The cockpit was so far down and back that their driver, Kenny Arnold, had to crane his neck

Emery Cook at the controls of Cliff Bedwell's Fenn TE-440. Compare the cockpit location to that of Ernie Hall's contemporary machine (page 47). (Courtesy National Hot Rod Assn.)

to see over the cowl, but a TE-440 probably had more weight concentrated on the rear tires than any other dragster. Fenn sold a lot of them and even more of a subsequent model he termed a K-88. Across the country in Pennsylvania Pat Bilbow's Lynwood Welding sold frank imitations to eastern racers.

An NHRA "Drag Safari" made several cross-country treks in the mid-1950s, encountering dragsters from coast to coast. There seemed to be scarcely any two alike. In 1955, in Oklahoma and Texas, they found Jack Moss sitting atop the rear axle, Mel Heath in front of it, and Ray Harrelson with his feet up against the front axle in a mid-engine car resembling Ollie Morris's. Mid-engines became more commonplace as one moved away from California: There were at least three at Oswego, Illinois, three at Woodbine, New Jersey, and the Arfons brothers from Ohio dominated a regional event at Fulton, New York. Nevertheless, by 1956 *Hot Rod*'s chief photographer, Eric "Rick" Rickman, believed that "a consistent pattern of construction" was emerging in the slingshot.[26] There were still significant exceptions, such as the Bustle Bomb and Henslee's roadster, now being driven by Emery Cook, a thirty-year-old San Diegan who was married to Joaquin Arnett's sister; over the 1956 Labor Day weekend at Lions Cook turned 157 miles an hour. But that same weekend, at NHRA's Nationals (moved from Great Bend to Kansas City for 1956), Bob Alsenz clocked 159 in a slingshot owned by Ken Lindley. Back on the West Coast, at the Southern California Winter Championships at Morrow Field in Colton, Calvin Rice broke into the 9s, as did two other drivers, Kenny Arnold and Walt Nicholls, all three in slingshots.

By 1957 Cook had parted ways with Henslee (who had become too involved with his business of making mechanical rabbits for dog racing) and joined forces with a bodyman named Cliff Bedwell, who owned one of the original Fenn TE-440 chassis. Their engine, one of the larger 354-cubic-inch Chryslers, had been prepared by Bruce Crower of San Diego using a six-carburetor manifold of his own design, which he was beginning to market as a "U-Fab." It also had a double-disc clutch designed by Paul Schiefer, who had been setting records on the dry lakes since 1934; as with the roadster, there was no gearbox. It had a pair of Bruce Alexander's slicks, then the best available. And it had an Iskenderian cam (later it was named the "Isky-U-Fab Special"). It had previously turned 153, and people knew that this was a machine with a great deal of potential.[27]

On February 3 Cook and Bedwell towed through the pit gate at Lions. Cook made six runs all told. First off he set a track ET record, 9.28, seven-tenths of a second quicker than the Henslee roadster. He also clocked 156.25. Although he did not better his ET, on the next two passes he turned speeds of 157 (equaling the track record he held in the roadster) and 159 (tying the world record held by Bob Alsenz). On the fourth he fell off a bit, but on the fifth he turned 165, and finally he clocked a speed of 166.97. That clocking was verified by every means possible, leaving little doubt about its credibility. Later both Mickey Thompson and Kenny Arnold drove this same dragster to speeds in excess of 156. Its performances did much to assure the ascendancy of the slingshot design; even though mid-engine machines would remain fairly commonplace into the 1960s, only a few times more before 1971 would one of them go to the final round at a major event.

More important, it got key people thinking that 167 miles an hour was *too* fast. Actually, 167 was just what some theoretician had already deduced was all that physical laws such as coefficient of friction would ever permit, but who could trust an abstract premise like that? Cook, or somebody else, was likely to go even faster: Wasn't that arguably a "purpose of the thing"? If so, then whatever could *Drag News* mean when it referred to "speed problems"? For the first time people began to realize that purpose was *truly* a matter of opinion.

A lot of racers had been recording big numbers—some of them richly experienced like Cook, some not; some in machines that were apparently pretty safe like Chassis Research TE-440s, some in cars that appeared to be dangerously flimsy or unstable. Fenn had offered to help: "If your car does not handle, if it skates, hops, or hunts, send us your problems." But some people were thinking about an entirely different purpose—about how to *slow dragsters down.* By far the readiest way was to exclude the use of fuel; Cook was reputed to have had a 90 percent load of nitro in his tank when he turned his 166.97. The very next day C.J. Hart announced that Santa Ana would place an immediate ban on any fuel but pump gasoline, citing the problems of getting stopped safely; some strips like Colton and Saugus had shutoff areas that were only marginally safe for 160-MPH speeds, at least before anyone had considered using parachutes. But Hart also talked about the "skyrocketing cost of participation" as well as "a general desire expressed by participants to return to gas."[28] Three weeks later management at five other California strips including Lions, Saugus, San Fernando, and Kingdon mandated "pump gas only." By the middle of March Pomona, Paradise Mesa, and Colton had joined in, and an NHRA representative had announced that the 1957 National Championships, scheduled for the Labor Day weekend in Oklahoma City, would operate under the same restriction.

There had always been racers who were opposed to fuel.[29] Fuel was "mere chemistry," they said; fuel permitted "buying" records. At most strips there had initially been "fuel" classes and "gas" classes, and one could choose a mode of competition accordingly. For dragsters, however, many promoters had adopted NHRA's classification system, which was essentially unlimited. There were no restrictions on displacement, on supercharging, or on fuel. Cars "specially constructed for all-out competition" ran in one class, "D" class. With off-the-shelf chassis available from people like Fenn, dragster competition was more inviting to more racers and undoubtedly many of them did not want to bear the extra expense of running fuel; they saw purpose as a matter of close competition, not getting from A to B as quickly as conceivable. Hence, there may actually have been some "general desire" for a gasoline-only policy.

Although NHRA did not instigate the "fuel ban," it became a staunch proponent with its gasoline-only policy for the Nationals—the biggest, best-publicized, and best-attended drag race of the year. NHRA was seeking to fend off "damaging attacks being launched against drag strips from outside sources," one of them being the National Safety Council. So, while noting that gasoline-only would afford "a better break for all contestants," it also stressed the safety angle, particularly the "need for better control" under initial acceleration.[30] Presumably, a car with less power would be a safer car.

Nevertheless, a dragster as well designed as Bedwell's TE-440 handled quite well, and one is inclined to wonder about the motives for punishing the sport's premier competitors. Fenn, a deliberately contentious man, suggested that the villain was General Petroleum; Socony-Mobil had sponsored the NHRA Nationals (as well as regional meets and the Safari) and, according to Fenn, wanted to make certain that all vehicles in competition were powered by the product it refined. "Wasn't gas shoved down everyone's throat?" he later asked in one of the rhetorical exercises that became his trademark.[31] NHRA sued. Though the case never went to court, Fenn's charge had lost credibility when Mobil and NHRA parted company. Indeed, there *was* something incongruous about a dragster running on a mixture of nitro and methanol while sporting a large decal of Mobilgas Flying Red Horse. But Wally Parks insisted that Socony-Mobil had a long-standing "experimental" interest in fuels other than gasoline. One thing was nonetheless obvious: Political forces had made a marked intrusion and people with fuelers were outraged. Racing was a technological activity, they said, and everybody knew that you could not stop technological progress. Progress, however, had not been stopped. In technical realms progress has many possible trajectories, and in this case it had merely been deflected.

Actually, the fuel burners were never banished altogether, not even where the move began, in southern California. Strips that were off the beaten track and needed some special performance to attract spectators continued to welcome fuelers; hence, Cook kept setting records at places such as Inyokern in the Mojave Desert. Saugus, Pomona, and Colton quickly withdrew their initial assent in the ban (at Colton a few months later Cook upped his speed record to 168.85 MPH), but the major racing venues in the Los Angeles area, particularly C.J. Hart's Santa Ana and Mickey Thompson's Lions, stood firm. Still, there were plenty of strips that NHRA had not enlisted and which did not prohibit fuel—strips that, as we shall soon see, were to become the stomping grounds for some very fast fuel dragsters from places like Michigan, Missouri, Illinois, Texas, and Florida which had never been seen in the West. While the fuel ban was partly aimed at warding off the incipient forces of "professionalization," it actually imparted momentum to these forces because people would pay just to see fuelers perform.[32]

As for the California fuel racers, some like Cook held out, but others decided to see what they could do with gasoline. At Santa Ana's first big all-gas show Calvin Rice, who had often run 150s, struggled to a 116 on his first pass on gasoline and finally managed a 137. Meantime, there seemed to be a lot of waffling about what the fuel ban was really for, safety or an attempt at parity:

> Drag racing used to be full of such stirring triumphs—until the bugaboo of $5 per gallon fuel with accompanying consumption of gallons per run split the sport right down the middle. The *haves* won the big trophies, while the *have nots* just had it. But hot rodding is right back in stride again, with the little guy . . . getting his biggest break in years with the announcement that "gasoline only" will be used for the forthcoming 3d National Championship Drags.[33]

NHRA was not the only organization, however, that could promote an event as the "National Championship." Wally Parks might call the others "outlaw" events, but they attracted competitors like Cook and Bedwell, Red Greth, and "a wiry little Armenian from Detroit"—Serop "Setto" Postoian, who had been barnstorming midwestern and eastern strips.[34]

Top eliminator at the 1957 NHRA Nationals was won by an Oldsmobile-powered slingshot from Tucson entered by Joe Dillon, Allen "Lefty" Mudersbach, and Buddy Sampson. It was quite a sophisticated piece of machinery, and its times, 141.50 in 10.42 seconds, were both exceptional for pump gas. But such numbers sounded a little pale compared to Cook's 168 at the ATAA's World Series and the 169 that the Speed Sport roadster turned late in 1957 at Davis-Monthan Air Force Base near Tucson. And they particularly paled in comparison to the news that Iskenderian broke in February 1958 in his customary full-page ad on the inside front cover of *Hot Rod*. There, below a fuzzy photo of a nondescript black slingshot, Isky reported on "the latest shocker from the Southeast circuit":

> Don G. Garlits, driving his stock bore and stroke 1957 Chrysler which was built in his own speed shop in Tampa, Fla., went well beyond the old record set by the Isky 5-Cycle equipped Cook-Bedwell Dragster to a new Top Time of 176.40 mph and an ET of 8.79 sec. This historic drag racing event took place at the Brooksville, Fla., concrete strip on November 10, and was checked and verified by the latest electronic clocks.

Don G. Garlits? Brooksville, Florida? Any claim made in an Isky ad? Hardly anyone in California believed it. But, if fuel dragsters really were capable of that sort of performance (and it turned out they really were), across the country they would draw crowds bigger than anyone had ever anticipated, and drag racing had a commercial potential bigger than anyone ever thought.

By 1958 the design paradigm was fairly well if not quite finally established: The Speed Sport roadster provided evidence that mid-engine machines could run as fast as slingshots—under optimum conditions, anyway. Yet *Drag News*'s Dan Roulston did not think that there were "two identical cars in the entire United States." Dragsters had moved out of the "experimental" phase after having found a "workable configuration," but there were many other questions that had to be addressed in seeking a best "particular instance" of dragster design.

With the most important race of the year limited to gasoline only, was there a performance potential in gasoline close to that of nitro? What about dragsters with huge airplane engines, which had always run gasoline—could one of those be designed for ET as well as MPH?

Maybe two automobile engines were the way to go after all, despite the complexity. With a new twin-engine dragster Jazzy Nelson was racking up impressive performances. Ever the nonconformist Nelson remained in the flathead camp. But what about twin Cadillacs, Don Jensen's setup? Or twin Chevys? Or Chryslers, a combination nobody had yet tried? And placement—both in front, front and rear, side by side?

How about supercharging? Ernie Hashim, Ken Lindley, and Calvin Rice had blowers; Cook, Setto, Garlits, and most others seemed to do fine without. But, clearly, supercharging was a route to more power without adding a great deal of weight. Inside the engine how about billet steel rods instead of castings, crankshafts with the throws 180 degrees apart rather than 90?

What about streamlining? Low aerodynamic drag was a chracteristic of the mid-engine form, as embodied in the Henslee and Fisher-Greth roadsters, but, as Mickey Thompson and others had shown, the aerodynamics of a slingshot could be cleaned up considerably too, if a designer thought this was worthwhile.

What about traction? Given the possibility that tires would rarely be all that racers might desire, maybe the secret really did lie in driving through four of them. On the other hand, did there actually need to be two wheels in front? Why not just one? With so very little weight on the front end, even if there were two wheels, couldn't they be as lightweight as bicycle wheels?

What about suspension? Would torsion bars be preferable to leaf springs? Did there have to be any suspension at all, front or rear?

What about gearboxes? Most dragsters ran two-speed transmissions, but Speed Sport and both of Emery Cook's record breakers had direct drive. Was direct drive, as Roulston suggested in *Drag News,* the beginning of a "whole new chapter for drag racing"? Was there anything to Romeo Palamides's idea of using a torque converter?

What about wheelbase? Everybody regarded a slingshot as a "lever," but some thought the lever arm should be long, notably Garlits; some thought it should be short. Fenn fostered a theory that a dragster's wheelbase should be equivalent to the circumference of the rear tires, 90 inches or less. Dissatisfied with a machine he had built with a 110-inch wheelbase, Palamides shortened it and clocked quicker elapsed times.

What about materials? Some chassis builders favored 4130 chrome moly tubing, but many swore by 1020 mild steel, Fenn among them. Some favored arc welding, but many dragsters, including Fenn's, were still oxy-acetylene welded. Quite a number of fast machines, such as Red Jones's, had been built using aluminum channel, not steel tubing.

These and hundreds of other questions of design and fabrication were yet to be worked out. They would not be worked out in the same sort of context which had prevailed for the first few years. In 1953, when one of the magazines asked, "Who is to dictate the shape of a drag machine?" the implied answer was obvious; the answer was nobody at all.[35] As people discovered that purpose was a matter of opinion, however, the context would encompass many other considerations: safety, for one, but also commercial considerations—actually, something broader than that, considerations best termed political. There were those, it turned out, who actually did have designs on dictating the shape of drag machines.

4

POWER

The NHRA plans to transform the hot rod movement from a disorganized, sporadic, rudderless activity into an integrated, regulated and supervised sport.

LEE RYAN, FOR THE NATIONAL
HOT ROD ASSOCIATION, 1952

quarter-century after C.J. Hart flagged off the first pairing at Orange County Airport a journalist named Rick Voegelin wrote an article titled "The Ten Most Powerful Men in Drag Racing." Only three were primarily racers—Don Garlits, Don Prudhomme, and Bill "Grumpy" Jenkins—and Voegelin saw their power only partly in terms of their victories in competition: "Being a successful racer does not necessarily make one a *powerful* racer," he said. Beyond his countless wins in fuel dragsters, Garlits had been instrumental in founding an organization that mounted a direct challenge to the NHRA's premier event, its national championships on Labor Day weekend. Prudhomme, the dominant force in funny cars, had been instrumental in selling potential sponsors on "the idea of buying space on the sides of racing cars." Jenkins, a master of stock-bodied machines, had the power to influence rules makers and had been largely responsible for persuading NHRA to institute the pro-stock class.

Lou Baney was one of the ten, too, but not because of his exploits with dragsters; rather, it was for what he had attained as managing director of the Specialty Equipment Manufacturers Association (SEMA), the organization that had assumed responsibility for standardizing specifications (specs) and, more important, for representing the automotive aftermarket in potentially hostile legislative councils.

The other six men likewise related to the industry or sport—the two descriptive terms were used interchangeably—in ways that had nothing to do with driving or sending a car they owned down any drag strip. Bob Duffy was a retailer in Red Bank, New Jersey, whose establishment functioned as a nationwide clearinghouse for racing information: "If NHRA announces a rules change which is billed as benefiting the sport, one call to Duffy will reveal who got hurt by the revision." Vince Piggins and Dick Maxwell worked for the Chevrolet Division of General Motors and the Chrysler Corporation, respectively. Even though Chevrolet maintained that it did not "support any racers," Piggins's Chevy Product Promotion saw to it that "the hardware required to win finds its way into competent hands." Maxwell's situation was different; he was responsible for a large-scale corporate program to supply loyal racers with parts and, sometimes, with cash.

Voegelin's powerful men included Bill Doner, who operated a string of NHRA tracks on the Pacific Coast, and Larry Carrier, who had founded the International Hot Rod Association (IHRA) a few years before as a rival to NHRA.

Finally, there was NHRA's Wally Parks, a man whose power clearly transcended that of Carrier and every one of the others. Since the early 1950s Parks's organization had indeed transformed drag racing into "an integrated, regulated and supervised sport." Actually, it was not so much a matter of transformation as of something more basic: Parks and NHRA "were the primary creators of the sport," Voegelin concluded.[1]

Wally Parks had been a racer himself, but he termed his exploits "nonspectacular." His forte lay elsewhere. Born in Goltry, Oklahoma, in 1913, he had come to California in his teens, and first visited Muroc in the early 1930s. He found himself pressed into duty as a spotter. "I got hooked on the details of running and operating such an event," he recalled.[2] In 1937 he helped found the Southern California Timing Association. After service in the South Pacific (he was a technical sergeant) Parks returned to his prewar job at the GM assembly plant in South Gate, but only briefly. Soon he had become SCTA's salaried general manager. "Always thinking," was how Garlits described the most successful racers. Parks was always thinking, too, but about "the organizational end of things"—about matters like regulation and supervision, about power.

In 1947 Parks met Robert E. "Pete" Petersen, a former MGM publicist who represented a concern called Hollywood Publicity Associates. His primary client was Earl "Mad Man" Muntz, already a local legend as a hustler, but he thought he might also be able to help SCTA ward off the threat of legislation that would ban automobile racing and make hot rods illegal. (Most lakes machines still did double duty as street transportation; indeed, double duty was central to Don Montgomery's definition of a hot

During the first NHRA National Championships an acrobatic starter gives the go to a pair of slingshots, including the eventual winner, Calvin Rice. Mickey Thompson's machine (right), running as the "Panorama City Special," had been much tidied up from its initial appearance. (Photograph by Bob D'Olivo, courtesy Petersen Publishing Co.)

rod.) Above all, Parks was concerned about the hysterical headlines, about hot rodding's image. He wanted to show that hot rodders were not simply some new breed of delinquent; rather, they were (most of them, anyway) responsible young men who were seriously dedicated to technological progress. He hoped to transform popular perceptions. Parks suggested to Petersen the possibility of producing an exposition:

> I said that we [the SCTA] had the vehicles and the personnel to produce the show, but we needed the publicity, promotion, and the personalities from the entertainment industry that could help make it a success. We felt we could serve a couple of purposes. We could paint our own picture of the types of cars and people we were to the public that attended and also raise some funds that would make it possible for us to put an active publicity campaign in effect to try and offset this proposed legislation.[3]

The Los Angeles Automotive Equipment Display and Hot Rod Exposition was staged at the National Guard Armory in Exposition Park in January 1948. The excellent turnout stirred Petersen's enthusiasm for another project: publication of a monthly magazine.

The first issue of *Hot Rod* also appeared in January 1948. A year later Parks became editor, and in 1951 he spearheaded founding of the National Hot Rod Association. The NHRA aimed to do "what the Golden Gloves have done for boxing by way of transforming street fights and alley brawls into supervised boxing contests in the clean atmosphere of a gymnasium."[4] "Order out of chaos" was a phrase that appeared time and again in NHRA literature. For fourteen years Parks worked tirelessly to promote "the Association" on the pages of *Hot Rod*.

Like Pappy Hart, Parks would often repeat that he "didn't invent drag racing." Indeed, NHRA initially focused its energies on shows, safety campaigns, gymkhanas, and "reliability runs" on the model devised by the Pasadena Roadster Club, an SCTA affiliate. As early as 1949, however, Parks had participated in acceleration tests under the combined aegis of *Hot Rod* and *Motor Trend* (Petersen's second magazine) on an airstrip at Ramona in San Diego County. (Also on hand was Fred Davies, the car dealer who later became the promoter at Paradise Mesa.) In 1950 Parks, the SCTA, and the American Motorcycle Association staged drag races at the Naval Lighter-than-Air Base (later the Marine Corps helicopter station) in Tustin, not far from the airport where Hart inaugurated the first commercial operation. And by 1953 NHRA itself was producing drag racing events. Its initial venture, the Southern California Championships at the Pomona Fairgrounds, provided the first clear-cut indication of drag racing's potential as a spectator sport. Parks glimpsed the future; activities of this sort were but the tip of an iceberg. "Wherever there's progress, there is always a leader," NHRA literature would boast. "Guiding light of all this progressive endeavor is the National Hot Rod Association," *Hot Rod* proclaimed.[5]

SCTA was still conducting time trials at El Mirage a few times a year and once a year at Bonneville. Drag races could be staged every weekend, all over. As *Hot Rod* took off, car clubs were being formed from coast to coast. There were small airports or decommissioned military fields almost everywhere, but gaining access legally was contingent upon posting liability insurance. Underwriters naturally demanded that there be properly enforced rules regarding the safety of spectators and participants. The situation seemed to cry out for some sort of organization with oversight authority, with the power to bestow sanction.

Enter NHRA's Drag Safari, the aim of which was to "show rodding clubs how to set up races and keep them safe." The safari hit the road for Caddo Mills, Texas, and points east on June 10, 1954, not returning to California until November. Heading it up as NHRA field director was Bud Coons, the onetime Pomona police sergeant who had been instrumental in establishing the strip on the fairgrounds. *Hot Rod*'s Rick Rickman went along on "special roving assignment" to make certain there was plenty of material for photo stories in Wally's magazine. The safari made cross-country treks three years running and provided a tremendous boost for NHRA. By the end of 1956 it had conducted regional events in at least twenty-four states besides California.

Although NHRA membership was open to any individual, Parks began to regard it primarily as a trade association. The benefits to anyone who ran a strip included advice on rules and matters of organization and safety; publicity, particularly in *Hot Rod;* and "eligibility for the finest insurance program available." NHRA had initially insured its events by means of a rider on a policy held by a Chicago group that staged dirt track races, but in 1954 it established a tie with Aetna Life and Casualty. "Some of the nation's strips have resorted to cheap cut-rate insurance," NHRA warned, "enabling them to pocket a few dollars but often ending up with unsatisfactory results when the claims are in."[6] NHRA's insurance program gave it a powerful lever. Soon Parks was convening meetings of strip operators, with formal addresses on "the growing business of drag racing," and NHRA itself had a growing staff. By the mid-1960s it could claim to have sanctioned ten thousand events.

The safari brought along everything from timing equipment to portable telephones to trophies, and Coons was warmly welcomed everywhere. NHRA made demands in return for its largesse, however, demands for allegiance which did not always sit well. Art Pillsbury, head of the American Automobile Association's (AAA) racing program, had warned Parks: "You're not going to win any popularity contests, you're going to be an S.O.B."[7] Parks recalled that this left him "kind of shaken," but he adapted to the reality of Pillsbury's forecast, even if he tended to overreact to criticism when he seemed imperious, as he sometimes did. He called NHRA's rules and classification system a "universal standard," for instance, when it was nothing of the sort. In 1960, three years before he left Petersen Publishing to become NHRA's chief executive officer, he founded a semiweekly (soon weekly) newspaper, *National Dragster,* as "the voice of the organized drag racing sport"—pointedly ignoring *Drag News,* then five years old and doing rather well under the editorship of Chet Herbert's sister Doris.

Articulate and impeccably groomed, usually in coat and tie, Parks was always prepared for an interview or photo opportunity. He was keenly intelligent, even charismatic. And there was something of a cult of personality about him. Loyal subordinates might honor Parks for doing "a fantastic amount of work" while seldom receiving "tangible recognition,"[8] but in reality his likeness appeared over and over in *Hot Rod* and shots of him bestowing honors were printed literally thousands of times in *National Dragster* over a thirty-year period.

People were bound to have strong feelings about Wally Parks, and, as early as 1956, NHRA spun off a rival, when three onetime members of its technical committee in Pennsylvania founded the American Hot Rod Association (AHRA). The AHRA set up headquarters in Kansas City and began staging a Labor Day weekend championship—at Great Bend, where the president, Nelson Pointer, resided. (NHRA had set a new venue for its own event, in Kansas City, coincidentally enough.) Although begun without anything like the safari and a tie to a thriving monthly magazine to foster nationwide visibility, the AHRA soon fell to the control of a man of considerable ambition, James Tice, a veteran of the air force and former racer, who was in the insurance business in Kansas City. Tice knew that he was admirably placed to capitalize on NHRA's decision concerning fuel. Dragsters registered comparatively tepid performances at the NHRA Nationals throughout the latter 1950s, far slower than the numbers run under AHRA auspices by fuel racers like Emery Cook and Setto Postoian. In 1958 Don Garlits won the Great Bend event, to begin a three-decade reign as drag racing's biggest star.

Fuel was one factor in giving Jim Tice's organization a means of competing with Parks's NHRA. The other factor was money. Tice and his associates were avowedly profit oriented and readily understood that motivation in others, whereas the confinement of racers to amateur status was fundamental to NHRA dogma. "Diligence of the sport's devotees, who compete for *enjoyment* rather than monetary rewards, has kept the sport wholesome, respected and successful," wrote Parks. "The guy to watch out for is the conniver who would exploit your sport by dangling prize-money inducements before the eyes of unwary contestants."9 He reiterated his opposition to "prize-money inducements" time and again. There was, however, a wonderful irony to NHRA's fuel policy: It greatly accelerated the collapse of this ideal. NHRA might claim that drag racing was "by choice an amateur's sport" and that it was "mandatory, in the sport's best interest, that cash awards be restricted from this type of competition."10 But there were bound to be those skeptics who asked "Whose choice?" and "Whose best interest?"

Fuelers had established themselves as the best draw, and Tice conceded that their performances ought to be worth cash, either purses for open competition or "appearance money" just to show up. As people like Garlits acquired a nationwide following, this naturally enhanced their monetary value to promoters, and, in turn, their own bargaining position. The emergence of a band of itinerant performers who competed for something besides "enjoyment" obviously marked a major turning point. NHRA claimed to be upholding traditional American values, but what value was more traditional than quid pro quo? Another competitive arena had been opened, another concept of purpose defined. There was competition on the drag strip, of course. There was competition among manufacturers and machine shops, especially the cam grinders. And there was competition among drag strips and now among organizations that bestowed sanction on drag strips. At one point Garlits was named president of the AHRA; the title was only honorary, and Tice actually held a closer rein on his organization than Parks did on his. But Parks, with his high-profile style and his order-from-chaos rhetoric, seemed intent on something much more ambitious.

Actually, there was yet another sanctioning body that had been involved with drag racing as long as NHRA, the Automobile Timing Association of America, backed by Arnold Maremont of Chicago's Maremont Automotive Products. ATAA's initial venue was Half Day Speedway, some twenty miles from Chicago, which had opened in 1953, one of the first drag strips in the Midwest. ATAA was organized as a nonprofit corporation, like NHRA. But it paid purses, and beginning in 1954 it staged one of the premier events for fuel dragsters, the World Series of Drag Racing, first at Lawrenceville, Illinois, and later at Quad City Drag Strip in Cordova.

ATAA's managing director, Jim Lamona, pointedly ignored the fuel ban, even as NHRA endorsed and promoted it. Periodicals sympathetic to ATAA suggested that NHRA's stance entailed hidden political agendas.[11] In 1956 ATAA had sanctioned only 70 events, about one-tenth as many as NHRA, but in 1957 it sanctioned 250 and even took California strips into its fold. Early the next year, however, Parks and Maremont negotiated a merger, which was widely interpreted as a defensive measure on NHRA's part.[12] Lamona reorganized the ATAA as the American Timing Activities Association and carried on with a nucleus of strips in Illinois and Michigan that paid cash purses and welcomed fuelers.

Small-time operators often needed the crowds fuelers could attract in order to survive. As even Parks would later concede, as many as two-thirds of the NHRA strips did not adhere rigidly to its "gasoline-only" strictures. While the rule book did not allow for open competition among fuelers, promoters could book them in for match races, best of three or best of five. But NHRA treated all times turned by fuelers as unofficial, the ban came to be regarded as a measure Parks had fathered, and he remained firmly committed to a purpose that implicitly precluded getting from A to B as quickly as possible.

NHRA staged its first gasoline-only Nationals in September 1957, having changed the site for the second year in a row, this time to Oklahoma City. The winner's marks were 141.50 and 10.42, both excellent for gasoline, and Art Arfons topped 150 with Allison power. But Cook and Bedwell were clocking 160s all over the West in the summer and fall of 1957, and Postoian won the ATAA's World Series with a 9.33. And in November Garlits turned his 176.40 in 8.79 seconds at Brooksville. Such fuel performances dominated the headlines in *Drag News* while rarely getting any mention at all in *Hot Rod,* except in the ads. Generally, the best fuelers were about 25 miles per hour faster and a second quicker. The gap would narrow considerably within a few years, especially as certain racers went to twin-engine setups to regain an edge in horsepower. Yet the added weight was a crucial trade-off that would assure the continued presence of a gap. Obviously, Parks had decided that pushing technology to the limit was not a fundamental priority. Enhancing drag racing's image and overall visibility was a more important purpose, and there were many possible ways to do that.

In January 1958, for example, Parks arranged for a pair of dragsters to make record runs under the certification procedures of the Fédération Internationale de l'Automobile (FIA), an organization headquartered in Paris. The target was the standing ⅝-mile mark, a record set in 1937 by Bernd Rosemayer in a German Auto Union. The dragsters selected to challenge the record were those of Calvin Rice, winner of the first NHRA Nationals but pretty much out of the limelight since then, and the stylish but relatively obscure "Glass Slipper" of Ed Cortopassi from Sacramento. Rosemayer's record was broken, to be sure, but Iskenderian led an outcry against NHRA's "secret and closed" session: "It's not clear why the top performing cars like Don Garlits, Cook-Bedwell, and others were by-passed. Garlits's time for the ¼-mile 176.40 mph (official) is actually better than Rice did in the ⅝ mile run, 2½ times as far. These cars undoubtedly would have passed well over the 200 mph mark in this distance." Parks countered in terms that bespoke a different sense of purpose. Rice and Cortopassi were both "well qualified in experience, capability and cooperation." Crewmen "met the requirements for dependability, efficiency and sincerity." And, finally, in an unmistakable dig at Isky, Parks addressed the matter of "integrity." "Since the basis for this entire project was the establishment of *genuine* records, no one could risk having irresponsible advertising or misleading claims jeopardize the end results."[13]

That issue was a bit clouded, since Iskenderian was one of two cam grinders who got advertising mileage out of Cortopassi. In light of Parks's assertion that both cars "fit the standards for stability at high speeds, performance potential, general construction, appearance and overall interest," one might also have argued that there were plenty of other cars that fit these standards as well or better. Rice's "Hot Rod Magazine Special" did not even look entirely safe: NHRA rules called for a "head high" roll bar, and his was not even close. All that was beside the point. Isky was saying that the purpose of the FIA record session should have been top performance in a technological sense. Parks saw its purpose as a matter of gaining "international recognition" for men whom *he* regarded as dependable, sincere, and cooperative.

There was just a tinge of paranoia in *Hot Rod*'s editorial style. Parks depicted his foes as "unscrupulous parties, posing behind the community-benefit nature of the sport." He warned darkly not only about "irresponsible advertising" but also about the designs of promoters who sought "to exploit every possible dollar from the drag racing sport."[14] Yet, even while maintaining his opposition to get-rich-quick schemers, Parks could readily agree that promotional success was tied to black ink. One response to this anomalous stance was inevitable. "Somebody once said this was a 'Non-Profit' sport," Scotty Fenn wrote in one of his memorable missives. "All we can ask is—Non-Profit for whom?????"[15]

Fenn, who needled NHRA relentlessly, was anathema to Parks and everyone close to him.[16] Yet Parks was well aware that he had put himself in a position to take a lot of heat for a long time to come. Even as NHRA eventually readmitted fuelers to its fold, even as it instituted cash purses, even as it adopted the sort of promotional ploys it had once decried, Parks remained at the center of controversy. "NHRA is like an octopus, reaching in every direction, trying to bring in as much as it can for itself," ran one refrain. "Where would drag racing be if it were not for the efforts of NHRA?" ran a counter-refrain.[17]

No matter how wildly irresponsible they might be in ascribing evil motives to the Association, its critics could always command a forum. Yet the commercial excess that later inundated drag racing tended to confirm many of Wally's warnings from his *Hot Rod* podium. The men who founded NHRA had the zeal of missionaries; whatever else they may have been, they were confirmed technological enthusiasts, and there was no mistaking the irrepressible affection they felt for high-performance machinery. In other parts of the country, however, different sorts of people saw drag racing in a different light. For the pioneer generation of California racers it might be a technical challenge pure and simple. For the NHRA cohort it might have become a challenge to that group's logistical ingenuity. But elsewhere it was strictly a matter of making money for a growing band of promoters who had, at best, passed quickly through the phase of technological enthusiasm. The most prominent were in the Midwest. Eventually, NHRA would call on some of these same men to promote its own events, but at first *midwestern* was something of a code word for unrestrained commercialism. Not that there were not plenty of people with similar attitudes in other parts of the country. Later some of the most wretched schemes would be fostered by Californians; then people who decried this turn of events might have recalled Wally Parks's conservative preachments.

Still, NHRA was certainly culpable for "trying to bring in as much as it can for itself." In 1960 there was no *Drag News* account of the NHRA Nationals in September, and a California strip official named Steve Gibbs complained to Doris Herbert. Herbert answered that, since Parks had started his own newspaper, he had prohibited her from selling hers or even giving it away at NHRA events. She said that he had also thwarted her efforts to get a story at the Nationals.[18] Even though less than half the tracks were affiliated with NHRA, it claimed to be "the single official sanctioning body for drag racing in the U.S.," and *National Dragster*'s Dick Wells explicitly restricted his coverage. NHRA usually treated rivals (others besides AHRA sprouted up from time to time) as if they did not even exist.

There was an exception, however: the National Association of Stock Car Auto Racing (NASCAR), founded in 1948 by Bill France, a man who was, if anything, even more passionate about order than Wally Parks. NASCAR quickly became a dominant power in closed-course racing, especially in the southeastern states. In the mid-1950s it moved into drag racing, sanctioning strips as far afield as New York state. Drags were also staged in conjunction with NASCAR's "Speed Week" in Daytona Beach in February, but in 1956 these degenerated into a widely publicized "hot-rod riot." Miamian Ernie Schorb, who became one of Parks's first two regional directors, saw an opportunity and opened negotiations with NASCAR vice president Ed Otto. Four years later NASCAR joined NHRA in presenting a week-long "Winter Nationals."

Schorb's hope was that the two organizations, "by showing a spirit of friendliness, co-operation, fellowship, and, most of all, good sportsmanship," would ensure "a long and successful future" for this event.[19] It was not to be. The relationship between Parks and France always seemed edgy; the establishment of NHRA's tie to the FIA was said to be partly a move to thwart NASCAR's drag racing ambitions.[20] And, among other untoward happenings, there was a memorable clash involving Parks and Don Garlits, a denizen of the NASCAR heartland, who would certainly have supported France's organization in any contest over power. Garlits had put in a few laps on gasoline that winter, and on February 7 he headed over to the Spruce Creek strip, along with his brother Ed and his boyhood chum Art Malone. Garlits and Brock Yates tell the story of what ensued in *King of the Dragsters.*

There were runoffs each evening, with the competitor who did best for the whole week to be named the overall winner. The fireworks took place on Wednesday, when Garlits lined up against Alabamian Lewis Carden for the final round. As the flagman signaled "Go!" Garlits powered away from the line. But Carden, who knew he was outclassed and wanted to get a jump, exercised his option of not accepting the start. Seeing what had happened, Garlits slowed, turned around, and drove back up the strip. As he and Malone were maneuvering the dragster into position for the usual push start, Parks walked over and said: "Let it sit, Don. I'm afraid you're through for the evening." It turned out that Garlits had incurred a disqualification for crossing the center line, and Carden was going to get to make a solo run for the win.

Garlits had quite a temper: "*Like hell he is.*" He raised his hand in the air, "waving Art to start pushing us forward."

"Wait a minute, you guys," Wally shouted and tried to step between the rear of the dragster and the push car. At that moment Malone floored the accelerator. "Wally had no choice but to make a wild leap onto the hood as it picked up speed. . . ." Garlits engaged the clutch, fired up, gunned his engine, and roared past Carden at half-track.[21]

Naturally, the disqualification stood. At the end of the week, Parks handled presentations gracefully. But Garlits recalled that he won the top prize, where *National Dragster* reported that Carden was the winner. The seeds of dragdom's most enduring contest for power had been sown.

NHRA never ran another meet in conjunction with NASCAR, and in his book *Drag Racing: Yesterday and Today* Parks never even mentioned this Florida event. Beginning in 1961, NHRA scheduled its Winternationals for the Pomona Fairgrounds in California, a venue that had always been profitable and would continue to be. To hear Wally tell it, that 1961 meet was the first Winternationals, but Don Garlits, for one, would remember the one before, the one produced jointly. NASCAR continued to stage drag races as an added attraction to Speed Week, and in 1965 it inaugurated a fuel-dragster circuit in the East and Southeast, headed by Walt Mentzer, one of the trio of NHRA dissidents who had founded the AHRA in the mid-1950s. But the circuit enjoyed only modest success, and eventually NASCAR left the straightaways to NHRA.

Actually, a joint venture between NHRA and NASCAR seems unlikely on the face of it, given the personalities involved. William France and Wally Parks were almost the same age and just the same in many other ways. Each took what was essentially a regional pursuit, something "rudderless," and made it into a highly structured national pastime, something "orderly." Each was reticent about sharing power. Each was accused of dictatorial tactics. In the words of Shav Glick of the *Los Angeles Times* France "created a sport and held it hostage."[22] Parks never dominated drag racing quite to that extent, but he came close. By the 1960s Garlits was far and away the sport's biggest name—"the only man in drag racing who can pack a place to capacity," *Sports Illustrated* called him—but he knew that NHRA did not need him. As he whistled down the Spruce Creek strip with Parks sprawled on the hood behind him, his gabardine overcoat flapping in the wind, he recalled thinking: "Now you've gone and done it, Garlits. That's the boss of the most important drag racing organization in the world who's holding on for dear life back there. This is your last NHRA race for a long, long time."[23]

For all its vocal opposition to what it called prize money inducements, NHRA had actually been authorizing substantial merchandise awards almost from the beginning. Several brand-new engines were presented to various winners at the first Nationals in 1955. The top eliminator in 1958, the Escondido, California, team of Ted Cyr and Bill Hopper, received a Chevrolet pickup. In 1959 it was a Chevy El Camino for Rodney Singer and Karol Miller of Houston.

And beyond the tangible rewards the NHRA Nationals were far better publicized than any other event, even though times fell short of fuel times elsewhere, elapsed times anyway: The Arfons brothers always turned big numbers, but no Green Monster could get off the line quickly enough to advance very far in elimination rounds. Nevertheless, there were those who felt that the Arfons dragsters had a lot more in them, including drag racing's resident engineering expert, Roger Huntington. In their years of

In this shot, taken on a chilly evening after the conclusion of the NHRA/NASCAR Winternationals in 1960, some of the expressions are evidence of what Don Garlits described as the "wild and wooly" events that came before. Posed with Don's car and his various trophies are, from left, Ed Garlits, Shelby Carswell, Garlits, Ed Otto of NASCAR, Ernie Schorb of NHRA, and a rather striking Wally Parks. (Courtesy Museum of Drag Racing)

Seen here at the 1958 NHRA Nationals at Oklahoma City is Art Arfons in the latest from the stable of airplane-engined "Green Monsters." Though never a winner at a big NHRA event, top speed of the meet belonged perennially to the Arfons brothers. (Photograph by Bob Pendergast, courtesy Petersen Publishing Co.)

experimentation with aircraft engines (they also used Rangers, Rolls Royces, and several other types), Huntington thought that the Arfons brothers had made technical contributions "fully as important as the California boys with their nitro fuel and sling-shot chassis." He also believed that their 1959 model Monster had the potential for elapsed times in the 7s as well as 200-MPH speeds.[24]

And possibly this was true. What the Arfons brothers liked best about aircraft engines was that they were so inexpensive (they boasted of having acquired a surplus Allison for $35, a $12,000 Ranger "still in the original packing crate" for $50) and this permitted a lot of ongoing experimentation with the overall combination. Traction was their holy grail. In early 1959 Art Arfons turned a reported 8.53 at Chester, a strip with better bite than most. At the NHRA Nationals he topped 172 miles per hour but again fell early in elimination rounds. Allison-, Ranger-, and Franklin-powered dragsters hit better elapsed times than ever before, however, and by the next summer there were at least seven such machines on the scene. Chances seemed excellent that one or another would begin to match Huntington's expectations. Actually, no Monster ever got another shot at the NHRA Nationals. Headquarters had ruled that aircraft engines were ineligible for competition, just like that.

What was behind this ruling? One answer, though not the only possible answer, is suggested by the new location of the Nationals, Detroit Dragway, a strip ten miles from Dearborn and twenty from downtown Detroit. As Parks had now demonstrated any number of times, he had his own purposes. He was very excited about the 1959 event, not so much about the racing itself as something else:

> What was undoubtedly the best *selling* job ever done in behalf of the hot rod sport, and drag racing in particular, was accomplished at Detroit this year when members of the National Hot Rod Association presented the 5th Annual National Championship Drag Races there.
>   . . . among the 80,000 or so who attended the five days and two nights of activity were hundreds of representatives from the various auto manufacturing companies. . . . Never before has so much high level attention been focused on the drags. . . . Top leaders of the auto industry stood shaking their heads, some stating "I'd never have believed it if I hadn't seen it!"[25]

Parks had long sought a tie-up with the nation's automotive establishment, and this "high level attention" was like a dream coming true.

Of course, not everyone saw things as Parks did. Earlier Garlits had won an expense-paid trip to Detroit; he opted instead to go to the AHRA meet in Great Bend, one of the "fictitious championships" that *Hot Rod* never mentioned, except to lament their very existence. There he not only set top time and low ET (an extraordinary 8.23) but also took home a considerable chunk of cash, as did Chris Karamesines, a Chicagoan a few years Garlits's elder, who was emerging as his closest competitor. Karamesines had a penchant for fuels even more exotic than nitro, such as hydrazine, an anhydrous compound capable of bursting into flame when merely spilled on oxidized metal. He would have scoffed at the idea of confining himself to gasoline when there was such "liquid horsepower" available with so much more performance potential.

Hot Rod's account of the Detroit Nationals conceded that the fuel restriction limited performances somewhat in one regard. In another, however, "drag racing was never more potent." NHRA had mailed hundreds of invitations and drew "an unprecedented number of interested guests representing the auto manufacturers, administrative officials, engineers, stylists, technicians, sales and advertising reps, public relations and agency people, and countless other associates." And, when the finalists began rolling up to the line, "it was FORD VS CHRYSLER VS GENERAL MOTORS. . . . Detroit and the automobile industries got their first real look at drag racing à la NHRA, and judging from all appearances, they *liked* it."[26]

Scotty Fenn continued to be a burr in Wally's saddle, accusing NHRA of having accepted fifty thousand dollars "to swing the Nationals from Oke City to Detroit." Parks might have asked, "From whom?" but instead he answered, somewhat unconvincingly, that this decision was not his to make.[27] He could not, however, deny the political utility of having a situation in which the finalists had engines from each of the so-called Big Three. Allisons had been a General Motors product, to be sure, but GM corporate executives were not likely to have been thrilled by Art and Walt Arfons. What if—as the brothers believed and Roger Huntington thought possible—a dragster with an aero engine would soon prove capable not just of setting top time but of actually winning the NHRA Nationals, defeating Chryslers, Lincolns, Pontiacs, Oldsmobiles, and Chevrolets? How would Detroit feel about NHRA drag racing if that happened?

As far back as 1953, Art Arfons had been paid appearance money by an Illinois promoter; in 1959 he toured the country from coast to coast, drawing big crowds everywhere, and less than three months before the Nationals Ed Eaton had posed for publicity shots in the Green Monster. Yet Parks could later write that "the aircraft powerplants did not suit the basic automotive appeal of drag racing."[28] Perhaps he meant to say that they did not suit *him* or suit the Detroit people who flocked to the 1959 Nationals and would do so again in 1960. Looking back, another Allison enthusiast, Jim Lytle, thought he knew the answer: "Art Arfons was entering the record books far too often with his Allison dragster. And Allisons do not sell new Fords or Chevrolets, nor do they sell speed equipment." When NHRA added "automobile engines only" to "gasoline only" in the rule book, again it cited safety considerations. Lytle believed that the reasons were "purely political and economic."[29]

*E*ven though NHRA might see fit, for its own reasons, to preclude certain technological choices, an organization whose motto was Ingenuity in Action could hardly be inhospitable to novelty per se. Ever since 1957 there had been talk about minimum weights and ceilings on engine displacement. Both were instituted eventually, but in this early period NHRA generally heeded pleas about not "limiting the tools" with which designers of "unlimited" dragsters had to work.[30] There was every reason to anticipate that the performance gap between fuel and gas dragsters could be narrowed substantially, and that prospect was naturally pleasing to the NHRA cohort. In any activity involving technological devices ingenuity has been most strongly elicited where the mandated handicaps weigh heaviest. Many 1990 top-fuelers could have come from a cookie cutter, whereas cars in classes with sweeping restrictions displayed rich variety. So it was with gas dragsters during the period that fuel was out of favor with the Association.

There were, for example, the sidewinders—dragsters with the engine mounted transversely. A few of these had showed up even before 1957. Bob Tennant, a member of the Akron contingent who favored airplane engines, raced one with a 460-cubic-inch Ranger for several years. Tennant used chain drive, but Bert Kessler and Dean Gammill of Mattoon, Illinois, went to the big ATAA and NHRA meets in 1956 with a Crossley-bodied sedan that had a transverse Oldsmobile engine geared to the rear axle. At about the same time a Pennsylvanian named Lowell Lister debuted a sidewinder dubbed "Crossfire," and in California one of C.J. Hart's original partners at Santa Ana, Creighton Hunter, built a flathead-powered sidewinder called the "Piece of Pie" that clocked 140s before it was demolished in a crash.

Except for Hunter's, sidewinders had mid-engines, a configuration that had been rapidly losing favor to slingshots because of handling difficulties that seemed inherent. But it might be different with the crankshaft spinning in the same plane as the axle. Dynamic forces would work to load the rear tires under acceleration. Moreover, because the power train did not have to negotiate a 90-degree angle, both rear tires would be loaded equally, and there would be no tendency for everything to torque over to the right, impairing both traction and control. And the engine's mass could be crowded up even closer to the rear tires than with a slingshot. In theory at least the mid-engine sidewinder heralded design progress that might compensate for the gasoline-only handicap.

The concept was pushed most deliberately in southern California, where Lions, Santa Ana, and other major tracks were unwavering in their support of the fuel ban and did not even bring in fuelers for exhibitions. In 1957 Paul Nicolini and his father-in-law, Harry Duncan, a wealthy Orange County contractor who had been active in drag racing since its Santa Ana beginnings, built a tube-framed sidewinder with a Chrysler engine and chain drive. Nicolini coaxed this machine to elapsed times in the high 9s but was plagued by bent axles; with so much of the weight concentrated

Bert Kessler and Dean Gammill show off the mounting of their Oldsmobile engine at the 1956 NHRA Nationals at Kansas City. (Photograph by Dick Day, courtesy Petersen Publishing Co.)

right next to the rear end and with no suspension, any wheel "hop" coming off the line was devastating. The chain drive itself was always troublesome. Moreover, even though the engine's attitude permitted locating the cockpit further from the front wheels than with most mid-engine cars, it was certainly "forward enough to require a sensitive feel in the seat of the pants to detect any inclination of the rear wheels to take off on a tangent during acceleration."[31] (Recall that drivers counted on "sighting" along what part of the car was in front of them to detect whether or not they were going straight.) Like almost all mid-engine dragsters, it soon gained a reputation for spooky handling.

Fed up with bent axles, broken chains, and aborted runs, Nicolini sold the chassis to Chuck Jones, a graphic artist from La Crescenta who had raced a fuel coupe at Santa Ana in the days of Hunter's Piece of Pie and who also worked as managing editor of *Drag News* for a time. Jones convened a team that included Joe Mailliard, owner of a Long Beach machine shop catering to drag racers, and Wayne Reed, a house mover who had been involved in various quarter-mile and Bonneville ventures. The driver was Art Chrisman's Uncle Jack. Born in Oklahoma, like many of the southern California drag racing crowd, Jack Chrisman operated a service station in Willowbrook. He had been involved with the drags almost as long as his nephew (they were close to the same age), gaining experience in all sorts of vehicles, from a Chrysler-powered Model-A sedan to a roadster of Reed's to a dragster of Duncan's. Skilled and versatile, he was destined for nation-wide exposure as one of drag racing's first "touring professionals."

Chrisman had driven mid-engine machines previously and quickly got the feel that eluded Nicolini. With Reed's supercharged Chrysler gulping air just over his head he began winning local meets with some frequency in 1959. In May, at Lions, Chrisman set an NHRA record of 9.11, which was only about a half-second off his nephew's fuel record, and he also won $1,100 worth of bonds. A week later C.J. Hart prepared to close up at Santa Ana by staging a grand finale. He drew his biggest crowd ever, and those on hand saw Chrisman win top eliminator with a 9.16, while nobody else in competition ran quicker than 9.40s.

In the hands of an adept chauffeur the sidewinder configuration showed a lot of potential, so much that Jones had Nicolini build him an improved version. Presaging a tactic others would try—cutting the odds by half—he entered both of these in the 1959 NHRA Nationals in Detroit. Chrisman came very close to winning in the newer machine, and it appeared that a few additional refinements might be all it would take. Weight could be pared, and there could be more wheelbase for better directional stability.

Jones went to the San Fernando Valley, to Kent Fuller, who was gaining repute as the most accomplished designer of dragster chassis. Fuller was ordinarily a partisan of chrome moly steel, so-called aircraft tubing. But the dragster that he and Jones planned was made from *magnesium* tubing—expensive and tricky to work with, but very light indeed. While this "Mag-winder" had a 113-inch wheelbase, longer than most slingshots, it measured only 30.5 inches to the top of the roll bar and weighed just 1,443 pounds ready to race, even with full magnesium and aluminum body panels, the work of Indy car builder Wayne Ewing.[32]

Jones, Fuller, and Chrisman debuted their elegant new machine at Lions during the 1960 holidays. It ran speeds in the 170s but was not an instant winner, and Jones competed only sporadically in 1961 and on into 1962, his dreams fading as his purse emptied. Sidewinders suffered from mechanical problems like bent axles, and they could be a handful to drive. One run might be straight as a string, the next a brush with disaster. The design would continue to prove seductive. Jim and Don Sivenpiper, brothers from New York state, built a chain drive, DeSoto–powered machine in the mid-1960s. Nearly twenty years later Chuck and Mike Sage, brothers from

The Jones-Reed-Mailliard-Chrisman blown Chrysler sidewinder, 1959. "5 Cycle" was a type of camshaft produced by Iskenderian. (Photograph by Donald Nickles, courtesy National Hot Rod Assn.)

The Chet Herbert twin-engine machine campaigned by Lefty Mudersbach is seen here in 1962. (Courtesy National Hot Rod Assn.)

Ohio, built one with a gear drive for Garlits, in which he installed a state-of-the-art fuel Chrysler. Art Malone tried a sidewinder, too, and Jack Chrisman even commissioned a Mustang-bodied "Funny Winder." But mastering a novel configuration requires tenacity, and nobody was ever willing to make the requisite investment of time, money, and energy. Although Garlits was overjoyed when he first took delivery of his sidewinder, he eventually concluded that it was "jinxed."

Nobody seemed to have had a feel for the breed like Chrisman, but his skills were much in demand by others besides Jones. In 1960 he took a slingshot on tour, for a series of bookings at strips in the South and Northeast; he did the same with another slingshot in 1961 and never drove for Jones again. The slingshot configuration was now perceived as "normal," and most racers were now searching not for a new paradigm but simply for large measures of brute horsepower.

Given NHRA's proscription on fuel and airplane engines, where could such power most readily be found? In multiple automotive engines, of course. Two gasoline engines could easily outmatch the power of one fuel burner. Some strips had instituted limitations on displacement; idealizing "ingenuity not inches," for example, San Gabriel Dragway put a 341-cubic-inch cap on blown dragsters. But NHRA resisted any urge to limit cubic inches or ban "twins," which would have simply amounted to handing the competition another card to play (AHRA rolled out the red carpet for airplane engines, just as it had done for fuelers). Two-engine setups had precedents, of course, notably the unorthodox Bustle Bomb campaigned by Lloyd Scott. With regard to configuration there were other unorthodox options—Lee Titus of Santa Monica even built a sidewinder designed for engines fore and aft the rear axle, à la Don Jensen's twin—but the obvious choices involved the tried-and-true slingshot, with the engines either side by side or in tandem.

By all odds the most successful of the side-by-side twins was designed and fabricated by Phil Johnson and Howard Johansen and sponsored by Johansen's company, Howards Racing Cams. Johansen, born in 1909 in Shelby, Nebraska, had settled in Los Angeles in 1941, opening an auto repair shop on South Main Street. Regrinding cams was a sideline at first, but gradually the repair business was phased out. Howard was known as a man who "could make anything." He got into producing forged aluminum connecting rods and a great many other engine components but always kept in direct touch with the racing scene. He himself had driven on midwestern ovals before the war, and in the late 1940s he was involved in the lively California dirt track circuit as well as in boat racing and the action at El Mirage. He helped dozens of drag racers, and from the 1950s through the 1970s there was always a top-flight machine running as the Howard Cam Special. When Howard died in 1988 he was dubbed "the quintessential hot rodder."[33]

Howard was enthusiastic about novelty. He had once outfitted a '35 Ford coupe with a Marmon V-16 in the rear, and he built a Bonneville streamliner using two aircraft drop tanks, one for the engine, the other, beside it, for the driver. There had been dragsters with side-by-side engines before but never one quite like Howard's. The engines, Chevrolets with GMC blowers, were installed with the one on the left set to run backwards (i.e., with the firing order reversed) and geared to the other via the flywheel ring gears. Power was transmitted from the right-hand engine to the rear axle via a single drive shaft. Howard debuted this machine in the summer of 1958, and he and his sons campaigned it for more than three years as the "Twin Bears," or just "the Bear." There were different drivers at different times, but usually it was the well-traveled Jack Chrisman. Although a broken axle spelled defeat at the 1958 NHRA Nationals, later that year the Bear clocked 166 at Famoso, and by the end of 1959 it had turned 8.75 at almost 178 miles per hour. On at least one occasion it won top eliminator against a field of fuelers.

Howard's cam-grinding rival, Chet Herbert, soon debuted a twin, too, but with a different configuration, the engines in tandem; each was nearly 100 cubic inches larger than the Bear's, but there were no superchargers. The builder and driver was Lefty Mudersbach, who had fabricated the Money Olds Special in which Buddy Sampson won the 1957 NHRA Nationals (the last machine without a supercharger ever to do so). And a young Hollywood actor named Tommy Ivo, with a chassis from Fuller, merged both approaches, big unblown engines (Buicks in his case, totaling 928 cubic inches) but arranged side by side.

As the 1960 NHRA Nationals approached, there was something nearing a consensus that a twin-engine slingshot would be the winner, or at least would clock the low elapsed time.[34] And certainly there were plenty of twins on hand in Detroit. Bill Tibboles of Ohio and Don Westerdale, a local boy, played variations on the side-by-side theme, as did Texan Jack Moss. Moss had actually pioneered the configuration along with Jazzy Jim Nelson, but, like Nelson, he faced both engines forward and used two separate drive shafts. Another Jim Nelson, this one from Oceanside, California, rather than Venice, played another variation. Nelson and his partner, Dode Martin, who had recently entered the ranks of commercial chassis fabricators with their Dragmaster Company, showed up with the "Two Thing." It had 354-cubic-inch Chevys geared to one another, but the blowers were in front, driven directly off the crankshaft and ducted to the engines. Two Thing turned the event's best time, 171 miles per hour, only fractionally short of Art Arfons's mark from the year before.

Despite expectations, the twin-engine legions ultimately fell to a conventional slingshot with one supercharged Olds, entered by Gene Adams and Ronnie Scrima and driven by Leonard Harris of Playa del Rey. Scrima had built one of the very first rail jobs ever, and Adams was one of the truly

Jack Chrisman poses with Howard Johansen's son Jerry after they won the 1961 NHRA Winternationals at Pomona. Note the aerodynamic device, a sheet of plywood; dragsters of this period often had very little bodywork. (Courtesy National Hot Rod Assn.)

keen minds in drag racing. Then twenty-five, he had started out at Santa Ana in his dad's '50 Olds Rocket 88, become involved in a fabled rivalry with Howard Johansen's souped-up '55 Chevy, and eventually moved up to a supercharged dragster in partnership with Scrima and cam grinder Jack Engle. After that car crashed in 1959 at Lions, Adams and Scrima purchased a Fenn K-88 chassis and turned to Harris, a muscular gymnast who was highly regarded for his natural prowess as a driver. Harris also arranged sponsorship from a local Oldsmobile dealer, and the Albertson Olds Special dominated Saturday night competition at Lions throughout the summer of 1960, going unbeaten for almost three months. To most competitors that was a feat almost beyond comprehension.

Back in California, after winning the Nationals, Harris picked up right where he left off, and Lions impresario Mickey Thompson knew a money-

At an event at Amarillo Dragway prior to the 1961 NHRA Nationals, Eddie Hill charges off the line in his twin-engine Pontiac. Hill was always an innovator; with this machine, in addition to everything else, the engines were integral with the frame. (Photograph by Don Brown, courtesy Museum of Drag Racing)

making promotion when he saw it. "The Match Race of the Century" between Howard's Bear and the Albertson Olds Special was scheduled for Saturday night, October 22, 1960. But that afternoon, while testing another dragster, the steering malfunctioned, and Harris was killed.

Leonard Harris was far from the first dragster driver to meet with a fatal accident, but he was the first star to do so. Actually, he was a rookie sensation. Fans, promoters, and drivers were reminded that racing dragsters was dangerous, and tears were shed. And then, of course, the show went on. Adams soon had a new driver, Tom McEwen, twenty-three, and kept right on winning. But the Bear won elsewhere just as convincingly, and Chrisman and McEwen emerged as the favorites for the NHRA's big winter meet, now slated for Pomona.

The Pomona pits were packed with twins. John Peters and Nye Frank debuted another variation on the theme, tandem Chevys with a crank-driven blower feeding into duct work. Jack Moss came out from Texas with a revamped twin-driveshaft machine. Mickey Thompson had four-wheel drive. The Two Thing from Oceanside turned 174.92, breaking Arfons's speed record, then another twin did it too. But the Bear finally had its day, defeating McEwen in the final.[35]

Chrisman, Nelson and Martin, Peters and Frank, Ivo, and Lefty kept up the momentum, and new twins kept appearing. Chet Herbert financed another machine similar to Lefty's, driven by Zane Shubert, who had worked in his shop since he was sixteen (Herbert himself was confined to a wheel-chair). In the fall of 1961 Thompson staged four straight meets at Lions worth one thousand dollars to the winner, three of which were won by twins—Shubert, Chrisman, and Lefty—the fourth by Adams and McEwen.

Although twins delivered more horsepower, they inevitably traded off weight. The Two Thing came in at almost 1,900 lbs., and, apropos of Mickey's twin, Jim Nelson remarked that "all 4-wheel drive twin-engine cars have ended up weighing 3,000 pounds." That was as much as the airplane-engined dragsters. Even if weight could be pared drastically, the twins were still twice as complex, and the risk of engine failure was exactly doubled. Mudersbach argued that two engines yielded sufficient horse-power without the addition of superchargers, and that was certainly a step back towards simplicity. But even Lefty's unblown machine weighed 1,660 lbs. Nelson felt that "it would be better with no twin engine cars. . . . It wouldn't hurt my feelings if they [the NHRA] cut them off."[36]

Nelson, whom Wally Parks liked as much as he loathed Scotty Fenn, was conceding the obvious—that the ultimate constraints on the way these machines would be set up would be political, not technological. But Parks was apparently still hoping that the twin-engine dragsters could overtake the fuelers, which were proving such a boon to Jim Tice. There was no longer any pretense that the gasoline-only ruling had anything to do with keeping costs down for competitors, or speeds either.

In 1961 NHRA moved the site of its Nationals for the fourth time, to Indianapolis Raceway Park (IRP), a new facility a few miles west of the venerable Brickyard. Texan Eddie Hill showed up at IRP with two blown 422-cubic-inch Pontiacs and quadruple rear tires. After literally tearing holes in the starting line asphalt, he tied with the Two Thing and two other cars for top time, 170.45. But the other two had single engines, and, as with the Albertson Olds machine in 1960, the event was again won by a carefully engineered single-engine lightweight.

This time the victor was Lew Russell "Pete" Robinson, twenty-seven, a virtual unknown, even though he had been coming up from Atlanta to compete at the Nationals in a '40 Ford coupe for several years. Pete Robin-son foretold a new breed. Like Eddie Hill, he had formal training in engi-neering, at Georgia Tech. He had a methodical way of working which re-

Pete Robinson poses with the machine he took to the 1961 Nationals, as cleverly engineered as Eddie Hill's but much lighter and less complex. (Photograph by Eric Rickman, courtesy Petersen Publishing Co.)

minded *Hot Rod*'s LeRoi Smith of "German race teams."[37] And he was always acutely conscious of weight. "Your engine has a certain amount of available power," he said, "and the less total weight it has to propel, the faster and quicker the vehicle will accelerate."[38]

What Pete brought in the gate at IRP appeared to be a conventional Nelson-Martin Dragmaster chassis with a supercharged 352-cubic-inch Chevy engine. But it weighed only 1,280 pounds, nearly 50 percent less than the Dragmaster "factory car," the Two Thing. (It also induced much consternation among the NHRA inspection crew, who feared that it was too flimsy to be safe and tapped every inch of the frame to determine whether Pete had substituted something very thin.) Jack Chrisman had again been one of the favorites, but, as Chrisman himself put it, Pete "just plain outran me" in the semifinals. And in the final round Robinson beat McEwen in the single-engine dragster of Gene Adams, a racer whose approach was much the same as Robinson's.[39]

Notwithstanding their own considerable success with a twin, for the 1962 Winternationals the Dragmaster team, Nelson and Martin, switched to a similar single-engine sprinter, a car that worked "on getting running room off the line and then holding on." Weight was the principal topic of conversation in the Pomona pits. Michigan racer Connie Kalitta was sporting aluminum heads on his Chrysler, "special little gems" that saved 95 lbs. and enabled him to come across the scales at only 1,250 lbs. But the Drag-master "Dart" weighed 1,114 and Robinson's car less than 1,000. With a longer and somewhat lighter frame the Bear did very well at Pomona, clocking top time, 176, and tying Robinson for low elapsed time, 8.50. A low ET in qualifying was never, however, any guarantee of a win. The contestants in the final round were Adams and McEwen, who had a new lightweight Kent Fuller car, and Jim Nelson.[40]

The winner was Nelson, but that was immaterial to most of the other racers, who were faced with a serious question indeed: What was the "best way" to design one of these things? Two engines and sacrifice on weight, or one engine and sacrifice horsepower? At the Winternationals the Bear had another driver, and Jack Chrisman was handling a Mickey Thompson machine that had one aluminum-block Buick and one aluminum Olds F-85 and was as light as any twin. In September they took this car to Indy, along with a single-engine dragster powered by a hybrid Pontiac engine with hemi heads à la Chrysler. It was in this latter machine that Chrisman set low ET and also won the event. He defeated Garlits, who now had a deal with the Dodge Division of Chrysler Corporation and had been helped by Dodge public relations to patch up his differences with Wally Parks.

The Best Engineered award went to a twin from Dayton, perhaps in an effort to shore up interest in the breed. But twins were definitely a waning fashion, even though the Peters and Frank machine did win at Pomona the next February. NHRA, starting to waver on its position regarding "the bug-aboo of fuel," elected to admit fuelers to the 1963 Winternationals on a "trial basis." Peters and Frank called their machine the "Freight Train," and it surely put on a stirring performance, coming within 1.5 miles per hour of the fastest fueler on the fairgrounds. But the fuel ban had been rationalized on the basis of safety and economy. Safety seemed like a moot point if twin-engine gas dragsters were going nearly as fast as fuelers. Economy was moot, too: The Freight Train was considerably more costly than a fuel machine, and yet it was still a quarter-second shy in the elapsed time department, 8.36 compared to 8.11 for the quickest fueler at Pomona. Some enthusiasts just liked the *idea* of two engines. But others were wondering about the whole point of the fuel ban. They were surely wondering in the NHRA boardroom.

Political constraints had largely sustained several years of intensive development with gas-burning dragsters: sidewinders, twins, and light-weights. When NHRA put the lid on special fuels in 1957 the best gas dragsters (with automobile engines) were running 140s in the mid-10s. Six years later they were up to 185, with elapsed times two seconds quicker; two engines or one, it did not seem to matter, but one engine was a lot less bother. Although it appeared that the major NHRA events would attract huge crowds no matter what, the machines that drew most attention at most strips operating week in and week out were, as Garlits said, "the faster, noisier, more spectacular fuelers."[41] Even though the disparity had been narrowed, under the right conditions fuelers could hit elapsed times no gas dragster could touch, times in the 7s as well as speeds in the 190s.

Mickey Thompson had been one of the instigators of the fuel ban, and the torrid level of competition at Lions had been one of the primary factors sustaining it. At the beginning of 1962, however, competition from nearby promoters who permitted fuel induced Thompson to institute separate "top fuel" and "top gas" categories. Soon Lions, like most other California strips, was paying more money to the fuelers. When NHRA "legalized" fuel for the 1963 Winternationals Ed Eaton termed this a "test" to measure "the amount of interest" in the fuelers.[42] There was plenty, and in 1964 they competed at the Indianapolis Nationals as well. That was also the year NHRA first paid prize money and the year that fuelers broke through to 200-MPH speeds. Within three years elapsed times would be in the 6s, speeds in the 220s. NHRA would continue to stage competition among gas dragsters until the early 1970s, but the spotlight remained on the fuelers. At least as far as dragsters were concerned; as we shall see, Jack Chrisman would pioneer an entirely different sort of machine, which proved just as popular as dragsters—more popular, in fact. Their performance would be measured theatrically, not technologically, and they would put NHRA in another quandary not unlike the one it had to face regarding "pump gas only." NHRA could alter evolutionary paths, but its power was never absolute.

To be sure, it fueled an ongoing debate by its very existence. Its partisans asked, "Did the [drag] strips just drop as manna because all the hot rodders were such good guys?"[43] Detractors played dirty tricks such as distributing laminated plastic tags that read "I Hate Wally," and the accusations of duplicity were endless. Yet there were people outside NHRA councils who felt that the organization's muscle was deficient. A *Drag News* correspondent named Al Caldwell agreed with a reader who wrote: "Drag Racing needs a NASCAR type organization to take over. NASCAR is better organized, better run, and has a better safety record." Caldwell added, however, that the outlook of the typical NASCAR competitor and the typical drag racer were "vastly dissimilar."[44]

While sidewinders, dual engines, blowers, chrome moly chassis, and a lot of other innovation came out of California, not all of it did by any means. In the late 1950s fuel dragster technology was pushed by intrepid pioneers from the Midwest, Texas, and Florida. The shrewdest innovator of all was from Tampa. Garlits was a technological innovator nonpareil, but more important was his innovation in the realm of entrepreneurship. Garlits was the first drag racer—not a promoter or a denizen of some sanctioning body, not a manufacturer or a supplier—the first *racer* to turn the activity to consistent profit. In so doing, he provided a role model for thousands of others; he made it possible for them to dream of being pros on the drag racing stage, too—and for many years he made it impossible for Wally Parks to control his domain in quite the way that Bill France controlled his.

5

FAME

The Californians were so smug about their drag racing accomplishments that they figured it was downright impossible for anyone from another section of the country—especially swampy ol' Florida—to outdo them.

DON GARLITS, 1967

ike all of the "car culture," drag racing flourished in California because of the year-round sunshine and because there was a critical mass of specialty equipment manufacturers and suppliers. The cam grinders; the foundries and machine shops that turned out cylinder heads, pistons, rods, flywheels, and intake manifolds; the firms that made ignitions, fuel injectors, exhaust headers, clutches, and other such equipment—in the 1950s virtually all of them were located in California. Men like Paul Schiefer and Vic Edelbrock had been denizens of Muroc, and the "world's first speed shop," Bell Auto Parts in Bellflower, started in business when Henry Ford was still making Model-Ts. Edelbrock began manufacturing dual-carburetor manifolds for flathead V-8s in 1938, later enhancing his line with many other products such as high-compression heads and kits for converting Ford Stromberg carburetors to nitro. Schiefer began making aluminum flywheels for flatheads in 1948, then branched out into clutches and magnetos. Moonlighting while employed as a machinist for Schiefer, Bruce Crower started to produce his U-Fab intake manifolds for Chryslers and other overhead valve engines. In 1955 he teamed up with Dave Schneider to challenge Ed Iskenderian, Howard Johansen, Chet Herbert, Jack Engle, Kenny Harman, Phil Weber, Clay Smith, Racer Brown, Dempsey Wilson, Chuck Potvin, and others in a very competitive camshaft business, which had been pioneered before the war by Ed Winfield, a mechanical genius who did R&D work for Detroit, held numerous patents, and served in a mentor's role to men like Isky.

In Long Beach Frank Venolia began casting racing pistons from permanent molds just after the war. A few years later, in Santa Fe Springs, Dean Moon began making aluminum fuel tanks and pumps, then expanded by taking over Potvin's shop. In Glendale Kent Enderle began competing against Stu Hilborn's Fuel Injection Engineering in Santa Monica, and Crower later joined the injector fray too. Bill Hays made clutches and flywheels; Jerry Jardine made exhaust headers. Harold Nicholson in Arcadia, Nelson Taylor in Whittier, and Bobby Meeks at Edelbrock's put together completely reworked flatheads, for which there were more than two hundred manufacturers producing special equipment as early as 1950. A new generation of engine builders—men like Ray Brown, Dave Zeuschel, Keith Black, Ed Donovan, Ed Pink, and Sid Waterman—favored the Chrysler, with its hemispherical heads. Machinists like Phil Weiand, who had initially made parts for Ford flatheads, were soon concentrating on hemis—manifolds and blower drives in Weiand's case.

In North Hollywood Bob Spar and Mort Shuman formed B&M Automotive in 1955, manufacturing beefed-up transmissions, and Don Clark and Clem Tebow formed C-T Automotive, specializing in hard-chromed crankshafts. When Milo Franklin and Don Alderson started Milodon Engineering in Van Nuys they mostly did aircraft work, but soon they were concentrating on main-bearing supports ("girdles") and other components used in drag racing engines. Muffler shops abounded. The energetic Mickey Thompson opened one as a sideline in the late 1950s; his M/T Enterprises in Long Beach was eventually involved in the manufacture or distribution of more than fourteen hundred products for dragsters and other sorts of race cars.

These names do not begin to exhaust the roll. There were Earl Evans, Tommy Thickstun, Bob Tattersfield, Frank Baron, Barney Navarro, Charles "Kong" Jackson, Eddie Meyer, Al Sharp, Joe Hunt, Don Alpenfels, Tom and Bill Spalding, Al Gonzales, and Wayne Horning; Offenhauser, Scott, Reath, Riley, McGurk, Nicson, Halibrand, Hildebrandt, and dozens more specialty manufacturers, not to mention the jobbers and retailers like Alex Xydias in Burbank, Al Hubbard in Hayward, and Don Blair in Pasadena. Firms like Hilborn's and Engle's, Isky's and Edelbrock's, cross-fertilized one another's R&D efforts. Pattern makers and machinists would apprentice with one shop then split off to form their own firm nearby—sometimes amicably, sometimes not. Later there would be complex litigation over patents and a wave of mergers and buy-outs. But some of the pioneering firms, such as Edelbrock, Iskenderian, and Crower, remained all in the family, if rarely in the same location at which they had started out; Orange and San Diego county addresses later became commonplace.

Not that *every* specialty equipment manufacturer was in California. Harvey Crane, for example, founded Crane Engineering in Hallandale, Florida, in 1953, although his name did not become widely known among drag racers until Pete Robinson won the 1961 NHRA Nationals with a Crane cam in his engine. In Lakewood, Ohio, Joe Schubeck began fabricating dragster frames commercially in the late 1950s, as did Pat Bilbow in Wilkes-Barre, Pennsylvania.[1]

But firms like Crane Engineering, Schubeck's Lakewood Industries, and Bilbow's Lynwood Welding were simply exceptions that proved the rule. Much of drag racing's intense competitive energy was imparted by the cam grinders, and very few of these were not located in California. Much of the impetus towards normalizing dragster design came from the commercial chassis makers, and all the major ones were in California: Scotty Fenn in Inglewood, Jim Nelson and Dode Martin's Dragmaster Company in Oceanside, Lefty Mudersbach in Pico Rivera, Kent Fuller in Van Nuys.[2] Bruce's Slicks were made in Oakland. Even though M&H tires came from Massachusetts, the primary distributor was Ernie Hashim in Bakersfield.

The leading monthly concerned with drag racing, *Hot Rod,* had its editorial offices on Hollywood Boulevard. *Drag News* started out in Long Beach, then, after Doris Herbert took over, moved to Santa Ana and later to Highland Park. The NHRA was headquartered on North Vermont Avenue in Los Angeles. The most famous drag strips were at Long Beach, Pomona, San Fernando, and San Gabriel in Los Angeles County; Santa Ana in Orange County; Famoso in Kern County; Riverside and Fontana out around San Bernardino; and Kingdon, Fremont, and Half Moon Bay near San Francisco. Even small operations such as Colton, Cotati, Santa Maria, Vaca Valley, and Inyokern attracted strong competition. Until 1957 every successive drag racing record was set by a California machine, the sole exception being the speeds attained by the Arfons brothers with their airplane engines. One can hardly blame the Californians for being "so smug" or wonder at their skepticism when a fuel dragster from "swampy ol' Florida" started making regular headlines in *Drag News.*[3]

Unless they had happened to notice a letter to the editor in the spring of 1957 complaining that there was "nothing about Florida" in *Hot Rod,* California drag racers first heard the name Don Garlits that fall when he defeated Emery Cook at the ATAA World Series in Cordova. Some of them confused Garlits with Art Arfons, somebody else from back East. (As *Drag News* remarked later, as far as the Californians were concerned, "back east was everything on the other side of Casa Grande and Tucson, Arizona.")[4] Actually, Garlits had been racing as long as almost any Californian, since 1950, on the streets of Tampa and St. Petersburg and on the rough-and-ready drag strips at Zephyrhills and Lake Wales.

Garlits was the son of a former Westinghouse engineer who had come to Florida in 1927 with his wife, Helen. Edward Garlits prospered raising oranges, then went broke in 1932, the year Don was born. Brother Ed was born two years later. The elder Garlits gradually recouped, opening a small welding and machine shop. He was not always a gentle man, and Helen divorced him in 1943, but not before the boys had grown to appreciate "the hard, precise feeling of metal" and their dad's "tough principles regarding workmanship" had been firmly imbued in them.[5]

Don and Ed's stepfather, Alex Weir, was not mechanically inclined. He raised cows, and tending them was up to the boys. They also earned some change repairing bicycles. When Don was a senior at Hillsborough High in Tampa he bought his first car, a '40 Ford sedan, and also took an elective in metal shop. The instructor introduced him to *Hot Rod,* and reading each issue of the magazine over and over reinforced his enthusiasm for tinkering with automobiles. After graduation he got a job in the accounting department of a Tampa department store. It was awful. When he asked his stepfather for advice Alex encouraged him to "take advantage of [his] mechanical ability." So he quit his bookkeeping job and went to work in a body shop. He also began to hang out with the local hot rodders, street racing in a '40 Ford convertible with a Cadillac engine. It was quite a wild passage in Don Garlits's life. Then he met Pat Bieger, a young woman from "a gentle middle-class family in Tampa" who worked as a secretary for the chamber of commerce. Love calmed Don down a lot. He and Pat married in 1953, and, while employed by the American Can Company, Don built a house in northern Tampa.[6]

Then, one Sunday afternoon, Don and Pat stopped by the Lake Wales Drag Strip. He entered their '50 Ford, won a trophy, and all the old enthusiasm was rekindled. Soon he was racing a '27-T roadster, which, in emulation of Mickey Thompson and Calvin Rice on the West Coast (and probably Bill Martin on the opposite coast of Florida), he reconfigured as a slingshot before the NHRA's Safari came to Florida in 1955. Even though he won the event at Lake City, *Hot Rod* never ran a photo of his machine, which was by his own admission "probably the ugliest dragster in the business."[7] Early in 1956 he sold it and built another one around a set of narrowed 1930 Chevy frame rails, which were torsionally strong and also had the correct "kick" over the rear axle. This dragster looked quite a bit like its contemporaries in California; later dubbed "Swamp Rat," it would stir strong passions among the racers there. The power plant was a 331-cubic-inch Chrysler hemi with a Crower U-Fab manifold, Isky cam, Harman and Collins magneto, and long header pipes that swept out past the rear tires. Running as much nitro as he dared, 25 percent, Garlits turned speeds around 135 at places like Amelia Earhart Field in Hialeah. Some of the Californians had already topped 150. But, if Garlits was a little behind performances being posted on the West Coast, he was way ahead on one score: In Florida, at least, he was in a class by himself, and already he was thinking in terms of operating "full-time in racing."[8]

After opening a garage specializing in "performance tuning," Don decided to enter an event in which he "could take on some really big-time competition." NHRA competition was gasoline only, and Garlits was hooked on the power of nitro. So it was that he trailered to the World Series in Cordova, at the time probably the most important event independent of the NHRA. And that was how he met Emery Cook, a man with a "gigantic reputation" who was nevertheless willing to show a novice how to modify his fuel system to run *lots* of nitro—95 percent, not 25. Garlits was able to pare two seconds off the elapsed times he clocked upon first arriving, and he proceeded to outrun Cook in one of the elimination rounds. Even though he subsequently lost to Setto Postoian—whose new slingshot took low ET of the meet at 9.26—word of this new "easterner" got back to California pronto.[9]

While Iskenderian was stirring the pot with his ads featuring the Don's Speed Shop Special from Tampa, Garlits set about upgrading his machinery in accord with the setups he had observed in Cordova. Taking a cue from Postoian, he installed a big 392-cubic-inch hemi (the kind Chrysler put in its New Yorker in 1957 and 1958) with an eight-carburetor Weiand "Drag Star" manifold. Figuring that he now had at least 650 horsepower at his disposal on 95 percent nitro, he converted to direct drive, as Cook had done more than a year before. On his first outing he turned 156 at Kissimmee, south of Orlando. After solving a fuel flow problem, he turned 176 at Brooksville, with an elapsed time of 8.79. And, with a set of experimental seven-inch tires from Marvin Rifchin—the first M&H slicks ever run on a dragster—he won an event at Chester, South Carolina, sanctioned by the International Timing Association (ITA), one even richer than the World Series. Isky played up events at Brooksville and Chester for all he was worth, as did Phil Weiand, and the reaction among the California racers was altogether predictable.

Garlits was dismissed as "some kind of a big phony—running against bad clocks with a second-rate car." Even if the Brooksville timers were accurate, that strip was concrete, and times were not to be compared with those turned on asphalt. As for winning events staged in places such as South Carolina, people asked, who besides Postoian had there been to give him a real run for the money?[10] When Garlits first got wind of this kind of talk he was furious, but, as he thought further, he could see that all the uproar was impelling him towards his goal of becoming a full-time drag racer. People were going to want to see him perform, if only to see him shown up. With Ed's help he prepared for the challenges he hoped were coming, and, indeed, the phone began to ring.

In May 1958 the Red River Timing Association, which operated a drag strip in Wichita Falls, offered Garlits $450 to come to Texas. This was his

The first of Don Garlits's "Swamp Rat" series is seen here at Sebring, Florida, in 1956, essentially as it was set up at the start. On Don's left is his wife, Pat, and his brother Ed. On his right is Bob Phillips, out of whose brother's Chrysler the engine had been taken. (Courtesy Museum of Drag Racing)

very first "appearance money," and, with the exception of payouts to Art Arfons and the Arizona Speed Sport team simply to make exhibition runs, almost certainly the first such deal ever made by any drag racer. Also on hand at Wichita Falls was the former Cook and Bedwell machine, still running strong with new owners from Kansas; the Speed Sport roadster, which had turned times in the 170s; Bobby Langley, whose "Scorpion" slingshot had been garnering a reputation around Texas similar to Garlits's in the Southeast; Mel Heath of Rush Springs, Oklahoma, who had won the NHRA Nationals the last time fuel was allowed, in 1956; and Jack Moss, with one of the earliest of the side-by-side twins. Garlits returned from Texas victorious, then went on to win the ITA meet in Chester once again, although his marks fell off from previously. Now Postoian was stirring the pot along with the Californians, announcing that Garlits's car was simply not capable of the performances that Isky had been claiming.

Again a call came from Texas, from Lynn Huiet, promoter for the Freeway Drag Strip near Houston. The first time Garlits had gone to Texas it was to race Texans and people from close by. This time it was different. Through Iskenderian and Chet Herbert, Huiet had arranged a "California Chal-

Besides the obvious differences, such as fuel injectors rather than carburetors, contrast the detailing of Romeo Palamides's machine with Garlits's (facing page). Here Pete Ogden is seen in the cockpit. (Courtesy Don Jensen)

lenge." The primary challenger was Jack Ewell from San Pedro, driving Jim Kamboor's "Jado Spl." Also on hand was Pete Ogden, the adept handler of Romeo Palamides's machine from Oakland. Kamboor was the owner of a Los Angeles catering business. Romeo had well-to-do patrons in San Francisco and was renowned for the style and glitter of his equipment. Langley was there too, as was John "Red" Case, in a tiny mid-engine machine that had turned speeds close to 170.

Garlits had some trouble getting things sorted out and was subject to taunts from the California contingent—"Hey, Garlits, where are those 176s? See what honest clocks can do, Garlits." But when time came to race he successively polished off Case, then Ewell, and then Ogden. He was particularly pleased about the final, not only because Ogden was known for his skill at driving off the line with maximum weight transfer—front wheels in the air—but also because Romeo's slingshot was probably the most costly machine of any in drag racing; there was no reason for doubt when he said that the total investment was more than five thousand dollars. Ever since his street-racing days around Tampa and St. Petersburg Garlits had relished taking on "the rich guys."[11]

With brother Ed along Don spent plenty of time on the road in 1958. At the World Series he picked up five hundred dollars for the fastest time. At Caddo Mills he won the "Texas State Championship," defeating Langley, and at Great Bend he won the AHRA's National Championship, also setting top time. At the NASCAR strip in Montgomery, New York, he clocked an 8.36 ET on new eight-inch M&Hs. While these tires yielded a definite edge, Montgomery used a brand of timers generally regarded as unreliable. Across the continent cries of "phoney times" went up. They grew louder when Garlits clocked 180 in a match race with Langley at Brooksville, despite the fact that NHRA's Ernie Schorb had attested to the timing equipment. "Get Garlits to California!" went the call. "Get him up against some real competition and some accurate clocks!"

Garlits could see the dollar signs, but in dragdom's capital a lot of other people were starting to have visions of cashing in, too. When the deal was finally put together to bring him west the principals included officials of the Smokers, the Bakersfield club that ran the strip at Famoso (which was called Bakersfield as often as Famoso, even though the two places were twenty miles apart). Also involved were Scotty Fenn and Ed Iskenderian, who seemingly had their own competition going over who could stir up the most controversy. The Smokers called their event the United States Fuel and Gas Championship, but, with the major southern California strips adhering to the fuel ban, the fuelers would naturally be the main draw, and Don Garlits in particular—people like Isky would make sure of that.

Early in 1959 Garlits contracted to appear at three events—the two-day Famoso meet at the beginning of March, another at Kingdon two weeks later, and a third a week after that at Chandler, Arizona, near Phoenix. He would receive a total of $5,000 in guarantees against winnings, $2,000 from the Smokers, and $1,500 from each of the other two strips. This was a world in which a California enthusiast would travel fifteen hundred miles just to compete: "No bucks, not one dime of tow money, nothing," Pete Ogden recalled of his Houston trip.[12] In California winning a major event rarely paid more than $100; more often it was just trophies and $25 bonds. In a world in which most drag racers still conceptualized their activity as a hobby, $5,000 was incredible. Isky delightedly broke the story in *Drag News:*

> Throughout the south, the eastern seaboard, the midwest and Texas, wherever he has run, the exploits of Garlits and his string of victories have drawn unreserved praise and open amazement. The only skepticism has remained on the West Coast, the hotbed of drag racing. Certain factions, though soundly beaten in a Texas showdown, have used various alibis, including weather conditions, to cover up their defeat and have spread the report that Garlits would more than meet his match "on the coast." This spring every top West Coast car will have an opportunity to match wheels with Garlits.

Garlits was not the only one who was coming west "to race the 'big ones' right in their own back yard." Iskenderian made sure that Postoian would be in California, too; indeed, Don and brother Ed met up with Setto in Houston, caravanning the rest of the way westward to Bakersfield.[13]

Judged against the cream of the California cars, Postoian's machine was not state-of-the-art, and Garlits's was even less so. Garlits still had his Chevy frame-rails, whereas most of the best California fuelers had custom-fabricated chassis. Garlits had indicated to Ernie Schorb that he had $2,500 invested in his machine, but, when he pulled into the pit gate at Famoso, one bystander is said to have remarked, "Hell, there isn't a thousand dollars there, if you threw in the trailer, both suitcases, and the gas money Garlits is saving for his return trip home."[14] The California fuelers were epitomized in the new "Hustler" of Art Chrisman and his wealthy partner, Frank Cannon. Chrisman, who had been setting records since the very earliest days at Santa Ana and Paradise Mesa, had recently turned almost 182 at Riverside, with an ET of 8.54. While his basic power plant was the same as Garlits's, a 392 Chrysler, Chrisman's engine had all sorts of "trick" items such as a hard-chromed stroker crankshaft, forged pistons, and, most notably, a big 6-71 GMC blower, along with a set of Hilborn fuel injectors fed by a high-volume pump.

Garlits knew about Chrysler fuelers with blowers driven directly off the crankshaft. There was Tony Waters, for example, who was famous for his attention to detail and could run in the 8s even though he favored a DeSoto engine, a down-scaled version of the Chrysler hemi. But mounting a Jimmie on top permitted overdriving it faster than the speed of the engine. And Chrisman was only one of several Californians to have this powerful combination; another was Gary Cagle, in the slingshot he drove for Chet Herbert. Moreover, while Garlits had recently been able to count on M&H tires to give him an edge, the Californians had now been introduced to M&Hs, too, via the peripatetic Texan Bobby Langley, who had appeared at Riverside and Famoso a few months before.

Besides the names with which Garlits was already familiar, there were dozens of other racers with cars that looked polished and costly. In Saturday time trials whenever one of them ran a big number the announcer would inquire, "How'd you like that one, Tampa Dan?" The crowd, none too hospitable at best, would go wild. And the crowd was huge, perhaps the largest ever to attend a drag race in California since its beginnings a decade before. At the spectator gate ticket sellers were stuffing the take into shopping bags in the bed of a pickup truck. Elsewhere there were scenes out of Wally Parks's worst nightmares. Technical inspections were cursory. Crowd control was virtually nil. Empty beer cans littered the staging area, and there were drunken fistfights in the grandstands. Time and again competitors flew into a rage over some imagined act of favoritism and attacked the hapless official charged with flag-starting the races. Jay Cheatham, who

Gary Cagle in Chet Herbert's slingshot at Famoso. GMC blowers, though seen occasionally atop flathead Fords (Tom Cobbs had a 4-71 like this one at Goleta in 1949), had been difficult to adapt to Chryslers because of problems with fuel flow, distribution, and pressure. This basic setup would become virtually standard for fuelers within a year or two (see page 98) and remain so into the 1990s, albeit with a toothed Gilmer-belt drive, rather than the V-belts seen here, and much larger blowers. (Courtesy Museum of Drag Racing)

campaigned one of the strongest gas dragsters in the land and had come within an ace of winning the NHRA Nationals a few months before, was killed when his Pearson Olds Special from Sunnyvale crashed on Saturday.

In the pits Don and Ed Garlits had their own troubles. Tuning a fuel dragster is largely a matter of adjusting to the available traction and the relative air density. Although both are affected by such factors as ambient temperature and time of day, every strip has a normal range. Racers like Chrisman, Cagle, and Waters had run at Famoso many times before and knew what to expect, but conditions there were entirely new to the Garlits brothers. "It takes time to learn a strip," Don observed. "You send any West Coast driver to Florida and his times will drop until he learns."[15] But this was the San Joaquin Valley in California—hot, dry, and dusty—not swampy ol' Florida. In time trials Garlits finally clocked 172 in nine seconds flat. While these would have been considered excellent marks a year before, they were outclassed by the likes of Art Chrisman. To make matters worse, Garlits ruined a cylinder wall and had to prepare a new short block for the elimination rounds.

After an all-night session at Ernie Hashim's shop, with Ed and Isky working at his side, Don made it back to the track on time on Sunday morning. But, as he staged for the first round, an insert bearing failed and a rod broke. He recalled that the crowd, which had turned positively nasty, "nearly jeered us out of the place when the announcer jubilantly gave them the news that we couldn't run."[16] Chrisman ended up taking the win, defeating Waters in an exciting final round that was run in near darkness, the only light being from spectators' headlights.

Amid rumors that he had deliberately sabotaged his engine to avoid getting humiliated, Garlits began preparing for the Kingdon event, working at Iskenderian's shop in Inglewood. Fortunately, he had money for new parts. The first order of business was to get a Jimmie blower. With Isky helping set up the belt drive unit (a component he was tooling up to manufacture), Garlits assembled a 454-cubic-inch engine with a hard-chromed crankshaft from C-T Automotive, forged pistons, and rods that had been reinforced by "boxing" the area between the flanges—costly items he had never used before. When he finally rolled his dragster onto the trailer and headed over the Ridge Route for Bob Cress's Kingdon strip, he felt reasonably confident.

Chrisman sat out that event. Cagle, a police officer, could not get off duty. But Waters was there, along with Bill Crossley, driving Hashim's machine, and Ed Cortopassi's Glass Slipper, one of the dragsters Wally Parks had selected to challenge FIA records—and Setto, of course, who had also installed a blower. Garlits outran Cortopassi, then Postoian, and in the final round he consigned Waters to his second runner-up finish in two weeks. He had saved face, but it could still be argued that the level of competition was simply not the same as at Famoso. In Arizona a week later, however, Garlits dominated a field that included not only Waters, Setto, and Langley but also Ted Cyr (who was equally skilled with gas and fuel and had won the 1958 NHRA Nationals), an up-and-comer from Pasadena named Chuck Gireth, and both Art Chrisman's Hustler and his Uncle Jack's Bear.[17]

At Chandler Garlits discovered a new driving technique: M&Hs dissipated heat so effectively that, rather than coming off the line at less than full throttle, he began charging away wide open, the tires pouring smoke. While Cyr had top speed, 178, nobody else was even close to Garlits's best ET of 8.43. "Lighting the tires" became the convention for driving top-fuelers, and there were those who believed that this was a way to get around the physical limitations imposed by frictional coefficients. Others would conclude, however, that slippage should be confined to the clutch and that the tires ought to be hooked up to the pavement, notably Keith Black, a name soon to be heard often. As it turned out, Black's way would win out, but this was a matter not to be resolved conclusively for nearly a decade.

All in all Don Garlits had reason to feel pretty good as he and his brother headed for home at the end of March 1959. They had proved themselves against "the California boys." And those boys for their part had come to an understanding that Garlits was for real, that he was driven by competitive

The final round at the 1959 U.S. Fuel and Gas Championship. Art Chrisman was beaten off the line by Tony Waters but won the race when Waters lost control. (Courtesy Museum of Drag Racing)

Garlits versus Setto Postoian at Kingdon in March 1959. Although both men had installed blowers, they set them up with Stromberg carburetors rather than Hilborn fuel injectors, such as several of the Californians were running. (Courtesy Museum of Drag Racing)

fires few of the Californians could even comprehend, and that he was a quick learner—more than that, eventually it would become clear, as the AHRA's Jim Tice put it, that he was "without question the smartest man in drag racing."[18]

Garlits regained his speed mark from Art Chrisman in short order. With the 180s, however, all the fuel racers would hit a wall. While they continued to chip away at ET records, they would gain only a few miles an hour in the next five years. Still, it must have seemed to Garlits that his own commercial prospects were boundless. In June 1959 a series of full-page ads began appearing in *Hot Rod* in which he endorsed Pennzoil Z-7. In June 1960 he made *Life.* With an offer of five thousand dollars a theretofore obscure machinist named Ray Giovannoni broke Don's allegiance to Iskenderian, and GIOVANNONI CAMS was lettered on the nose of his dragster where Isky's name had long been featured.

And Garlits was getting plenty of bookings, even as the whole concept of "appearance money" was changing drag racing's commercial profile. When, for example, *Drag News* arranged a seven-week series of appearances for Gary Cagle, Don signed to compete against him at two of his own favorite venues, Houston and Chester. Cagle promised to be a good draw, but he never got further than his first appearance, at Great Bend on Sunday, June 7. On his second run he got off the track after clocking 169 and flipped several times. While the roll bar held, he was thrown out the underside of the frame and critically injured.[19] His bookings were salvaged when Chet Herbert arranged to put his engine in Bob Sullivan's "Pandemonium" from Kansas City. Sullivan had been gaining something of a reputation in his own right at places like Chester, but Garlits defeated him handily at Houston the next weekend, turning 182 in 8.48 seconds. Texans began pinning the nickname "Daddy" on Garlits, which soon gave him a name recognition value no other drag racer could approach.[20]

After Houston "Daddy Don" (he would not become "Big Daddy" until the early 1960s) headed for a second confrontation with Sullivan the following Sunday at Chester. Whether or not Garlits thought much about such things, Cagle's accident, the death of Jay Cheatham three months before at Famoso, and the death of Red Case at Vaca Valley signaled dark days ahead. Harold Nicholson would die at San Fernando. Yeiji Toyota would die at Fremont, Mickey Brown at Lions, Hank Vincent at Fremont—all of them topflight drivers. In 1960, before a hometown audience in Detroit, Setto Postoian crashed when one of his new spoked wheels collapsed. He was hospitalized with a broken back, broken legs, and other grave injuries, and he never drove a fueler again. Over the next few years men would die by the dozens in dragsters, as lightweight construction was taken to absurd extremes, as overstressed engine and driveline components disintegrated, and as strips wide enough or long enough for 130 MPH proved quite inadequate for 180s.

When Garlits raced Sullivan at Houston he was using a Hilborn fuel injector for the first time. His performance was way ahead of the rest of the field, and he anticipated improving even more at Chester. But in practice on Saturday, June 20, with a 70-percent mixture of nitro in the tank, he confronted an awful new peril: "Racing supercharged fuel dragsters was a whole new world and nobody had ever considered the possibility of explosion and fire." Garlits tells the story in searing detail in his autobiography:

> Suddenly there was a loud explosion and I was immersed in flame. I sat there for a split second, roaring along at 170 mph, not even responding to the heat. No pain had started to register and it was like a dream, a nasty sensation that would all go away in a second. But it didn't go away and great licks of murderous fire lashed back from the engine and enveloped the cockpit. The manifold had ruptured and the tremendous blower pressure had popped open a gaping hole three feet away from my face— from which flames were belching like a giant blow torch. The pain was unbearable. . . . Stunned by the explosion, I had kept my right foot partially on the throttle, which in turn kept the butterfly of the supercharger open and continued to pump raw fuel into the holocaust. . . . The thought flitted through my mind that I was about to die.[21]

Garlits managed to get stopped and was taken to the local hospital, critically burned on his face and hands. One physician wanted to amputate his left hand. His survival was almost certainly due to the fact that he was wearing a leather jacket Pat had given him just a week earlier; before that he ordinarily drove wearing just a cotton T-shirt. Recovery was going to take a long time, however, and, with unlimited hours to assess his options while hospitalized in Tampa, he thought to himself, "I've run my last drag race." He told his brother Ed to sell everything.[22]

One day his old chum Art Malone dropped in to visit, along with his wife, Lorraine, and Garlits announced that he was through for good. "Too bad," Malone answered, "just when you were getting a name."

"Right," Don agreed. "I had some pretty good contracts to appear at strips up north."

"Listen, Don," Art said, "I'll lay it on the line. I'd like to take a crack at running your dragster."[23]

Garlits thought it over. Malone had never driven a fueler, but he had a lot of natural ability and had been fairly successful in stock car racing while doing construction work around Clearwater. Malone remained an unknown, but Garlits knew that his own fame was worth money. A lot of top competitors limited themselves strictly to the mechanical end of things— Gene Adams, for instance. Others such as Tom McEwen concentrated solely on driving. For the first generation of touring fuel racers, however, the idea of having to divvy up the income was not particularly appealing. These men usually fabricated their own chassis, assembled their own engines, and drove, taking some youngster along to races to serve as a "donkey" in return for travel expenses only. Still, Garlits had worked himself into a position from which he could command more appearance money than anyone else in drag racing. And he did need money; not long after getting out of the hospital, he found out that the promoter at Chester had failed to pay his insurance premium, and his hospital expenses were not covered.[24] If anybody could make a partnership work in financial terms, he might be the one. And so he agreed to split the gross with Malone, while he himself would cover operating and maintenance expenses.

Their first booking was in Sanford, Maine, a strip that paid well enough to attract the best touring fuelers. Although Malone's starting line reactions needed honing, he proved to be a quick study otherwise. By the time he and Garlits arrived at Great Bend for the AHRA Nationals in September 1959 he had developed into a skilled "shoe," making a pass in the 8.20s during preliminaries. Yet, in the elimination rounds, Malone lost to Chris Karamesines, running out of Al's Speed Shop in Aurora, Illinois.

"The Greek," he was called. Garlits described him as "a friendly bushy-haired guy," and a *Sports Illustrated* writer later called him "a mustached, sleepy-eyed man who looks like a character out of Steinbeck's *Tortilla Flat*."[25] This was his first season in a fueler, but already he owed Garlits a great deal. While he still held down a regular job, his dream was to follow the trail that Garlits had blazed, racing full-time. He and his sidekick, Don Maynard (a former partner of Fisher and Greth), had recently converted to a supercharged hemi set up like Garlits's. For fuel racers back East, anyway, Garlits was the man to emulate in technological realms. Even more important for some of them was his conceptualization of drag racing as a business.

Karamesines had a keen flair for the theatrical. His penchant for "keeping his foot in it" rather than aborting a crossed-up run quickly earned him a revised sobriquet, "the Crazy Greek." Daredevil antics would quite literally be the death of a good many drivers in the 1960s, but luck was with Karamesines; in more than thirty years he would never be badly injured (he was still at it in the 1990s), and he acquired an avid and loyal following. "Chris could get a car down the track as good as anyone I ever saw drive," remarked one competitor.[26]

The Greek ran Iskenderian components, and Isky made a lot of hay out of his exploits, particularly when he was credited with turning a speed of 204.54 miles per hour at Alton Dragway near St. Louis in April 1960. This was no doubt a fluke, but nothing like that ever stopped Isky's ad writers, nor was the Greek inclined to admit that there was anything dubious: "It definitely felt faster to me than any previous pass," he deadpanned.[27] Ultimately, after Karamesines bought a chassis from Rod Stuckey, the Kansas City fabricator who built Sullivan's Pandemonium, his "204" machine was enshrined in the Chicago Museum of Science and Industry. As a folk hero, the Greek was Garlits's peer. In two-out-of-three match races—the touring fuel racer's staple performance—he would prove more successful against Garlits than any other competitor he ever faced. But Karamesines knew as well as anybody that there would have been no band of professional fuel racers had it not been for two other men. It would not have happened had there been no Wally Parks and no NHRA gasoline-only rule. And, more important, it would not have happened had there been no Daddy Don to serve as a role model and to assume the status of "top gun" for others to challenge.

Sometimes Garlits had to put other people in the cockpit—first Malone, later a young Okie named Connie Swingle, who had formerly made his living, such as it was, street racing for money. Like the Greek, Swingle could electrify a crowd. Indeed, there might have been better pure showmen than Garlits, but nobody had his magic fame, and nobody had a keener business sense. His confrontation with Wally Parks provided priceless publicity. So did a feud that developed when Postoian, abetted by Iskenderian, went after "Swamp Rat" Garlits for striking his bargain with Malone: "You're in the sport for what you can get . . . you don't care who you hurt as long as you can make a buck at it . . . you shoot off your mouth about safety and then you take a green kid and put him in the biggest blown fueler you can possibly make.[28] When under fire Garlits's humor sometimes failed him, and the implication that he was being careless with another man's life might have been an outrage, especially in view of the fact that there were car owners on the West Coast who were actually guilty of this.[29] But Garlits and Malone both saw Setto's blast as a blessing. Malone had THE GREEN KID lettered on his helmet at once, and when the two of them hit the road for Riverside Raceway in California in December 1959 (for one thousand dollars—by then Garlits's standard guarantee) the Swamp Rat moniker was being used by the promoter, Rex Gilbert, to stir up excitement.

The whole idea was to get people to take notice; whether or not they actually *liked* Don Garlits was immaterial. And, indeed, in California it was now his very successes that were thrown up to him. "You know, I've never seen anything original originate from the south since the Civil War," wrote a *Drag News* correspondent named Bill Friend, referring to Garlits's switch to the Jimmie-Hilborn combination the preceding spring. "Come on . . . come up with some ideas of your own. Quit copying ours."[30] While Friend was hardly Garlits's only foe, he was the nastiest one to get his views in print regularly. Time and again he lashed out at Garlits (or at "Mr. Garlits-Iskenderian"—attacks on Isky, an Armenian who was making a lot of money from drag racing, had a distinctly racist flavor). Friend also harped on the appearance of Garlits's car, with its "clobbered welds." "If I owned a car like his and received the same publicity," he wrote, "I'd be too ashamed to admit it."[31]

On the other hand, there were admonitions to "face up to the facts."

With guys like Garlits, Sullivan, Langley, and the Greek coming from all parts of the country, California had simply lost its edge. And to call Garlits's car unsafe was to impugn the integrity of technical inspectors everywhere as being men "easily swayed by a celebrity." "Just because his car is not all chrome and candy apple red, that doesn't mean it isn't safe and fast."[32] Still, one had to concede that a promoter who had paid a performer just to show up was not likely to turn him away because a tech crew expressed concerns. Even Dean Brown, *Drag News*'s founder, complained about Garlits's guarantees and the preferential treatment he got from promoters such as Riverside's Rex Gilbert and Ron Lawrence of Fremont.

Whether pro or con, controversy did nothing but increase Garlits's commercial value. And all of his critics seemed to miss one crucial point: He was simply taking advantage of that sine qua non of American capitalism, the law of supply and demand. As he himself explained:

> the reason that the Garlits-Malone Dragster and other dragsters like mine are paid to appear at drag strips is because the big-name dragsters, by traveling all over the United States in their quest for competition, have built up spectator interest in drag racing within the last several years to make drag racing a profitable business for drag strips. Therefore, they are now willing to pay those cars who have the popularity to draw spectators . . . mine being one of them.[33]

At Riverside, on December 20, a reported twenty thousand spectators were on hand for a meet pitting the newly dubbed Swamp Rat of Garlits and Malone against fifteen other fuel racers, including Karamesines (who was on the coast for the first time) and Art Chrisman. It was becoming harder and harder to claim that somehow Garlits's machine did not perform as well as claimed. For his seven runs Malone averaged 178, 8.70. In the semifinal round he beat Chrisman (whom Garlits himself had never raced) with the low ET of the meet, 8.50, and in the final he beat the Greek.

Garlits and Malone returned to Florida, running gasoline for a couple of events, including the one at which the set-to with Wally Parks occurred. Then it was back to the coast for the second annual Famoso meet, followed by appearances at Kingdon and Fremont. They almost always excelled, yet Malone was as headstrong as Garlits, and he was getting restless. Garlits was edgy, too. When the break came ostensibly the issue was a matter of technological choice: Malone thought Garlits should build a lighter machine; Garlits, "the wallet," disagreed. Malone went to Wichita to build a fueler of his own in the shop of Al Williams, known for his weight-saving tricks, and soon he joined the ranks of touring fuelers on his own.[34]

When Malone left, Garlits had dates to fulfill in New York, Pennsylvania, New Jersey, and Wisconsin. Even though his burns were still not fully healed, he strapped in behind a blown fuel-burning engine once again. In August 1960 he headed for Alton and Great Bend, after having constructed

In September 1960 five of the nation's best-traveled fuel racers pose at the AHRA Nationals in Kansas City. From right, Chris Karamesines, Don Garlits, Bob Sullivan, Bobby Langley, and, in Al Williams's machine, Art Malone. (Courtesy Museum of Drag Racing)

a new car, his first made from chrome moly tubing, Swamp Rat III (Swamp Rat II was a gas dragster campaigned by Ed). Then it was on to the coast again, to Riverside, where it was "'Swamp Rat' Don vs. Racers of the U.S." The California appearances had become particularly important to him "because successes there invariably got into the drag racing press and led to contracts." Promoters from other parts of the country would often sign up the pros right on the spot.[35]

Actually, Garlits was in a slump, and, as often happens, he started fooling with his combination. He had constructed Swamp Rat III with a 128-inch wheelbase, almost two feet longer than Swamp Rat I. Given current engine and tire technology, this was actually a smart move, but there was still a strong school of thought on the West Coast that favored a short wheelbase for maximum weight transfer: Palamides had lopped more than 10 inches off a dragster that was short to begin with. Garlits lost confidence in his own judgment, cut back to 105 inches, then was plagued with handling problems.

Back in Florida, he went to the Golden Triangle strip in Oldsmar to try to sort things out. There, at 185 miles per hour, he had another fire, this one fueled by engine oil spraying out through a hole poked in the pan by a broken rod. He was wearing an aluminized suit, gloves, and a mask, which afforded some protection, and he was able to get up and shuffle around his shop within a few days. He was nonetheless convinced that "driving a fuel dragster was a shortcut to disaster." And, as always, he had contracts to fulfill.[36] Swingle, his cocky and eager helper from Oklahoma, was like most young guys in the sport: "He simply couldn't conceive of anything happening to him." And so Swingle climbed into the cockpit of Swamp Rat III in the summer of 1961.

Garlits got a lot of satisfaction out of seeing his new protégé beat Malone at Brooksville in his first outing, but soon, just like Malone had done, Swingle was complaining that Don's car was too heavy. This was confirmed by a visit to Pete Robinson, the acknowledged master of lightweight design, and Connie and Don succeeded in paring 1,800 pounds down to 1,350. "Now, I'll run you *real* times," Connie promised. Sure enough, Swingle clocked 190 at Saginaw, Michigan, and again at Atco, New Jersey, but then at Emporia, Virginia, his parachute failed to open, and he crashed into a pine grove. He only had cracked ribs, but the car was junk.[37]

In one week they had a date at York, Pennsylvania, against a new eastern hot shoe, Dick Belfatti, and somehow they managed to get a car together and fulfill the commitment after a nonstop drive. Then, in an exhibition at the Reading fairgrounds three days later, Swingle crashed again. And later he crashed yet again. Both times the car was reparable, and Swingle went on to

An inspector checks Al Williams's "Hypersonic I," a slingshot with a frame fabricated from rectangular aluminum tubing throughout (note the roll cage). Williams's next machine was conventional chrome moly. (Photograph by Union Electric, courtesy Museum of Drag Racing)

win the World Series at Cordova and almost take the AHRA Nationals. Then it was on to California, to a 185-MPH run at Lions (which reinstated fuel in January 1962), then to the Smokers meet at Famoso, with eighty-seven fuelers on hand and a huge crowd. Garlits and Swingle qualified on Saturday afternoon, hightailed it over the Ridge Route to race Texan Vance Hunt at San Gabriel that evening, then rushed back to Bakersfield for the Sunday elimination rounds. Swingle lost to Tom McEwen, and the ultimate winner was a budding star named Don Prudhomme, but aside from that quite an amazing thing happened: The loudest cheers came for Art Chrisman—and for Don Garlits. The Swamp Rat not only had won fame but also favor.

This surely pleased Garlits, and surely he was doing well financially. But chaos seemed to have become a way of life—frenzied dashes across vast distances in the middle of the night, not to mention fires, obstreperous drivers, crashes, various and sundry incidents. Garlits received a concussion in a Houston parking lot when a jack slipped out from under his GMC Carry-All, pinning his head under the front suspension. Not long afterwards Swingle saved his life when the same GMC plunged into a murky ditch at 2:00 A.M., on the road between a Saturday date in Columbia and a Sunday date in Savannah. Again, the dragster was a write-off. Don was a family man now with two little girls, Gay Lyn and Donna. He and Ed were starting to do well fabricating chassis and selling them to other racers—Vance Hunt, for one. In April 1962 Don put Swamp Rat III up for sale for eight thousand dollars and announced that he was temporarily retiring from active competition. He had negotiated a nice deal with the Chrysler Corporation and was being urged to curtail his "outlaw" activities and return to the good graces of the National Hot Rod Association to campaign under the Dodge banner.

NHRA competition, orderly, regulated, and comparatively safe—one rationale for gasoline only, recall—seemed like an attractive proposition even to Don Garlits. And so it was that after a summer-long layoff he went to the NHRA Nationals at Indianapolis Raceway Park under the wing of a Detroit automaker as well as Pennzoil, Wynn's Friction Proofing, and Champion Spark Plugs. He had a driver slated for Swamp Rat IV, a new gas dragster, and another for a Dodge stocker as well. Don could never be an establishment man, but he could certainly seek a day-to-day routine that was somewhat less hectic and perilous than his had become.

In the four years since Garlits signed his first contract to appear for a guarantee, drag racing had changed enormously. It was not just that the cars were faster, the crowds bigger, corporate America more deeply involved. What Garlits had done was establish an alternative to Wally Parks's vision of the future. Parks thought in terms of an amateur pastime uncorrupted by money—money for the performers, anyway. A big winner would reap lots

After losing in the final round to Connie Kalitta the year before, Garlits showed up at Famoso in 1965 with three different entries, one of them driven by Connie Swingle and outfitted with an experimental sprung rear end. Here Garlits attempts to prevent Swingle from creeping into the ET beam on the starting line and spoiling a qualifying run. Fans loved this sort of drama. (Courtesy Museum of Drag Racing)

of glory, maybe even leave the Nationals in a new car, but then he would return to his regular job, whatever it might be. Even a big winner would be out-of-pocket. Garlits had a different vision, one much more in tune with the American entrepreneurial spirit. He wanted to make drag racing pay, not cost. Bass fishers, bowlers, and beach volleyball players (among many other enthusiasts whose activities would air on television by 1990 as "professional" sports) might all have had the same dream, but they could not conceivably fulfill it then. For a drag racer, however, fulfillment might be feasible because that activity was based on automobility, technological innovation, and human derring-do—matters that fascinated the American public.

Not that Garlits had brought drag racing much nearer to social respectability; he knew that as well as anybody. "Mention drag racing," he remarked, "and right away people look at you kind of queer, and you know what they're thinking: 'Poor boy. He was promised a pony one year for Christmas, and he didn't get it.'"[38] Maybe there had been no pony, but

Garlits had succeeded in inventing a profession that nobody had dreamed of in 1950: professional drag racer.

Don Garlits was an enthusiast but also a man of exceptional purposiveness. One may be certain that, had he not succeeded in his aim of becoming a professional drag racer, he would have made his fortune some other way. As a racer, he was both typical and unique. He was unique in that he became, and he remained for nearly three decades, "the most innovative guy in the sport. . . . The focal point, the center of attention" (the words are those of none other than Wally Parks).[39] Those he inspired mostly came and went. Setto retired in 1960; Langley a few years later; Malone, Swingle, and Sullivan a few years after that. But others like the Greek kept on, perpetual enthusiasts. The event that Garlits "made," the U.S. Fuel and Gas Championship, made other pros as well. It was a turning point for Don Prudhomme when he won in 1962 and Connie Kalitta when he won in 1964. (Malone won in 1963, and after seven years Garlits finally won it himself in 1965.)

Prudhomme wanted nothing more than to be able to quit his job as a car painter and go racing, as Malone had quit construction. Likewise, Setto's protégé, Kalitta, wanted to be able to punch out of the bottled-gas plant where he worked and never return. When he finally did so it was touch and go—until he won at Famoso: "That Bakersfield win gave me money, lots of publicity, respect among the racers as a competitor, and most of all . . . confidence in myself."[40]

By the early 1960s there were significant numbers of fuelers in the Midwest, Northwest, Southwest, and South; indeed, machines were showing up all over the country, even in the eastern states, the last region in which they caught on. Locals like Dick Belfatti, Pat Bilbow, Red Lang, Fred Forkner, Tex Randall, Phil and Bub Reese, Jo Nocentino, Mike Sforza, Al Czerniac, Bittie Winward, Bernie Schacker, Ron Still, Joe Jacono, Ron Abbott, Ray Marsh, and the team of John Gaines and Fausto Marino could draw crowds to strips in Pennsylvania, Maryland, Delaware, New York, and New Jersey. But they all raced for posted purses, no guarantees, and even the purses were not much. Few if any of them even came close to breaking even.[41]

By far the biggest concentration of fuelers was in California. There they numbered in the hundreds, and, moreover, there were at least a dozen drag strips that drew substantial crowds week in, week out. But supply and demand made California perhaps the hardest place of all to make money, and California racers took to performing in other parts of the country, just as eastern racers came west. LeRoi Smith of *Hot Rod* could correctly observe, however, that "most fellows who do a lot of cross-country driving to and from drag meets wonder if all the effort is worth it. . . . Many liken it to show business, where the only real satisfaction is in the participation rather than the results."[42] When one considers the big picture Wally Parks did not need to lose too much sleep; drag racing was still a hobby for the vast majority of people whose primary involvement was as competitors. True, it was a hobby with its own "show business" satisfactions. Nevertheless, as the demands of staying competitive necessitated more and more expensive equipment, some of these enthusiasts wanted some more tangible reward.

6

FORTUNE

I feel that anyone who races a fueler, if they were to put as much time, effort, and diligence into a job as they put into their car, would end up with fifty times as much money as they end up with from racing.

LOU BANEY, 1966

By 1990, according to what one read in *National Dragster* and heard on the television programs that featured drag racing, each event was more exciting than the one before, every year better than last year, and progress was what drag racing was all about. It just got better and better. But a historian could ask no more telling a question than one the irascible Scotty Fenn liked to pose: "Progress for whom?"

Of course, there was progress for those corporations involved in sponsorships that were getting increasing exposure on cable channels such as ESPN and TNN and even on network TV; they were well served by Diamond P Sports Productions, an enterprise tightly interlocked with the National Hot Rod Association, which did a weekly show called "NHRA Today" as well as a ninety-minute feature about every event on the NHRA tour. Events sanctioned by the rival International Hot Rod Association could be seen on TV as well. There was progress for "the two primary underwriters of motor racing in America,"[1] the beer companies and the tobacco companies; with ads for tobacco banned from the broadcast media RJR/Nabisco got what was far and away its most significant television exposure for Winston cigarettes through its sponsorship of motor sports, NHRA and NASCAR being the prime vehicles.

There was progress, too, for people who sold equipment to the racers, for those involved in the vast automotive aftermarket.[2] Progress for people who operated tracks that hosted championship events. For certain well-connected publicists, for TV personalities, for men and women in sports marketing. For purveyors of pricey souvenir items, from caps and T-shirts to cloisonné pins. For those racers who had managed to put together sponsorship programs with brewers, oil companies, and automakers. (And for their agents, who reaped the customary 10 percent.) For the corporate types and sundry groupies who munched catered tasties in the swank VIP suites that were now de rigueur at the major venues.

Certainly, things got better every year for legions of spectators who thrived on the sensual delights of drag racing; who relished the sound and the fury, the color, the dazzling machinery; who crowded around the eighteen-wheelers lined up in the pits—mobile, fully outfitted command centers—to get a glimpse of drivers and crew chiefs close up; and who filled the stands to feel the thrill of close competition at nearly 300 miles per hour.

But turn the coin over. There was no progress if one happened to feel that beer sponsorships fostered "a lethal link between drinking and driving" or, as Dr. Louis W. Sullivan of the Department of Health and Human Services felt, that tobacco's growing involvement in sporting events amounted to "blood money."[3]

There was no progress for the owners of small-time strips driven to the wall as a result of the overwhelming emphasis on NHRA's Winston Championship Series, which took in nineteen events in seventeen different states (plus Quebec). Fans jaded by one four-second pass after another, either on television or in person at a supertrack NHRA itself might own (such as Indianapolis Raceway Park or Gainesville International Raceway), were less likely to flock to a local track to sit in rickety bleachers and watch the kids race their Camaros.[4]

There was no progress for enthusiasts of one of the various types of machine NHRA simply dropped from the rule book. And there was no progress for someone who dreamed of competing at the top level, in a top-fueler, but could not conceivably do so in the big-bucks milieu fostered by the marriage of television and tobacco.

Once, a zealot like Fenn could declaim drag racing as "a wonderful hobby . . . the last outpost where the individual can stand on his own two feet." Even in 1990 drag racing remained a wonderful hobby for tens of thousands of men (and more and more women) who raced their Camaros, fiberglass replica Model-Ts, and even stock-engined dragsters at places like Sumerduck Dragway in Culpeper, Virginia, Rock Falls Raceway in Eau Claire, Wisconsin, and Samoa Dragstrip in Eureka, California. But racing a top-fueler could bankrupt even a wealthy man or woman unable to woo

In this 1987 view of the Texas Motorplex, south of Dallas, eighteen-wheel trans-
porters are seen lined up in the pits behind the packed grandstands at left. VIP
suites look directly down the track. Paved with concrete, the Motorplex was the
site of the first elapsed time quicker than five seconds. (Courtesy Billy Meyer)

corporate sponsorship, which might have looked plentiful because its
overt manifestations were so obvious (visibility was the whole idea, after
all) but, in actuality, was a luxury almost impossible to acquire.

In the halcyon days of the 1960s racers who campaigned top-fuelers
lived in virtually every state of the union; nationwide, there were hundreds
of them, and certain "low-bucks" teams were quite capable of running with
the best. Skill, ingenuity, and boundless enthusiasm counted for a lot. They
still did in 1990, yet there were no more than a couple of dozen top-fuelers
that could conceivably win a Winston Championship Series event and only
six drivers who actually did win. Fuelers had been absent from the overwhelming majority of drag strips for many years, and in some regions there
were no such machines at all.[5]

Despite the prominence that touring pros like Garlits and the Greek had
attained by the early 1960s, the majority of racers were Californians who
seldom if ever competed outside their state. There were far more places for
dragsters to run in California than anywhere else. Certain strips operated
year-round, and a few could attract several thousand spectators almost any

Saturday night. Results inevitably got big play in *Drag News,* which claimed a circulation of thirty thousand by 1960. Moreover, many promoters paid prize money. Riverside, Lions, and Kingdon were occasionally putting up as much as $500 for top eliminator by 1959, and soon $1,000 payouts were not terribly unusual. The norm was still much less, however, $25 or $50, and, when Fontana Raceway's "Super Strip" had its grand opening in October 1960, Ted Cyr and his driver, Chuck Gireth, took away $100 for besting the field.

Even with larger payouts the image of California prize money was deceptive. Unless the finalists arranged privately in advance to split the winner's purse—an impromptu form of insurance that became fairly common—the situation was winner-take-all. Until well into the 1960s no promoter paid even a token sum for runner-up, let alone for losers in preliminary rounds. So, while the possibility of winning money did provide a tangible incentive, it does little to explain why so many Californians started getting drawn into dragster racing in the early 1960s. To a much greater extent the inducements lay in the unmatched opportunities to perform, the roar of the crowd, the technical challenge itself. Drag racing's takeoff into sustained growth provides a fine exemplification of Eugene Ferguson's observation about "plumb[ing] the murky depths of human motivation with measuring rods precisely calibrated in economic terms"; such measuring rods leave a lot unexplained. Don Garlits might not have raced very long out of sheer enthusiasm, but in that respect Garlits stood apart from most of the others.

As the Eisenhower years drew to a close, there was only a handful of drag racers whose names would have meant anything at all to a typical sportswriter: Garlits, perhaps Mickey Thompson, the Arfons brothers, the Chrismans. These men had come out of total obscurity. But such was not the case with Tommy Ivo, who emerged into the spotlight just as Garlits was taking up his first California challenges in the fall of 1958 and the spring of 1959. Ivo was born in Denver in 1936. During the war his family moved to Burbank, and, as a teenager, he started taking tap dancing lessons. He recalls: "Somebody thought I looked like Dennis O'Keefe's son, and because I could dance I ultimately acted in 100 films and 200 TV shows." His debut came as Cousin Arne in *I Remember Mama.* Then there were "a bunch of jungle movies," the Mickey Mouse Club, "The Donna Reed Show," and "Margie," an ABC series in which he played a dopey adolescent named Heywood Botts.[6]

Ivo had time on his hands between acting stints, and he had money. One evening, at Bob's drive-in in Glendale (the original home of the "Big Boy"), he saw Norm Grabowski's Model-T roadster, an archetypal hot rod of the period.[7] He had to have one like it. After getting in trouble for racing on the street, he started going to Santa Ana—which indeed seemed to be serving

Actor Tommy Ivo poses with his unique new dragster shortly after it was completed in 1960. The two engines on the right drove the front axle; the other two drove the rear axle. (Courtesy Tommy Ivo)

its function of social control. Soon Ivo craved more speed, and in 1958 he took the Buick V-8 out of the roadster and built a dragster, with the help of Kent Fuller. With all the best of equipment Ivo's dragster became an instant winner, but now he craved still more speed. Switching to nitro was not an option because most of his favorite strips still operated under NHRA's gasoline-only rule, so he and Fuller cooked up a twin-engine machine, still using Buicks.

Eventually, Ivo had Fuller build him a machine with *four* Buick engines and four-wheel drive. At a reported thirteen thousand dollars it was undoubtedly the costliest dragster in existence.[8] But Mickey Thompson and others experimenting with four-wheel drive were starting to suspect that this setup was actually detrimental to performance, and four engines put Ivo's machine at a weight disadvantage that no amount of horsepower could offset. Four engines could not match the times of Ivo's twin-engine machine, nor could they match the performance of the best single-engine lightweights, such as Pete Robinson's and Gene Adams's. That was missing something important, however; that was considering performance simply in a *technical* sense. In a show business sense the "Showboat" (for so Ivo dubbed it) offered an absolutely marvelous performance—boiling smoke from all four tires, then stopping by means of a huge red, white, and blue ring-slot parachute. The Showboat was boffo. On occasion Ivo drove it himself, but more often he sent it around the circuit in the charge of hired troupers.[9]

The boredom of studio work eventually wore him down. Ivo had booked a cross-country tour in 1960, trailing his twin-engine dragster behind a Cadillac limousine. Wherever he appeared people already knew who he was. Eventually, he felt confident that he "could afford to walk away from acting." But he still remained onstage, a showman at heart, performing stunts like racing against the fearsome first-generation jets.[10] Yet he also had a strong desire to compete with the dragster pros on their own terms. When the fuel ban was lifted he switched to a blown Chrysler, trekking coast to coast year after year, match racing Garlits and Karamesines, Don Prudhomme and (later) Shirley Muldowney, time and again. "TV Tommy" matched Garlits in fame, but when victories were tallied Garlits held the upper hand. He appreciated Ivo for his drawing power, but he loved to put him away, too, for Ivo epitomized his image of the California racer; Ivo was a "rich guy."

Ivo was not the first Californian to arrange appearances in the East. Gary Cagle had several dates in the summer of 1959, though he never got past his calamitous debut at Great Bend. In 1960 Jack Chrisman headed east to take on local and visiting talent at places like Sanford, Maine; Roanoke, Virginia; Concord, North Carolina; Jacksonville, Florida; Bainbridge, Georgia; and Biloxi, Mississippi. Lyle Fisher likewise was paid to appear at Biloxi in the Speed Sport roadster. Ernie Hashim and Bill Crossley of Bakersfield appeared at York, Tony Waters at Gary. Pondering this new turn of events, Doris Herbert of *Drag News* called 1960 drag racing's "payola year." Certain Californians were said to have received as much as fifteen hundred dollars for their performances, though the usual sum was much less. More to the point, for most Californians "going on tour" was out of the question for practical reasons, or it was a distant dream at best. Yet more and more dragsters were showing up on the strips scattered along the length of the state, and more and more racers were coming to the fore whose names would become legend in local annals.

Perhaps the strongest fueler on the California scene in the latter part of

1959 was built and driven by Chuck Gireth, a salesman for a Pasadena electronics firm. In partnership with machinist Joe Oliphant, Gireth clocked outstanding 8.20s at Fremont in August, then took the money at a big meet at Riverside. Later Gireth teamed with Ted Cyr, then with cam grinder Dave Carpenter. Still later he drove briefly for Keith Black, who would become known as the smartest engine specialist in the business. Rod Stuckey, a Kansan who spent a lot of time racing in California, also drove for Black and his backer, Tommy Greer, but their most talented and ultimately most famous driver was Prudhomme, a friend of Ivo's, who went along as his helper when he toured in 1960.

After Ivo debuted his twin, Prudhomme bought his old dragster and campaigned it locally every week, occasionally winning top eliminator at some small-time strip such as Colton. In late 1960 he teamed up with a youthful Burbank engine builder named Dave Zeuschel (Zeuschel was twenty, Prudhomme nineteen), running a blown Chrysler on fuel, and next year the two of them joined forces with Fuller, who provided a state-of-the-art chassis. They set a strip record at Fontana the first time out in September, then got everybody's attention with runs in the 8.10s at Half Moon Bay in October. After Prudhomme won the U.S. Fuel and Gas Championship in 1962 he was hired to drive by Black and Greer, while Ivo teamed up with Zeuschel for initiation into the mysteries of nitro.

Ivo sometimes drove another Fuller car owned by Ernie Alvarado, the proprietor of a Glendale camera shop. Alvarado's "Shudderbug" had an aluminum body fashioned by Fuller's associate Wayne Ewing, with a tail section that faired rakishly around the parachute pack. Ivo himself quickly adopted this eye-catching "swoopy" design, which would become one of the conventions of the 1960s. Another of Alvarado's shoes was Bill Alexander, who would later drive for Jim Brissette of Eagle Rock, one of the first racers to top 200 miles per hour using a chassis fabricated by Fuller's competitor, Gar Wood "Woody" Gilmore of Long Beach, who had learned his skills in the aircraft industry. In Pomona Jim Fox and Dennis Holding teamed with Bill Adair, who built both engines and chassis and had been around drag racing since Goleta days. Fox, Holding, and Adair entered sixty-three events in 1961, all of them in California, but later Fox and Holding joined up with Norman Weekly and Ron Rivero and took to the national circuit. Ed Donovan, a creative machinist with a burgeoning business in dragster components, fielded a winning new fueler in the fall of 1961, a team effort in concert with Leonard Van Luven, an employee of Iskenderian's, and Bob Tapia, a computer programmer. These would all be names with which to reckon for many years to come, yet the ranks of southern California fuel racers had scarcely begun to gather.

By late 1961 fuel burners had returned to most California strips, the only major holdout being Mickey Thompson's Lions. Even though restricted to gasoline, dragster competition at Lions was as intense as anywhere in the

This style of fairing the bodywork into the parachute pack, pioneered in the machines of Ernie Alvarado and Tommy Ivo, was copied hundreds of times in the 1960s. Here Californian Larry Stellings (wrenches in hand) is seen at work in the pits at Amarillo, Texas, while a crewman changes tires. Note disc brakes, simple push bar, and weights for balancing the wheels. (Photograph by Don Brown, courtesy Museum of Drag Racing)

land, with the likes of Adams and McEwen, Howard's Bear, Lefty, and the Freight Train sharing the spotlight with visitors such as Pete Robinson, Connie Kalitta, and Kalitta's sidekick from Portsmouth, Ohio, Gordon Collett. Mickey was never stumped for ideas about drawing crowds; when back East racers were due at Lions he posted an extra $1,000 bond for any such challenger who could defeat the locals. Iskenderian kept things stirred up, as usual. According to Isky, what people like Collett lacked in funds was "more than compensated for by the zealousness and hot competitive spirit that still prevails in the East and Mid-West":

> We have always maintained that a predominance of factory-works cars, such as are prevalent in the west [i.e., dragsters running with direct support from Isky's competitors like Howard and Herbert], has stifled

and all but obliterated individual enterprise in the top classes. Hot Rodding was intended to foster the spirit of ingenuity, and in spite of outlandish western budgets, seems once again headed in that direction.[11]

Lions had opened in October 1955, then faltered, but, since the first time Mickey staged races under lights on Saturday night, it had attracted more paying customers than any other drag strip anywhere.[12] Two rivals, San Gabriel Dragway and Santa Ana, had closed in the late 1950s, and Riverside was too far away to put much of a dent in Lions's attendance, as was Dean Brown's new Fontana strip on Route 66 near San Bernardino. Thompson scored a publicity coup on September 16, 1961: ninety minutes of live television time on local channel KTTV. But that was just before the San Gabriel strip reopened. Located at Live Oak and Rivergrade Road, right off the San Bernardino Freeway, "at the very center of drag racing," San Gabe welcomed fuelers. The promoters, Jack Tice (no relation to AHRA's Jim Tice), his brother Will, and Jack Minnock, were soon on a roll.

For Lions the handwriting was on the wall, and on January 28, 1962, Thompson reinstituted fuel competition after a five-year hiatus. The racing that first night was about as fast and close as anyone had ever seen, except perhaps at Famoso. Art Chrisman, who emerged victorious, turned 180, 8.48. But along the way Fuller, Zeuschel, and Prudhomme turned 180, 8.49; Jack Chrisman, driving for Gene Mooneyham, turned 185, 8.53; and Chuck Gireth, driving for Keith Black, had an 8.53 ET as well. The return of fuel racing to Lions signaled a transformation on the California scene. Thompson and most other strip managers would continue staging competition among gas dragsters (collectively called "Blackie Carbon") but with smaller posted purses. Gasoline stalwarts like Gene Adams would soon switch to fuel, and a whole bevy of new enthusiasts would enter the fray. As if to provide a mark to shoot for, Garlits made his first Lions appearance in February 1962, with Connie Swingle driving his car to 8.20s; this record stood for a few months but was broken that summer in the Greer-Black machine by Prudhomme, who was emerging as a racer who might well be Garlits's equal. Although Greer, Black, and Prudhomme were hardly ever seen outside California, performances on the order of a 7.77 at San Gabriel on January 19, 1963, enabled them to command appearance money right at home. Few other locals could ever wangle guarantees, although open competition everywhere in California was being staged amid an escalating ballyhoo about cash purses.

"Several years ago monetary prizes were unheard of in the sport of drag racing," Doris Herbert remarked at the beginning of 1962. "Then slowly they started giving money, first $5, then $10 and so forth. Now it has almost [come] to the point where some strips are giving away $500 every week."[13] The Tice brothers were masterful at exploiting the tension when competitors such as the venerable Don Yates went on winning streaks. Mickey

Thompson was quite as adept with schemes like "balloon payouts." As word got around about the good show to be seen at the drags, spectators poured in. By 1963 Lions and San Gabriel were both posting one thousand dollars for top eliminator more often than not. There were always enough fuelers on hand to fill a strong sixteen-car field, and the names of many new enthusiasts began to appear in *Drag News* headlines.

There was John Wenderski, Sid Waterman, Joe Winters, Roger Wolford, Neil Leffler, Earl Canavan, Bill Scott. There were euphoniously named duos, trios, and even quartets: Adams brothers and Stewart; Brissette and Alexander; Babler and Clark; Baber and Cassidy; Blair and Hanna; Briggs and Cagle; Brooks and Rapp; Batto, Valente, and Bing; Cope brothers and Cook; Crossley, Williams, and Swan; Danylo, Cochrum, and Shipley; Doss, Clayton, and King; Safford, Gaide, and Ratican; Scrima, Bacilek, and Milodon; Skinner, Jobe, and Sorokin; Zeuschel, Moody, and Fuller; Stellings and Hampshire; Shubert and Herbert; Harbert and Tapia; Porter and Reis; Lechein and Drake; Masters and Richter; Gotelli and Milani; Warren, Coburn, and Holloway; Weekly, Rivero, Fox, and Holding; Mooneyham, Ferguson, Jackson, and Faust; Sandoval brothers, Madden, and Gabelich; Baltes, Croshier, Leavitt, and Lovato; Broussard, Davis, Garrison, and Ongais. By 1964 it was a rare Saturday night at Lions or Fontana (San Gabriel was closed again, but Irwindale Raceway would soon open on a nearby site) when there was not at least one brand-new fuel dragster, "first time out."

Purses at several strips were sufficient for a local team on a hot streak to do rather nicely. From late 1963 through the spring of 1964, for example, Jerry Baltes won $3,800 at San Diego Raceway in Ramona, as well as a $1,500 meet at Fontana and another $1,000 in Arizona.[14] In ten months Gene Adams and Rick "The Iceman" Stewart won $5,500, while Zane Shubert and Chet Herbert won nearly $8,000, though suffering much higher attrition. Ted Gotelli and Denny Milani won over and over at Fremont. Larry Stellings and "Jeep" Hampshire were often in the winner's circle at Lions, as were Dick Lechein and Lee Drake at Ramona. Don Moody won top-fuel eighteen times at a half-dozen different strips between August 1963 and August 1964. Driving Woody Gilmore's "factory works car," Paul Sutherland won twelve substantial purses in just four months and more than $10,000 all told in 1964.

But those were the big winners. By contrast, there were others who ran and ran, every Saturday and sometimes every Sunday too, yet rarely, if ever, went to the winner's circle. Terry Gall and Roy Thode, Kansas racers who had come west to get in on the action, won one purse in a whole year of trying, $500 at Fontana. Herb Reis won $500 at Pomona; Ray Ayres won $500 at the new Carlsbad strip near Oceanside; Bob Brooks and Ronnie Rapp won $350 at Ramona; various people won one or two of San Fernando's $250 Sunday purses. Others struck out.

By the mid-1960s it cost $6,000 to $8,000 to field a state-of-the-art fueler, depending on how much of the parts chasing and machine work one did oneself (Woody Gilmore charged $3,500 for a rolling chassis, complete except for engine, paint, and tires, or $1,900 for a "bare" chassis). One could try competing with secondhand equipment, but the routine costs were formidable under any circumstances. Tires, $155 a pair, lasted a dozen runs at most. Just one run consumed four or five gallons of nitro. Engine oil needed changing continually because fuel got past the piston rings and corrupted it. Beyond that there was always the danger of ruining parts, and even the smallest such incident—breaking a blower drive belt, for instance—could be a major setback for a competitor on a tight budget.

In November 1965 Terry Cook, staff writer for a new weekly called *Drag World,* observed that "a surprisingly large percentage" of the California fuel racers "live, or try to live, off their racing incomes alone." It was true; a lot of them did try. But Cook knew as well as anybody that most of them could not make anything like a living, nor could they realistically aspire to do so: "They just like to race."[15] It was a hand-to-mouth existence. Mike Doherty, Cook's editor, was pushing a scheme for ranking drag racers according to their winnings, on the order of professional golf and stock car racing. By 1965 it may have been, as Doherty put it, that promoters were offering "a small fortune . . . in cash prizes to fortunate competitors,"[16] but the ranks of the "fortunate" were pretty slim. The word itself had so much cachet that racers often claimed to be "pros," when actually they worked as wage earners much of the time. Quite a few of them had skills they could market readily—pattern maker, machinist, welder—and there was constant movement in and out of the shops whose vitality derived from the popularity of hot rodding and drag racing.

The folklore was rich in tales of racers who had a streak of luck and then "quit their job to go professional." Later a wonderful image of "touring" was fostered in *Funny Car Summer,* a documentary by Bruce Brown, who ordinarily made sensuous surfing films.[17] But Jim Dunn, who starred in that movie along with his young son, Mike, had simply taken time off his regular job as a firefighter in La Mirada. As it usually turned out when stagestruck Californians "went pro," touring was simply a diversion from regular employment.

Take the case of one of the most popular teams of the mid-1960s—three young men from Santa Monica, who had been collectively dubbed "The Surfers" because of their endearing demeanor, their denims, shaggy hair, and sneakers. Driver Mike Sorokin, born in 1939, had been competing for several years, beginning in a '41 Ford, then moving up through a Willys, two Fiat coupes, a modified roadster, and an unblown dragster. Bob Skinner and Tom Jobe, both slightly younger than Sorokin, had been street racers and "professional spectators" before deciding to jump in right at the top with a blown fueler in 1963. They tried a couple of other drivers, including "Lotus John" Morton, a seasoned sports car racer, before teaming with

In typical Saturday night final-round action at Fontana, Mike Sorokin, with a lead off the line, would defeat Herb Reis's 8.24 ET with a slower 8.26. Bob Skinner is seen by the door of the Surfers' '55 Chevy. (Courtesy Museum of Drag Racing)

Sorokin and becoming consistent winners at Fontana in the latter part of 1964 and 1965. Jobe, who had been an engineering student at Santa Monica City College, proved adept at running straight nitro (actually 98 percent, allowing for 2 percent benzol as an antioxidant) while keeping their Chrysler engine "living like Methuselah," as Sorokin put it.[18] Racers had generally run "straight can" only in final-round desperation, as it was so likely to result in damage.

Though open and affable like Tommy Ivo, the Surfers were, otherwise, quite the opposite. They initially financed their dragster "on a Bank of America furniture loan." They towed it behind a well-worn '55 Chevy with a lucky "Fontana branch" stuck in the hood ornament, not a Cadillac. They watched their pennies. Yet, even though they won at least a dozen top-fuel purses in 1964—at Lions, San Fernando, and Riverside as well as Fontana—they fell short of covering their $6,000 expenses. In 1965 they banked $15,000, including $2,500 from winning a special "200 MPH Club" meet staged by Mickey Thompson at Fontana (which he took over after leaving Lions), but they still ran slightly in the red. Then in March 1966 the Surfers won the U.S. Fuel and Gas Championship at Famoso, the toughest dragster race of all; there had been 102 top-fuelers entered. They collected $5,650. For the first time ever Skinner and Jobe were able to pay off all their bills, including their initial loan: "We've been trying to figure out a way to pay for the car for 2½ years," Jobe exclaimed from the winner's circle.

In "the lights" (or "the eyes") at Riverside in 1966. The fueler at right has just crossed the finish line, breaking the ET beam and stopping a clock that started when it crossed the starting line, 1,320 feet back. The narrow line across the track, 66 feet further on, denotes the end of the speed trap, one of a pair of beams (the first being located 66 feet before the finish line, where the cones are); the time it takes each car to cover the distance from one to the other will be read in terms of miles per hour. Mike Sorokin (left) is still pulling hard and will actually clock a faster speed than the winner, even though he has already lost the race. Drag races were often this close by the mid-1960s, and fielding an entry that could win with any consistency was becoming an expensive proposition. (Courtesy Museum of Drag Racing)

"None of us work and we all live off the race car's winnings."[19]

Truly, the Surfers were on the crest of a wave. They were immensely popular. They had clocked a 7.34 ET, the best ever, with an average for the entire weekend in the 7.50s. They were the first Californians to win Bakersfield in four years, coming through just as the locals were about to cement a reputation for not having the same instincts as people like Garlits, Kalitta, and Malone (who won in 1965, 1964, and 1963, respectively). The Famoso strip had been repaved, and the Surfers had the insight to tune their engine for all the horsepower it could deliver. When asked the "secret" of running 98 percent, Jobe made light of the question. They were always breaking the glass hydrometers used to check fuel mixtures, so they went to straight nitro to avoid having to buy "any more $17 equipment." (Recall Joaquin Arnett's answer in the early 1950s when asked about running 50 percent: "Easier to mix . . . a gallon of this and a gallon of that.") Jobe continued to embellish: They had come upon a "trick anti-detonate" through "years of engine explosion research." Actually, they simply understood the necessity of running an extremely rich air-fuel ratio in order to cool the engine internally. When asked about the motor that had powered him to the biggest win of his life Sorokin answered: "There has been lots of print about the pro engine builders who have the money and equipment to work with, and it seems like Skinner and Jobe do an awful lot with the small finances they have to work with."[20]

For all the talk about small finances, however, their machine was as well equipped as any. Skinner and Jobe's total investment was six thousand five hundred dollars, on the low side of the average, but only because they avoided frills. Now, all in one weekend, they had won almost as much as their total initial investment. They were a hot property for sure, and eastern promoters flocked around to sign them up. Still in the flush of victory, they began making plans for a showy new dragster, with swoopy bodywork, metallic paint, and lots of chrome—something like what Tommy Ivo had.

Towing their dragster behind their Chevy with its Fontana branch, the Surfers traveled coast to coast in the summer of 1966. They compiled an impressive string of victories over the strongest fuelers in the land, such as the B&M Automotive "Torkmaster," which Don Prudhomme was touring simultaneously. Back in Santa Monica that fall, however, Skinner and Jobe put their entire operation up for sale; they took delivery on their new car but turned right around and sold it to Ritchie Bandel from Brooklyn, who mostly toured it on the car show circuit, racing it very little. And Skinner and Jobe never again raced at the drags.[21]

What had happened? Either they could see that the P&L sheet was not great, even under the best of circumstances, or else they felt content to have had their day. No doubt, it was a little of both. Jobe regarded drag racing as "a big technical game"; Skinner thought it not only a game but also an "education." They had played, and they had learned. But they had not become addicted (a term that comes up over and over in conversations with drag racers), and so they were able to move on.[22]

As for Mike Sorokin, he married his sweetheart, Robyn, and soon they had a son, Adam. For a while Mike drove for Keith Black. He also drove briefly for Black's chief competitor, Ed Pink of Van Nuys, as well as for Tony Waters, the veteran from Bakersfield. He began making plans to campaign his own fueler, but for the time being he was beyond the spotlight, just one of dozens of journeymen drivers who populated the West Coast scene.[23]

All told, Skinner and Jobe's touring days were fairly typical in terms of duration, be it for Californians headed east for a series of paid match races or an easterner headed for the land of big weekly purses as well as matches. People dreamed of emulating Garlits, the Greek, Ivo. Yet very few others managed to put themselves in a position to garner a reasonable income for any extended period of time. Of course, there were exceptions, the most notable being Prudhomme, who first went east as a hired shoe, then went on his own and kept at it, with only one year off, into the 1990s. But for every Don Prudhomme there were a dozen more who tried it for a year, maybe two, had problems or ran low on funds, and retreated to the local scene or else dropped out altogether, even while dreaming of a comeback someday.

One of the first West Coast racers to hit the road was Bob Haines of Seattle, who headed for the Midwest with his partner Jack Cross in the summer of 1960. In the opening five weeks of their tour they split a $510 gross. Haines was the father of four small children. He had no job to return to. He had a $5,000 insurance policy but no double indemnity. Back home his wife JoAnn was finally moved to write an anguished letter to Doris Herbert:

> My husband has always raced because he enjoyed himself. The faster a machine would go, the better he liked it. . . . Now drag racing is out of the category of a hobby—to keep a car running in top shape is a full time job. . . . Consider what personal danger there is to the driver of these fast machines . . . manufacturers keep pushing to have their cars go faster, but you never see them up at the line behind the wheel. I feel like these drivers are just pawns, but it's of their own choosing too. Like I tell Bob, what is it gaining him? Up on top today—tomorrow you're either dead or have gotten older and have quit the racing game, and who really gives a darn one way or the other?[24]

Bob Haines would probably have appreciated a quip made by another driver, Richard Tharp, many years later: "Racin' might not be much, but workin' is nuthin'." Haines eventually drove some top-notch cars, put his name in the record book, managed to survive—and then, indeed, he quit.

By contrast, there was Joe "The Jet" Jackson, from the opposite side of the country, Jefferson, Maine. The first time he saw Don Garlits and Setto Postoian perform at Sanford, Jackson fell in love with fuel dragsters. He saved up, buying used parts here and there, and eventually fielded a fueler himself. As owner of the only such machine Down East, Jackson picked up a few match races and was nominally successful. In 1965, in the dead of winter—there would be no racing in Maine for months to come—Jackson pushed his antiquated slingshot into his trailer and towed it to California behind an aged hearse, which served him both as bedroom and workshop. He entered Mickey Thompson's big Fontana meet but failed to make the cut. He scuffled around the local tracks for several months, aiming mainly to pick up the twenty-five dollars that was by then the customary compensation for first-round losers. Open competition in California was much tougher than Jackson ever imagined, and he had to run his tired equipment much too hard.

Eventually, he found himself flat broke, with no funds even to get back to Maine. But he was a likable young man, and he had some friends. One February afternoon at Lions he agreed to make a checkout pass in a dragster from Huntington Beach which was giving the regular driver trouble. He smoked strongly off the line in the left lane, then at mid-track veered right, crossed the other lane and the grass beyond, rolled over, and crashed into a chain-link fence, striking his face on one of the poles. He died three and a half hours later at Harbor General Hospital, three thousand miles from home and without a dollar in his pocket.[25]

Joe Jackson's pathetic fate was hardly typical, of course. He had come west with no assurance of any compensation at all, whereas touring racers ordinarily sought to fill up their schedule with advance bookings. For a regular on the California scene, after having faced the stiffest competition in the land week in and week out, what could be more alluring than the prospect of "hitting two tracks per weekend, at $750 per day, making three singles or races at each plant, and then lounging away the week in a plush motel?" Terry Cook, who had begun his journalistic career in New Jersey doing a weekly column for *Drag News,* knew better. First, only a few racers could get two dates per week. Second, the payout was not always as much as $750. Third, touring entailed taking along many spare parts, because a contract to make three passes meant just that, no excuses. In the event of a rain-out promoters usually paid half, sometimes less. Motels, meals, gasoline, and tolls (double with a trailer) could eat up an awful lot. Then there was the race car itself: fuel, tires, and heaven forbid that you should run a match race hard enough to break anything (yet a racer who got a reputation as a "stroke" who took it too easy would find booking opportunities dwindling). And then there were dangerous tracks—too short, too narrow, badly lighted—and fly-by-night promoters, thugs, bounced checks. "Suddenly touring becomes a matter of survival," Cook wrote, "not security."[26]

The necessity for economy dictated going on the road with as few people along as possible—going alone if one had the temperament for it—and renting a room somewhere central to one's bookings rather than moving in and out of motels. "Sure, there's gold in those eastern tracks," Cook concluded, "but it takes a lot of hard work to dig it out." California enthusiasts would continue their quest, but fortune was elusive. In the mid-1970s a journalist colleague of Cook's, Dave Wallace, Jr., noted what had in fact been the case right along: "All but a handful of our digger heroes are working folks in between races."[27]

There were plenty of ideas for cutting overhead. Banning nitro became a perennial. But everyone recalled that when nitro *had* been banned it simply spawned an elite band of "outlaws." When asked how *he* kept costs down, Garlits said that he avoided any situation from which he risked emerging in the red. "Never, repeat *never*," he warned, "deliberately lean on your engine to get by one round."[28] Garlits was famous for his frugal ways, but others were clearly guilty of "trying to support 'champagne tastes' in what remains basically a 'beer sport,'" as a Pennsylvania promoter put it.[29] In the 1960s drag racing had very few Daddy Warbucks types (indeed, it had only a few in the 1990s, notwithstanding the infusion of millions in corporate sponsorships). Industrial machinery magnate Tom Greer came along in 1961 with the aim of fielding a fueler that was indomitable, and he pretty well succeeded. By 1964, however, Greer was bankrupt. Lou Baney headed several auto agencies in the Los Angeles area, and for several years he was in a position to retain top talent full-time. His dragsters won a lot of money, but he never came close to showing a profit and was quite direct about his actual motivations: "I would rather race than do anything."[30] For Baney, and for a lot of others, racing was neither a hobby nor a business. It was more like an obsession.

One did not have to spend money quite like Lou Baney did in order to win consistently. Gene Adams, who raced top-notch fuelers in the 1960s—almost always in open competition, hardly ever for guaranteed money (of which there was very little in southern California, anyway)—took pride in his balance sheet. "I know it's hard to believe," he said, "but my cars always manage to pay for themselves." In 1964, with Rick Stewart driving, Adams grossed $5,550, about the same as the Surfers, and, like Skinner and Jobe, he about broke even. Later, with John "Zookeeper" Mulligan in the cockpit, he recorded some spectacular performances, including the first elapsed time in the 6s, but expenses went far higher than ever before. In truth, the key role was played by his partner, car dealer Jack Wayre. Wayre was no mechanic, no driver, but he apparently did know how to exploit a drag racing operation as a tax write-off.[31]

Adams knew as well as anybody that the cost of running a fuel dragster on a weekly basis was "out of sight unless the car ha[d] a complete sponsorship or receive[d] many guaranteed match races." California's most consistent winner in 1966, the machine campaigned by Lee Sixt with Tim and Dave Beebe, brothers from Garden Grove, grossed $30,000 and netted 13 percent. Pete Robinson, who had switched to fuel by 1966, averaged out his costs at $175 *per run,* even though he was a master at minimizing parts breakage. Robinson kept careful books. What his figures indicated was that an unsponsored racer running for $750 or $1,000 guarantees might do all right. A racer running open competition for $750 or $1,000 purses was in trouble, even if he won half the events he entered, which was quite impossible for anyone to do for any extended period. By 1966 losers in open competition were usually paid $50 per round; that is, someone in a sixteen-car field who went down in the quarterfinals would get $150. Most racing was without guarantees. While the number of dragsters that consistently showed a profit may have been more than five, the figure Garlits was fond of throwing out, Cook felt certain that only about two dozen of them were breaking even.[32]

The touring match racers were attuned to profits, and so they conducted solo acts—perhaps paying a helper but avoiding partnerships. Garlits was the archetype. After Don Maynard was killed in a 1963 highway accident the Greek became a solo act. Kalitta was a solo, and so was Pete Robinson. The California racers, more sensitive to costs than profits, developed a regional style of organization, having two, three, or even four partners among whom to apportion the red ink. Racers who had been part of teams in California—both Prudhomme and Ivo with Fuller and Zeuschel, for example—hit the road as solo acts. There were very few exceptions, the most notable being Weekly, Rivero, Fox, and Holding, who toured together as the "Frantic Four." Of necessity the solo performers had to be versatile. Ordinarily, the California teams not only divvied up costs; they also developed specialized expertise and responsibilities: the engine, the chassis, the driving.

Drag racing was a technological activity whose ultimate constraints were political. But this is not to say that there might not have been some sort of "elemental force . . . within the technology itself," as Brooke Hindle puts it.[33] How else to explain Bob Haines, Joe Jackson, even Lou Baney? And how else to explain the technological "barriers" that somehow always gave way?

Fuel dragsters were running 180s by 1959. Racers could not get into the 190s regularly until 1963, but they broke past two hundred miles per hour only a year later. There was no one best way. The weight of the cars that did it ranged from less than 1,200 to 1,400 pounds. Garlits used a camshaft from Bruce Crower; others did it with Engle, Herbert, or Isky cams. Jim Brissette used a 331-cubic-inch Chrysler, Paul Sutherland's "Charger" had a 354. Garlits did it with a 392, as did Art Chrisman's onetime partner Frank Cannon in his "Hustler V." Garlits, Brissette, and Sutherland built their own engines, each with its own peculiarities; Dave Zeuschel built Cannon's. Cannon also used a two-speed Torkmaster from B&M Automotive. Everyone ran at least 80 percent nitro, Garlits and Sutherland more than 90. While the Goodyear Tire and Rubber Company was getting involved in drag racing, all the first 200s were turned on 10.50 × 16 M&Hs. But some used the "H" series, others the newer "J" series with a different cord angle; that depended on "the way the car was before," said Woody Gilmore, who built several of the first 200-MPH chassis. Brissette's "Woody" car had the engine mounted only 20.25 inches from the rear end; Cannon's was out nearly a foot further.[34]

All the 200-MPH cars had so-called zoom headers, a design Chet

Herbert had tried and discarded before it was reinvented by Gilmore. "Zoomies" directed the exhaust over the top of the slicks, cleaning up airflow and improving traction by blowing away bits of rubber on the surface of the tires. Garlits, who first saw zoomies in a *Drag News* photo of Cannon's Hustler, knew at once that they were "one of the keys to 200." Working at Kalitta's shop in Detroit, he welded up some of his own and ran the first official 200 shortly afterward.[35]

Whence 200? "You can't just think of one thing and give people 10 MPH in one jump," said Woody. "If you could, you'd be a millionaire in a week."[36] As Garlits noted, however, once the barrier was broken, racers began running 200 miles per hour "like nothing." Was it something like the four-minute mile? Maybe. It definitely was something that certain theoreticians had pronounced impossible—the racers never ceased taking delight in the shortcomings of abstraction—and the technology of a 200-MPH fueler of the mid-1960s deserves a close look. We happen to have a very thorough description of the dragster that Skinner and Jobe campaigned from 1963 to 1966; one or two features were unusual, but mostly it was quite typical. At the time Mike Sorokin won the U.S. Fuel and Gas Championship its best speed was 210.76; a few fuelers had gone a little faster, but only a few and only a little.[37]

It weighed 1,430 pounds ready to run, somewhat heavier than most California cars, though on a par with dragsters that toured, which needed to be more rugged in order to withstand the rigors of trailering long distances and running on poorly surfaced strips. The wheelbase measured 152 inches; it had been shorter initially but was stretched out when better-biting tires caused problems with wheelstands. The chassis was fabricated from 4130 chrome moly tubing and weighed 75 pounds. It had been designed and heliarc welded (i.e., welded in the cocoon of inert gas to prevent slag formation and possible occlusions) by Frank Huszar, who operated a firm called Race Car Specialties in Tarzana; the frame itself was 1¼-inch tubing with a wall thickness of .049, the roll cage was 1¾-inch with a .125 wall thickness, and there were various braces of ½- and ⅝-inch tubing.

The frame was rigid, a design favored by Huszar and several others, including Fuller and Rod Stuckey, over the "flexible flyer" type, whose chief proponent was Gilmore. This was a vexed question. One authority contended that a rigid chassis was preferable whenever there was good traction, "because the bite forces the engine torque to go back into the frame, causing misalignment in a flexible car." Flexible cars worked on slippery strips because torque was "pacified" by wheelspin, but cars without sufficient power to break the tires loose properly suffered from frame misalignment and tended "to go sideways right out of the chute." Gilmore contended that the overriding factor was a flexible frame's capacity for transferring weight while keeping the front wheels on the ground, and he positively bristled at the suggestion that his chassis were flimsy, pointing

On the starting line at Irwindale Raceway driver Jim Davis of Walnut Creek, California, peers between two sheets of flame from his "zoom" exhaust headers. (Photograph by Leslie Lovett, courtesy National Hot Rod Assn.)

out that he used 1⅜″ × .060 tubing for main frame rails, heavier than what Fuller or Huszar used.

Up front the spindles were from an Anglia, the torsion bar from a Volkswagen. Tires were 2.50 × 18 Pirellis run at 30 pounds pressure and mounted on spoked wheels that Jobe himself had laced up. The front axle was 1½″ × .125 chrome moly, the radius rods ¾″ × .049 with ⅜-inch Heim joints. Geometry was set with zero camber, ⅛-inch toe-in, and 45 degrees caster. The drag link, 1-inch 4130 tubing with a .049 wall thickness, connected to a Ross Crosley steering. A 20-pound weight was clamped under the axle, and a 7½ × 24 inverted NACA [National Advisory Committee for Aeronautics] Clark Y airfoil was mounted about a foot above. Such airfoils, standard equipment by this time, were usually fabricated from aluminum or sometimes acrylic sheet; a nice touch with the Surfers was that their airfoil was a piece of "Danish Modern" polished walnut.

The rear end housing was a narrowed '50 Olds, the axles and carrier were from a '56. The gears were 3.23:1 with stock lash. Differential gears were functional. The spider shaft was produced by Ed Donovan, as were the drive shaft, drive shaft cover, coupler, and "can" shielding the clutch and flywheel. Brake drums were '50 Olds, linings metallic, and brakes were activated by a hand lever linked to a .700 inside-diameter Girling master cylinder. There were two twelve-foot parachutes from Bill Simpson's shop in Torrance. Rear wheels were Halibrand, tires were 11.00 × 16 Goodyears run at 18 pounds pressure. Jobe had fabricated the abbreviated body out of 3003 H14 aluminum, .049 thick. He also stitched up the Naugahyde upholstery.

The engine had originally reposed under the hood of a Chrysler New Yorker. By 1966 better-biting tire compounds—Goodyear and M&H were locked in an intense competitive struggle—had enabled most holdouts to switch from 331- or 354-cubic-inch Chryslers to the more powerful 392s. (Some still preferred the 354, however, sometimes with a stroked crankshaft that brought the displacement up to 420 cubic inches.) Jobe had cleaned up the bore himself but took the crankshaft to Joe Reath's in Long Beach, where Henry Velasco had grooved it and ground it for clearances— .004 on the main bearings, .003 on the rods. Standard practice was to replace the crankshaft when it began to show cracks, usually after about sixty runs, along with the bearings. Although many engine specialists favored a bottom-end girdle of the sort manufactured by Milodon, Jobe felt that cold-rolled steel straps under each of the three center main caps sufficed for support.

The aluminum flywheel was manufactured by Paul Schiefer, along with the eleven-inch clutch that had two sintered-iron discs separated by a steel floater. Mickey Thompson manufactured the aluminum rods and pistons as well as the hard chrome rings and the wrist pins and locks. Compression was 7.5:1, piston clearance was .014. There was an AC filter, and the oiling system was pressurized at 100 pounds. The camshaft, roller tappets, push rods, springs, and keepers were all manufactured by Jack Engle, a neighbor of Skinner and Jobe's. Valve springs needed changing every three or four weeks, the timing chain every six months. Like Schiefer, Engle provided the Surfers with parts free of charge. A local distributor furnished their Pennzoil, but otherwise they paid for what they needed.

Jobe had ground and polished the ports. The valves, Ed Donovan's, were 2-inch intakes and 2 1/16-inch exhausts. The exhaust rockers were also made by Donovan, but the intake rockers were stock Chrysler. The hard chrome shafts were Milodon. To seal compression the heads had copper O-rings pressed into machined grooves; copper gaskets were changed each time the heads were removed. Exhaust headers were 2 1/2-inch mild steel, and each pipe was approximately 21 inches long measured from the port. The rear header on each side was cut on a bias and aimed over each slick at about two o'clock.

The GMC blower was turned 22 percent faster than the crankshaft by a 3-inch Gilmer belt drive from Cragar Industries, a division of Bell Auto Parts, the first of the California specialty shops. Santa Monica's Stu Hilborn had manufactured the fuel injector; Jobe added extra nozzles in the Cragar manifold which dumped raw fuel directly into the ports. Jobe also machined the drive for the Hilborn pump. Fuel lines were 7/8-inch inside diameter. The magneto was a Schiefer, and Jobe normally set the spark timing to a 36-degree advance. While the engine had never been on a dynamometer, horsepower was somewhere in the vicinity of fourteen hundred, meaning that the power-to-weight ratio was around 1:1. This was far better than any other type of racing car, though it needs be noted that the term *endurance* had a different meaning in most other forms of competition.

It is also worth noting how many of the components were reworked stock automotive parts from Detroit and from England and Germany—small items from Anglias, Crosleys, and Volkswagens, big ones from Chryslers, Oldsmobiles, and GMCs. A decade later, while there would still be many components deriving from automotive *forms*, virtually everything would be the product of a specialty manufacturer. In the 1960s, however, even a state-of-the-art fueler exhibited its hot rod lineage, and there were plenty of machines raced at the drags which had a lot more stock parts than a dragster had. Some of them even *looked* like stock automobiles, and some of those were almost as popular as dragsters.

Dragsters had never been all there was to drag racing—far from it. There had always been other classes. And some of these were becoming increasingly popular among spectators. Fuelers came into the 1960s running in the high 8s and 180s; they went into the 1970s running in the low 6s and 230s. Even as they were recording dramatic technical progress, however, different sorts of machines were stealing the spotlight and, with it, a good chunk of the kitty that promoters chose to make available for purses. While they were neither as quick nor as fast, they were often perceived as superior performers in a theatrical sense.

7

COMPETITION

I really enjoy watching a fueler smoke all the way through the eyes, but it gets boring quick.

DON NICHOLSON, 1966

Remember, it's the law of supply and demand.

STEVE GIBBS, 1967

Drag racing had its roots in hot rodding. Once the quintessential hot rod had been a roadster, a Model-T, Model-A, or '32 Ford; indeed, the Southern California Timing Association had stipulated "roadsters only" at the lakes, and the very name *hot rod* was a derivation of *hot roadster*. By the time drag racing began getting organized, however, roadsters were getting scarce. On an average Sunday afternoon at a typical strip there might be only a few roadsters on hand, perhaps just a couple of dragsters, but there would be plenty of prewar Ford coupes and even some hopped-up sedans. To people who were hot rodders before they were drag racers the appeal of a rail job, however fleet, might not be as strong as the appeal of a machine whose Detroit lineage was perfectly obvious. Spectators who had been wandering around the pits would run to watch when they heard the telltale fury of a fuel dragster firing up, but they would also run to see a coupe that was perhaps not so fast but had more, well, panache. And, considering an inevitable technical handicap in terms of weight and frontal area, some of these cars performed extremely well. Take Don Montgomery's '32.

Montgomery, who would become hot rodding's Boswell many years later, had begun street racing in the 1940s, but not in a Ford; his parents feared that, if he had a Ford, "it would become a hot rod." So he took his dad's '41 Hudson sedan and made *that* into a hot rod, installing a Buick straight-8 and fitting it with six carburetors. Later, as a member of the Glendale Coupe and Roadster Club, Montgomery ran a '37 Cord with a GMC engine at El Mirage, but after C. J. Hart got things started at Santa Ana nearly everyone in his club switched to the drags.[1] While the Cord's aerodynamics were advantageous at the lakes, its overall weight and particularly the weight distribution were not good for acceleration. The ideal machine for drag racing, Montgomery felt, would be a Ford coupe with a fuel-burning engine that was set partway back on the frame in order to enhance traction.

Many hot rodders believed that a "deuce" had the most pleasing lines of any Ford ever made, indeed any automobile ever made. Montgomery's '32 was stock to all outward appearances, except for a four-inch "chop" in the window area and the removal of a few nonessentials such as bumpers and headlights. The windshield had been replaced with a sheet of acrylic. Inside there was a war surplus bucket seat, and, to accommodate the engine setback, the firewall and dashboard had been cut away and replaced with aluminum. The rest of the interior metalwork was "gutted," bringing the total weight down to about 2,400 pounds. That was too heavy to match a dragster's charge off the starting line, but until well into the 1950s the best such machines could run almost as fast. Montgomery turned 120s with GMC power and 130s after he switched to a Chrysler with a top-mounted blower and fuel injection, a combination he managed to make work long before it became standard for fuel dragsters.[2]

During its heyday, the middle 1950s, Montgomery's '32 put on an exciting performance and was as popular with most spectators as any dragster. It was an exceptional machine, but far from unique. On October 14, 1956, Montgomery convened all the fastest full-fendered, fuel-burning coupes at Santa Ana. C. J.'s rules separated '34s and older from '35s and newer because the latter were heavier and thus at a disadvantage, even though they appeared to have had cleaner aerodynamics. Top time, 132.11, went to a '34 coupe owned by Gene Mooneyham and sporting a decal of cam grinder Clay Smith's "Woody Woodpecker" emblem."[3] Its name was its number, "554." Subsequent modifications made this more of a pure race car than Montgomery's '32; in its final evolution the fenders had been removed, the rear axle was narrowed, and the driver sat with his head up against the back window, steering through a 1929 Franklin unit of a type frequently adapted by circle track racers. But it remained readily identifiable as a '34 Ford. With a 392 Chrysler, a GMC blower, and a fuel injector made by Stu Hilborn's new rival, Kent Enderle, it topped 150 early in 1960 and eventually turned almost 180 running a set of one-off aluminum heads cast and machined by Mooneyham's partner, Al Sharp.[4] By then Mooneyham had

In the late 1940s a hot rod was, by definition, a fenderless roadster. Here a '32 Ford is seen parked in front of Karl Orr's, one of the pioneer southern California speed shops. (Photograph by Walt Woron, courtesy Museum of Drag Racing)

The chopped top aside, the body of Don Montgomery's '32 coupe was stock Ford. But the opening in the cowl for the injector indicates an engine setback aimed at improving weight distribution. (Courtesy Don Montgomery)

moved on to dragsters, but there were still plenty of enthusiasts who loved the idea of racing hot rods.

Ford coupes and sedans could be cut down to little more than four feet in overall height by chopping the top radically, repositioning the body lower on the frame—a procedure called channeling—and changing the suspension in any number of ways. Purists felt that it was important not to transform the stock lines too much, and one could botch the job entirely. The matter of aesthetics was subtle. In the mid-1960s Jim Lytle, a transplanted Texan, campaigned a '34 sedan whose one-piece fiberglass body had windows that were just an inch high (the windows were only simulated, and Lytle drove with his head protruding through the roof). Yet Lytle somehow managed not to violate the spirit of a '34's original configuration, and the attraction of his machine was not diminished by its 160 MPH times via an Allison engine, one of only a couple in drag racing after Akron's Arfons brothers switched to jets.[5]

Later, taking a cue from Tommy Ivo, Lytle dreamed up "Quad Al," with *four* Allison engines and a tiny Fiat body, a creation meant to be appreciated solely in an anything-goes theatrical sense. This was not true of many other machines using bodies from the ancestors of what would later

The Mooneyham-Sharp coupe, fenderless but still with an essentially stock body, is seen here at San Gabriel in 1962. (Photograph by Bill Turney)

With engines virtually identical to those in top-fuelers but drastic restrictions on overall configuration, machines like this were very dramatic. With its Fiat body the Dave Campos machine (far lane) was typical of fuel altereds; Gabby Bleeker's Austin (near lane) was unusual. The scene here is Famoso in 1964. (Photograph by Bill Turney)

During a technical inspection of his Chrysler-powered Willys in 1964, Tim Woods (left) clarifies a point to an NHRA official; note the decal on the injector scoop cover. (Courtesy National Hot Rod Assn.)

be dubbed "compact" cars: not only Fiats but also Bantams, American Austins, and Crosleys, even Messerschmitts, Isettas, and Nash Metropolitans. But the most popular was the Topolino, a body style made by Fiat from 1937 to 1948. The first Fiat-bodied drag racing machine of note was Jazzy Jim Nelson's, debuted in 1953 and campaigned into 1956. Nelson set it up with a weight distribution close to that of contemporary dragsters; indeed, this car was classified as a "competition coupe" and expected to compete against dragsters. Racers later put Fiat bodies, or sometimes Austins, on long-wheelbase slingshot frames. Many of these machines had blown Chryslers, and (like modified roadsters such as Tony Waters's) they were dragsters in any but a nominal sense.

Topolino bodies were also prevalent among so-called altered coupes, which were restricted by the rules in terms of wheelbase and the location of the engine and cockpit. An altered coupe had once been almost inevitably a Ford like the 554, but Fiat bodies were much lighter and aerodynamic. In the late 1950s a part-time movie stuntman named Tex Collins started manufacturing fiberglass Topolino replicas, thus ending "frantic junkyard searches" for a type of car that was never commonplace in the United States to begin with.

After NHRA mandated that altered coupes and roadsters had to run together in a single class, a more prevalent style of altered became a fiberglass replica of a '39 Bantam roadster. Howard "Ike" Eichenhofer and Jim Brissette's brother Bob were to the Bantam roadster what Jazzy Jim Nelson was to the Fiat coupe. Brissette and Ike had been using a '32 Ford frame with a Model-A body but were having trouble with the frame twisting, the axle walking out of alignment, and the car drifting sideways as it came off the line. Switching to a custom-fabricated tube chassis and a fiberglass Bantam body solved the handling problem and also saved some 500 pounds. With a blown Chrysler Windsor engine (354 cubic inches) Brissette had clocked 160s and high 9s by 1960.[6] Fuel-burning altereds later ran 8s, then 7s, 6s, and even peeked into the high 5s. That kind of performance required the same engine as a fuel dragster, the same risk of breaking parts, while the short wheelbase (normally around ninety inches) made altereds very unstable and certainly no match for a dragster. After there was some chance of winning a little money, many racers who had been running fuel-burning engines in machines other than dragsters switched over. And yet many did not, choosing instead to define the purpose of their endeavor differently.[7] "Fuel altereds forever" became the solemn pledge of a hardy band of enthusiasts.

Because an altered was a real handful for the driver, such machines owed much of their popularity to their technical *limitations*. Among fans there was hardly any direct identification with Fiats and Bantams, and in their more extreme permutations these did not look much more like a hot rod than a slingshot did. Some racers tried harder to keep the hot rodder's

With fenders, headlights, and sometimes even windshield wipers, supercharged gas coupes aimed to foster an illusion that they could be driven to town for groceries. Here George Montgomery's '33 Willys is seen coming out of the gate at Pomona in February 1967. (Photograph by Ron Lahr, courtesy National Hot Rod Assn.)

Junior Thompson's Opel GT. It was quite a trick to shoehorn a supercharged Chrysler into the engine compartment of such a tiny car. (Courtesy Museum of Drag Racing)

faith, however, and ultimately they were rewarded with a popularity—and, for a time, a match-race market value—second only to fuel dragsters. NHRA called their machines gas coupes, gassers for short, and strips that did not adhere to NHRA's classification scheme always had some equivalent. The basic idea was to have everything be "street legal"—headlights, windshield wipers, starter motor, everything. The engine had to be under the hood; fuel was prohibited. After a while no gasser could *really* be driven on the street (or driven very far, anyway), and all of them arrived at the drags on trailers, like dragsters. But the aim was to perpetuate a resemblance to cars that *might* be driven on the street, and the illusion worked.[8]

The first gassers were mostly prewar Fords, but there was plenty of variety. Gene Adams had a '50 Olds fastback. After Chevrolet came out with its V-8 engine '55 and '56 Chevy two-doors proved attractive, and Chevy V-8s were installed in older cars of all makes. Junior Thompson (that was his name) started out in a '34 Ford but later found that a '41 Studebaker was extremely light and transferred weight well—an important consideration with so little static weight on the rear tires—and soon others had Studebakers too. Then racers discovered the Willys, another oddball make like Studebaker, but even better for their purposes; for one thing they looked a lot like Fords. Some favored the 1940 body style; others preferred the '33, particularly George Montgomery of Dayton, Ohio, the first competitor from back East to make a mark in a gas coupe. Eventually, nearly all the faster machines had automatic transmissions to help control wheelspin and blown Chrysler engines identical to what powered many gas dragsters. Beginning in 1960 they were classified by NHRA as A/Gas Supercharged (A/GS).[9]

The most famous A/GS Willys coupe debuted in 1961. It was owned by Fred Stone and Tim Woods and driven by Jesse Douglas Cook, Doug Cook, born in Little Rock in 1932 but an Angeleno since he was sixteen. Both Stone and Cook worked for Woods, who owned a Compton construction firm and put up most of the money. Woods, born in 1917, had raced at Santa Ana almost from the start, just as Cook had. He was an African-American, as was Stone. Blacks had always been part of organized drag racing, but there had never before been one who owned a high-dollar machine; the Stone-Woods-Cook "Swindler" represented a larger investment than almost any dragster. At Lions, which was only a few miles south of Watts, a predominantly black section of Los Angeles, African-American spectators turned out in force every Saturday night and among other things were responsible for sustaining a spirited betting scene in the bleachers. Blacks competed at Lions, too, but mostly they competed in the twilight zone of street racing, where gambling was the acknowledged raison d'être.[10]

One key to success as a street racer was concealment of a car's potential; the less it looked like it might be fast, the better to lure someone into a wager. Don Montgomery relished the tale of how he once talked Jack Ewell into a race against his Buick-powered Hudson. Ewell's Ford coupe was top dog on the streets of San Pedro, but Montgomery "blew right on by him." Nobody had ever thought of a Hudson as fast, and that was the whole idea.[11]

A Willys had just about the best imaginable "sleeper" appeal, for in stock form no automobile was as tinny or as grossly underpowered; Willys had been the last American manufacturer to market a car for less than five hundred dollars.[12] Of course, like a pool hustler, the owner of a sleeper could spring only a limited number of surprises. Everyone knew that Tim Woods had a very fast Willys. They also knew how few of the mechanical parts had actually been made by Willys. But illusion was becoming a big part of drag racing, and the *idea* of a fast Willys—a Willys!—seemed very exciting. The excitement was enhanced by cam grinders' hype. That spiral had started even in the 1950s, but Cook took it to new heights in 1960, while he was driving a '40 Willys owned by Howard Johansen. Having attracted attention at the NHRA Nationals, he advertised in *Drag News* that strip owners were "WANTED" to book his "real crowd pleaser." He also took a dig at Iskenderian, who, he said, was "so adept at propaganda that he ought to be in the employ of our Government." Mike Marinoff, another gas coupe enthusiast, from Milwaukee, likewise cast aspersions on "loud-talking fruit pickers from California."[13]

That was the beginning of what came to be called "the gasser wars." The principals were Stone, Woods, and Cook, who were aligned with Engle; John Mazmanian (with various drivers), who was aligned with Isken-derian; and George Montgomery, who generated controversy partly on the basis of where he was from, much like Garlits. Mazmanian owned a fleet of trash trucks and was a wealthy man, like Tim Woods. He had been running a Corvette in a class for modified sports cars, but when he noticed that "Stone-Woods-Cook didn't have any real competition" he decided to finance the building of a '40 Willys with a blown Chrysler. Woods had begun racing with a '50 Oldsmobile—a teenage Tom McEwen drove briefly—and ran an Olds engine in his Willys at first but switched to a Chrysler in 1964. After graduating from flatheads, Montgomery went from a 331-cubic-inch Cadillac engine to a 389 Chevrolet to a 427 single-overhead-cam (SOHC) Ford.[14]

By 1964 all three gassers were turning times close to 150, and they had all been rechristened in volley after volley of *Drag News* broadsides: "Pebble, Pulp, and Chef," "Big June," "Ohio George." In a 1987 interview Mazmanian recalled what the hoopla was all about: "Drag racing needed more color during the mid '60s. With that and more controversy, we could generate more spectator and sponsor interest. And with more attention we could promote more match race money."[15] And, indeed, there was plenty of interest, plenty of match race action, particularly a series of well-advertised tours booked by promoter Ben Christ's Gold Agency in Evanston, Illinois.

A stalwart of those Christ tours was Junior Thompson, yet another enthusiast who had first raced at Santa Ana as a teenager shortly after it opened. While his friends were dreaming of dragsters, Thompson always preferred coupes. In search of better weight transfer he had switched from Ford to Studebaker to Willys. In 1966 he hit on the idea of using an (English) Austin, which had a ninety-two-inch wheelbase, eight inches shorter than a Willys.[16] Other California A/GS competitors switched to Anglias, which were even shorter and lighter, and several racers including Thompson eventually campaigned Opel GTs. In Ohio George Montgomery had also made a switch. He had been running his SOHC engine in collaboration with the Ford Motor Company and Connie Kalitta and Pete Robinson, both of whom had "cammers" in their dragsters. (He was also in business with Robinson, manufacturing and marketing magnesium blower housings and drive units.) Then one day someone important at Ford told him, "If you're going to run Fords, then use a Ford body." Montgomery could not refuse, and in 1967 he debuted a fiberglass replica of a fastback Mustang.[17]

At first he was delighted; with its superior aerodynamics the Mustang turned 160 and dipped into the 8s on its first all-out run. He seldom lost in 1967, though others were able to nudge Austins and Anglias into the 8s and 160s, too. But Mustangs and their ilk were actually the beginning of the end for the supercharged gassers. George Montgomery put his finger on one reason: "When we made the switch to that Mustang, it all went away. The new cars just didn't have that hot rod look."[18] A '67 Ford, that is, was no '32 Ford, no '33 Willys. For some racers, and some spectators, the "hot rod look" had a timeless appeal, and that is why Don Montgomery still drove a '32 coupe on the street in the 1990s and why the 554 still raced on the "nostalgia" circuit, as did clones of several Willys gassers from the 1960s. But the hot rod look meant a lot less to a new generation of spectators, who had never had any personal experience with prewar cars.

Technologically, George Montgomery's Mustang was a step forward. Yet, at this very same time, fuel-burning funny cars were sweeping into drag racing, tapping several resonant chords at once. First, they were skittish and unpredictable, like gassers. Once initial problems of aerodynamic instability had been sorted out, however, they were a lot faster. A fuel-burning funny car might not look much different from George Montgomery's gasser, but it made a lot more smoke and thunder. And, while no funny car had the hot rod look either, people born in the 1940s (as most spectators of the 1960s were) did not seem to care. The crucial consideration was product identity, a factor succinctly defined in a letter to *Drag News* in 1962: "Do you know what the majority of spectators and participants go to see? They go to see cars like theirs. . . . I wonder how long the strips would stay open if only dragsters were allowed to run?"[19]

As fast as they were, fuel dragsters lacked "excitement," said promoters. "It seems like you've seen one you've seen 'em all," echoed one racer, none other than Don Nicholson.[20] In so many words, these people were indicating a feeling that drag racing was about something other than being quickest from point A to point B. Actually, funny cars were "like" the cars that spectators drove only in the flimsiest sense. But they were enough like them. More to the point, if there had been any remaining doubt that high performance was no longer being defined solely or even essentially in technological terms, there was no doubt now.

Competition among stock automobiles (more or less as delivered from the manufacturer) had been the promoter's bread and butter from the very beginning. Better than anybody, Wally Parks understood the potential advertising value of stockers to Detroit automakers, and encouraging "factory participation" had become a cornerstone of NHRA policy. Detroit got the message when the Nationals were first staged there in 1959, after which NHRA devised an elaborate class system to equalize competition among different makes and models. Soon, in addition to stockers, there were "super-stock" (S/S) classes to accommodate all the manufacturers' performance options that could be loaded on—the biggest engines, heavy-duty suspension, four-speed manual transmissions. Don Nicholson began making a reputation in super-stock as early as 1960. With a new 409-cubic-inch Chevy engine he broke into the 12s in 1961, and soon the Beach Boys were singing a song about the 409. Even Don Garlits tried a Dodge Ramcharger.

The stakes kept escalating, and by 1964 NHRA had a classification called "factory experimental" (F/X); here it was Pontiac against Ford against Dodge against Plymouth, with drivers receiving paychecks directly from Detroit. As Dean Brown, now editor of NHRA's *National Dragster,* explained it:

> The name gives clear indication of how the game here is played. It's expensive, true, far more than racing in the other stock classes, and is precisely why NHRA set it apart from competition by the individual. Experimentation is always expensive, but the Factory Experimental players understand the stakes. They also understand the possible reward and we should all hope that a factory benefits when their product does well.[21]

NHRA did its best to encourage factory competition, but, as Dave Wallace, Jr., later wrote, it soon found itself being gnawed "by a monster of its own creation."[22] In a neat little case study in technological momentum, things went something like this: Chevrolet and Dodge started the ball rolling; Chevy already had "youth appeal," but Dodge was seeking to transcend an "older image."[23] By 1962 one could buy a 2,800-pound Chevy II with a 360 HP Corvette engine, or a Lancer rated at 410 HP. In 1963 there were Pontiacs with 420 HP that were some 350 pounds lighter overall than the year before, by virtue of a hole-sawed frame and numerous aluminum components such as front fenders, hood, exhaust headers, and bellhousing. Ford Galaxies were available with 427-cubic-inch engines and with fiberglass body panels and aluminum bumpers to match the racing weight of Pontiacs and Chevys. The Dodge "Max Wedge" performance package had comparable goodies. In 1964 General Motors lowered its racing profile, but Chrysler cast off any restraint, "releasing all-out racing cars right from the factory."[24]

The idea was still to convey the illusion of a showroom stock automobile, so the trickery remained fairly subtle. There were steel bodies that were "chemically cleaned" (i.e., dipped in acid), aluminum front ends and doors, plexiglass windows, magnesium wheels, spartan interiors. And Chrysler brought back the hemispherical head engine, out of production since the late 1950s, this time in a 426-cubic-inch, 425 HP version designed solely for racing. The Chrysler F/X program was the responsibility of a dozen company engineers, the Ramchargers Maximum Performance Corporation. President of the Ramchargers was James F. Thornton, twenty-six, high school valedictorian, first in his class at the University of Missouri, recipient of a master's degree in automotive engineering. There had not been many drag racers like Jim Thornton. One of the first "Hemi Commando" Plymouths was delivered to Hayden Proffitt, veteran of just about every kind of drag racing machine there was, and within a couple of weeks he was running low 11s at close to 130 miles per hour.

Ford countered with a fiberglass-bodied 427 Fairlane called a "Thunderbolt," assembled by an outside contractor, Holman & Moody, and resembling nothing that Ford had in production.

The Southern California Dodge Dealers Association commissioned a trio of "Dodge Chargers" from Jim Nelson and Dode Martin's Dragmaster shop. They had 480-cubic-inch engines with superchargers. Although one was wrecked during a testing session, Oklahomans Jimmy Nix and Jim Johnson toured the other two as full-time Chrysler employees (two hundred dollars a week plus expenses), their itinerary determined by the interest expressed by local dealers.[25] Because these machines did not conform to any existing class rules, NHRA created a new class, "super factory experimental" (S/FX). But their numbers, 135 miles per hour with 10.80 elapsed times, were not much better than unblown F/Xers, and back in California a wily veteran had an idea: If superchargers, Jack Chrisman asked, why not fuel?

Chrisman had not taken a ride in a slingshot since suffering severe injuries at Pomona when a rear axle "spun" (i.e., tore loose from its mounts); he was hospitalized for six weeks. Instead, he was managing a "Hi-Performance Center" for Helen Sachs, who owned a Downey Lincoln-Mercury dealership and had been sponsoring super-stockers and F/Xers since 1962. Chrisman's engine specialist was Gene Mooneyham. With help from Bill Stroppe, who had been responsible for Mercury's brief venture

Strong competition for favor among drag racing audiences was signaled by the debut of Jack Chrisman's supercharged Mercury Comet in 1964. Here Chrisman services the engine at Indianapolis Raceway Park. (Photograph by Bob D'Olivo, courtesy Petersen Publishing Co.)

In the first showroom-fresh stocker ever outfitted with a supercharged, fuel-burning engine, Chrisman performs with his Comet at the 1964 NHRA Nationals. (Courtesy National Hot Rod Assn.)

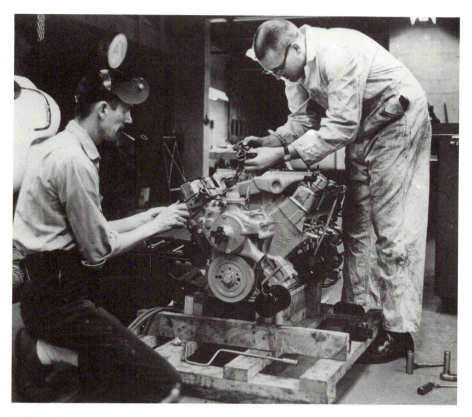

At Ramchargers headquarters in Roseville, Michigan, a stylishly shod Jim Thornton (right), *with Gene Meyers, prepares to drop a wedge-head Dodge engine into a '54 super-stocker. (Courtesy Ramchargers Maximum Performance Corp.)*

into NASCAR racing, Chrisman and Mooneyham put a blown, fuel-burning 427-cubic-inch engine in a 1964 Mercury Comet. Although much of the body was fiberglass and the windows plexiglass, the appearance was very close to showroom stock. With the engine in the original location traction was problematic, but Chrisman's Comet could still clock low 10s and speeds over 150. Jack hit the road, and, as *Hot Rod*'s Eric Rickman put it, his was "one of the best shows in quarter-mile competition."[26] Indeed, the "Super Cyclone" was a very popular act but not really "in competition," as NHRA would only classify a machine on fuel as a fuel dragster.

That winter Chrisman installed a Ford SOHC engine, relocated some two feet behind the stock mounts, and he gutted the interior entirely, cutting the weight down to 2,500 pounds. With this setup he was able to get well into the 9s at more than 160 miles per hour. While that was not as good as, say, Mooneyham's old '34, a car that looked like 554 was *supposed* to be fast; a Mercury Comet was not. Hence, Chrisman tapped the deep well of affection for sleepers, tapped it better than anyone ever had done before. Together with the Ford's high-performance honcho Fran Hernandez—who had been one of the featured competitors in his deuce coupe "the day drag racing began" at Goleta in 1949—Chrisman eventually designed a 1,650-pound tube-framed Comet. By that time, however, he had plenty of company in the exhibition stocker ranks, both newcomers and old pros like Bob Sullivan of Kansas City, who was campaigning a Plymouth Barracuda with a blown Chrysler hemi.[27]

The supercharged exhibition stocker, as a concept, dated to the Dodge Chargers and Chrisman's Comet, but the name *funny car* derived from a trick that Chrysler and its Ramchargers group tried in 1965. Performance of the Dodge Chargers was an embarrassment compared to Chrisman's Comet. But when Chrisman moved the engine back he had to sit up against the rear window, which was hardly conducive to the stock car illusion. There was another way to get more weight on the rear wheels: Racers had long known about the sort of tricks that could be played by shifting the relationship of body, frame, and axles. At the 1965 AHRA Winter Nationals at Bee Line, near Phoenix, Chrysler debuted ten hemi-powered Dodge and Plymouth A/FXers, built, like the Ford Thunderbolts, by an outside contractor, Ambelwagon, in Troy, Michigan. Some of these had the front wheels moved forward ten inches and the rear fifteen, shifting the bulk of the weight to the

In 1955, when Vince Polloccia combined the body from a '32 Ford coupe with a Model-A frame, juggling normal relationships in the interest of better weight distribution, things definitely looked odd, one might even say funny. (Courtesy National Hot Rod Assn.)

The cars are both Plymouths, but the wheelbase alterations to Ronnie Sox's (far side) are readily apparent. Sox's was one of the machines from Chrysler which gave rise to the appellation "funny car" (but compare with photo at left ten years before). (Photograph by Bill Turney)

Don Nicholson poses with one of the machines ordered in 1966 from the Logghe chassis shop in Michigan. This "flip top" means of access to the engine and cockpit inspired the sobriquet "flopper." (Courtesy Lincoln-Mercury Division, Ford Motor Co.)

slicks and yielding traction and elapsed times that no other F/Xer could touch. To the unpracticed eye these cars looked stock—pretty much. Actually, they looked "sorta funny"; they looked like funny cars.[28]

These cars were "illegal" by NHRA rules, which permitted altering F/X wheelbases no more than 2 percent. Ford ordered its factory-sponsored drivers not to compete against them at all but reconsidered when that closed off match race action. Ford went to altered wheelbases, too. Chrysler went to fuel injectors, then to nitro, and soon the Ramchargers had dipped into the 8s at Cecil County, Maryland. There was no longer any attempt to conform to NHRA's rules for stock-bodied cars, or AHRA's, or anybody else's.

The tube-chassied machines that Fran Hernandez ordered from Gene and Ron Logghe's fabrication shop near Detroit were outfitted with fiberglass Comet bodies, one-piece bodies with nonfunctional doors and hood. For access to the engine and cockpit they were hinged at the rear of the frame; flip-top bodies, they were called, or floppers, flops. One was slated for Nicholson, who had moved from California to Georgia because there was plenty of money for his kind of racing in that part of the country— even Richard Petty had campaigned a Plymouth F/Xer after NASCAR's decision to oust the 426 hemi, the engine that had given him his first Daytona 500 win.[29] Nicholson's Comet immediately clocked elapsed times in the 7s. Chrisman's supercharged version topped 185.

NHRA had no class for such machines, but it did "allow them to exist."[30] When Texan Gene Snow showed at the Nationals in 1966 in a fuel-burning altered-wheelbase Dodge Dart, he had to run as a dragster, like Chrisman's Comet had two years before. By 1967, however, there was official "super experimental stock" competition for funny cars. Doug Thorley, a manufacturer of exhaust headers in California, won in a Chevy, without any overt factory help, and soon "hundreds of veteran Super Stock machines were [being] updated in varying degrees and dozens of specials were completed to join the most exciting, most rewarding new facet of drag racing."[31]

Many of the names were unfamiliar—Tom Sturm, Arnie Beswick, Steve Bovan, Pete Seaton, Bobby Wood, Don Gay, Dick Branstner, Dick Landy, Malcolm Durham, Bruce Larson, Gene Snow—though some of these men had been racing in quiet obscurity for a long time. Snow, for example, had begun with a '58 Chevy Impala stocker, then moved up to super-stock and A/FX before commissioning his tube-framed Dart from Don Hardy, another Texan. Larson, from Harrisburg, had begun at the age of sixteen in a '32 Chevy coupe and continued competing in gassers and a Ford Cobra until building a fiberglass-bodied Chevrolet flopper in late 1965 (actually nosing out the Nicholson and Chrisman Comets as the first of the breed). Durham, who owned a speed shop in Hyattsville, Maryland, was a longtime Chevrolet devotee, who had earned quite a reputation around Washington, D.C., as "the Cassius Clay of drag racing."[32]

Several veteran A/GS campaigners like Woods and Mazmanian switched over to the floppers; Ron Scrima and Pat Foster built a Mustang-bodied funny car for Woods, and Doug Cook ran 174, 8.40, right off the bat.[33] But Cook was critically injured when the car flipped over backwards at a midwestern track. There was a lot to be learned about how aerodynamic forces affected a full-bodied car at high speeds, there were a lot of funny cars that were badly designed and poorly constructed, and there were a lot of mishaps. Nevertheless, the momentum seemed unstoppable. There was a sense that rules makers "had been unable to keep up with the headlong pace of evolution."[34] However mistaken that may sound to anyone who does not believe that technology "evolves," it was obvious that all design conventions had come loose. Several racers even draped automobile bodies rather awkwardly over slingshot dragster frames. Garlits teamed up with Emery Cook and mounted a fiberglass Dodge Dart on an old Swamp Rat. Cook turned 200 miles per hour, but Garlits dropped the idea when other racers claimed that this was going *too* far.[35]

That did not really matter to Garlits, but for most people who raced dragsters, particularly in southern California, funny cars represented anything but the "most exciting, most rewarding new facet of drag racing." There was a great irony: The floppers appeared onstage just when an attempt was being made to force a confrontation with strip management over purses. In 1963 Tom McEwen and Douglas Kruse had founded an organization called the United Drag Racers Association (UDRA). McEwen was, of course, a star driver; Kruse was a graduate mechanical engineer (University of Maryland), who had once managed a drag strip in Manassas, Virginia. In 1960, at age twenty-four, he headed for California and served an apprenticeship in one of the specialty machine shops, before founding Kruse Engineering in Long Beach. Later he became involved in all sorts of alternative energy projects—solar, steam, battery, even pedal power—but for several years his primary occupation was fabricating aluminum bodywork for dragsters.[36]

Kruse and McEwen were both self-declared "promoters," and one of the UDRA's stipulated purposes was "to foster and/or promote competitive racing events." But its primary aim was "to promote better relations and understanding between owners and drivers of drag racers and managers of drag strips." "Better relations" meant more money. While drag racing journalists often stood in awe of the participants, not all of them sided with racers automatically; the sympathies of Al Caldwell, a *Drag News* mainstay, were usually with management. The UDRA acquired a tireless supporter, however, in the person of *Drag World*'s Terry Cook:

> Without a doubt, there is a distinct parallel between the labor union movement of the early 1900s and the formation of the UDRA. Much as the sweat-shop laborers worked long hours in poor conditions for little pay, the drag racer of a few years ago raced at poor strips for little or no

prize money. As business prospered in the early years of this century, so has present drag racing. So shall we then not refer to the UDRA as a racer's union.[37]

Many racers prided themselves on their strong streak of independence. Most of them moved in the world of the small entrepreneur, and most had a knee-jerk aversion to the concept of collective bargaining. Yet Cook was quite right about the UDRA: It was a racer's union.

There had been mutual-benefit organizations previously, notably Drag Races [later, Racers] Incorporated (DRI), founded in Los Angeles in 1953. DRI staged six or eight events per year, with substantial purses for the time; as early as 1956, it put up a $1,000 bond for top eliminator and $500 for top time at its World Championship at Lions, attracting thirty dragsters on the same holiday weekend as the NHRA Nationals in Kansas City. (Top time was clocked by Emery Cook in the Henslee roadster, and top eliminator was won by Maurice Richer, in one of the first dragsters ever to carry a non-automotive sponsor's name, the "Nesbitt's Orange Special.") Under the leadership of Jack Ewell and Lou Baney DRI continued to stage money meets and maintain a modest hospital drawing account for members

Doug Kruse is seen here fitting the body to a dragster; typically, the material was .040 or .050 half-hard aluminum, attached to tabs on the frame by means of aircraft Dzus fasteners. (Smithsonian Institution collection)

throughout the 1950s, but it never became a major political force.[38]

Although many drag racers were anything but leaderly, Baney was a notable exception. In the 1940s he had been president of an association that staged dry lakes events, and he had operated a drag strip in the 1950s. In the early 1960s he became general manager at Bob Yeakel Plymouth in Downey, California. After Yeakel's death Baney acquired control of that agency as well as interests in several others in the Los Angeles area. He always managed to live well, but his close friends were racers, and many racers felt certain that they were not getting a fair shake from promoters.[39] Even though purses were growing, the perception was that they were not growing nearly as fast as gate receipts. This was a situation that clearly invited some kind of concerted action. Jack Ewell had been UDRA's first president, but when he declined a second term Baney looked like the ideal man for the job, and he accepted.

In 1965 a group began to coalesce within UDRA ranks which insisted that strip management should pay purses equal to a fixed percentage of gate receipts. The committee appointed to represent this so-called 40 percent group was headed by Dawes Wafer, who was Ed Donovan's partner in the Donovan Engineering fueler. The others included Don Alderson, owner of Milodon Engineering, who had been campaigning a fueler in partnership with Ronnie Scrima and George Bacilek; Bob Downey, who drove a fueler sponsored by Frank Hedge's A&W Root Beer stand in Venice, California; John Rasmussen, Gene Adams's partner in a machine driven by Rick Stewart; Frank Cannon of Hustler fame; and Tom McEwen, driver of Baney's Yeakel Plymouth Special.[40]

One of the most outspoken members of the 40 percent group was Roy "Goober" Tuller, a seasoned driver at twenty-nine—his rides had included the Surfers' and Milodon Engineering machines—who worked as a machinist for Jack Engle. Tuller had grown up in Oklahoma and once worked for Scotty Fenn. He had imbibed a lot of Fenn's ways. Asked by Terry Cook whether he raced "primarily for excitement," he responded that this had once been true, but no more. Money was important, too:

> Only a small percentage of the fuel racers are making enough money to get by on, but the majority of the strip owners are making money hand over fist. I realize that not all drag strips are profitable, but on the whole they are making a bundle, and, you might say, exploiting the racers. It seems as if everyone associated with drag racing and the speed equipment industry is making good money except the racers.[41]

When pressed most of the 40 percent group conceded that their prime target was Lions, which had been managed since Mickey Thompson's departure by none other than C. J. Hart.[42] The Lions books were closed, but Harry Hibler, a general contractor who operated San Fernando as a sideline, did not mind discussing finances openly: The average Sunday gross at San

Fernando (sometimes known as "The Pond," in contrast to Lions, "The Beach") was $3,700; normal operating expenses were $1,800; a $10,000 paving bill was being amortized at about $100 per week; management gave $5,000 to charity every year as a goodwill gesture; and $8,000 went for windbreakers, which were presented in lieu of trophies. That left $600 for purse money each Sunday, $250 of which went to the top-fuel eliminator. All things considered, Hibler said, "one week we lose, and the next week we win." But, under the best of circumstances, paying 40 percent of the gross would not leave "enough money to operate the strip, let alone even think of making any possible profit."[43]

The racers tended to give Hibler credence, especially in view of the fact that similar operations at Ramona and Palmdale had recently folded, and Fontana seemed shaky. But Lions was something else; Lions was without question the most profitable drag strip on earth. For top-fuelers C. J. ordi-

In 1965 Lou Baney was in a position to garner frequent guarantees at Lions and rarely raced out of state. But he did enter the AHRA Winter Nationals at Bee Line (near Phoenix), and here, wearing a UDRA cap, he poses with his driver, Tom McEwen. (Photograph by Eric Rickman, courtesy Petersen Publishing Co.)

narily paid a total of $2,100: $500 to the winner; $250 to the runner-up; and $150, $100, and $50 to third-, second-, and first-round losers, respectively. Most of the 40 percent group would have settled for a payout only $750 higher, with $750 to the winner, $400 for runner-up, and $1,500 in round money. But there were a lot of angry words. When C. J. reported that Lions had paid out $219,000 in cash in 1964 Donovan snapped back, "C. J. Hart didn't make the racers, they made him. I think that he should be thankful."[44]

Donovan's blunt manner when it came to money would later render him anathema to Wally Parks as well as to C. J. By contrast, Baney was witty and tactful, yet he was unable to stem the anger or mend a schizm between the racers of blown fuelers and enthusiasts for other types of machines, who also wanted a bigger share of the gate. Wafer declared that all the other classes had "been riding" on the blown fuelers. John Harbert added that "the 'little guys' don't try to live off of racing." John Garrison, Baney's engine specialist, said that there were "only two reasons" why anyone would run any other class: "Either they race for fun or they can't compete." Gary Cagle, who was not racing a top-fueler at the time, countered sarcastically that "perhaps the big guys should get together and work out some sort of program for themselves." "UDRA is supposed to be an association for the racer, and I feel that this should mean all racers," said Bob Tapia, Donovan's levelheaded driver.[45] There were few voices of conciliation, however, and the situation was not improved by intimations that McEwen, one of UDRA's two paid employees, was using his position to feather his own nest, and Baney's.

McEwen affected a flashy life-style and seemed rather proud of telling people that, except for a few weeks, he had "never worked for a living." In southern California, match races (i.e., guarantees) for fuelers were very scarce, yet of the few there were the Yeakel Plymouth Special seemed to be involved in most of them. Moreover, even though the factory experimentals and funny cars were commanding more and more of the match race action, Baney did not seem alarmed. Everyone knew that he was directly involved with such machines through the "Yeakel 426 Club" for super-stockers as well as an F/X Plymouth that Yeakel sponsored for Hayden Proffitt and the Southern California Plymouth Dealers Association's "Hemi-Cuda," an exhibition stocker that McEwen drove. Fed up with the innuendos and dissension, Baney resigned from the UDRA presidency and from the board as well. Both Ewell and Kruse declined nomination for the top spot; then who should the UDRA elect president but Tom McEwen![46]

Baney had always sought to mute the explicit threats, but shortly after McEwen became president, on December 4, 1965, there was a wildcat boycott at Lions. Hart had previously announced that, for one week only, the top-fuel prize would be cut back to $300 because he was committed to a $1,000 payout to each of the contestants in a match race—none other than

McEwen and Don Prudhomme. Woody Gilmore, who had built dozens of chassis for local fuelers (including Baney's), got on the phone to suggest that the racers stay away that Saturday. He did not reach everybody, and a few he did talk to showed up anyway. So, Woody made the rounds of the pits, along with Frank Cannon, asking people to trailer their cars. Most of them did so, including James Warren and Roger Coburn from Bakersfield, consistent winners everywhere they raced. But others would not comply— mostly men like Joe Jackson, who saw a chance, for once, to pocket some money. "Finks," Cannon called them. Naturally, the boycott unraveled.[47]

Everybody involved was left with a bad taste. Woody had felt it was "time for the racers to stand up and be counted . . . showing the management that we mean business." But the racers would not stand together. Doug Kruse, UDRA's only other paid member besides McEwen, "correctly diagnosed a severe case of infighting as terminal" and resigned to develop plans for a new organization with a structure that would not be vulnerable to factionalism. Hart announced he was quitting his job at Lions. He changed his mind but never entirely forgave UDRA activists. A lot of people were upset with McEwen. Asked his opinion of the situation, Don Garlits said, "UDRA could have every big-name driver behind them, from fuelers to super stocks"—but not, he said, without "a Jimmy Hoffa, an organizer."[48]

And Terry Cook summarized: "All members of the UDRA should all realize that they are all hurting. Of all the various members of this sport, be it speed equipment manufacturer, speed shop owner, sanctioning body official, strip owner, or someone connected with an automotive publication, the racers are getting the smallest proportional return for their investment of both time and money." The UDRA could count some significant accomplishments, including a drivers' licensing program and stronger safety regulations. Even had factionalism not weakened and eventually wrecked it, however, attempts at collective bargaining would have fallen in the face of management's perception that there were good draws besides fuel dragsters. Interviewed while Woody was fomenting his boycott, Hart said: "It's restraint of trade. I'll just have to find some other type of cars to run in the future if the fuelers don't want to race here."[49]

Ben Christ, the newest owner at Fontana, had already made it clear that he would build his normal shows around stockers and gassers, with just one special attraction each week, perhaps a wheelstander, whose performance consisted of running the entire course with the front end high in the air. Both Fontana and Lions were now sanctioned by the American Hot Rod Association, and AHRA held its national championships at Lions in September 1965, running two consecutive weekends. The first featured dragsters, the second super-stockers and F/Xers, and those machines drew a much larger crowd. In October Lions fans got their first taste of fuel-burning exhibition stockers using resin for traction, with racing on the order of what spectators in other parts of the country had been seeing most of the year. A couple of dragsters were on hand, making test runs, and they actually met with catcalls from the bleachers.[50]

Southern Californians also got a look at a new breed of professional racer, men in the direct employ of Detroit automakers, who took their cars from strip to strip in "gigantic air-conditioned tow rigs" and who tended to resemble "an ivy-League college junior who is a marketing major":

> They look like the sons of a bunch of oil magnates frolicking about, keeping their snappy duds impeccably spotless, regardless of how much work they do. Mid-week leisure-time activities for these guys include golf, rather than grease pits. When you see this group, you have no trouble relating them immediately to their thick Detroit financial ties, as they present an "image" which has been heretofore unknown in the sport.

It was enough to leave old-timers feeling "chilled to the bone."[51]

Most of the "factory" racers were from other parts of the country, and some of them had not done much racing previously. An exception in both respects was Dick Landy, from the San Fernando Valley. Born in 1937, Landy was the same age as Tom McEwen and had started racing at about the same age, nineteen. McEwen had graduated from his mother's '54 Oldsmobile into faster and faster machines: a Crosley, then a Fiat altered, then (with some coaching from Art Chrisman) gas and fuel dragsters. Landy had started in a '56 Ford pickup, then moved into super-stockers with his boss, Andy Andrews, who owned a Ford agency. After Andrews switched to a Plymouth dealership Landy was successful enough with a Plymouth super-stocker to get "a small amount of help" from the local dealers' association. Meanwhile, Chrysler was gearing up for its massive F/X assault, and in 1965 Landy was hired by the factory and given a $20,000 Dodge, and he joined drag racing's Bermuda-shorts jet set.

How did Landy manage such a deal? "He was two people," said one journalist, "a smooth hard-working businessman, really a salesman, and a hard-charging racer." While McEwen affected the life of a "semi-bachelor playboy," Landy had been married since graduating from Notre Dame High School in Sherman Oaks. After selling cars for several years, he had opened his own high-performance shop, Dick Landy's Automotive Research in Van Nuys. Amid all the dragster enthusiasts Landy had been a UDRA activist, trying to advise the others that they "should look more to putting on a show for the spectators, a diversified show." Why, he was asked, were cars like his becoming so popular?

> I think the biggest factor is the manufacturer's name associated directly with the cars. This direct factory sponsorship draws a lot of people on its own strength . . . people who otherwise wouldn't come.

But would that last?

> We know that personalities are the real big thing. Today most of the dragsters look alike. It's very hard to tell who is who when they come to the starting line. If the spectator can't identify with something on the car or with the people who own or drive the car, they are going to lose interest in it. It's for this reason that the [dragster] racers need personalities.

Would that be enough?

> There are so darn many of them. . . . They have flooded their particular market, and as a result there is no demand for them.

Landy also spoke confidently of a day when all racing operations would be like his, when there would not be a lot of "boys" trying to run with the "men," like there were among the fuel dragster racers.[52] Here he was not entirely correct.

As people with funny cars devised new means for "putting on a show for the spectators," they built a good part of their drawing power around various antics that preceded the actual run down the strip. There were a lot of would-be performers who might not have been so sharp mechanically and might not have looked like marketing majors but who knew a lot about antics.

Dragster drivers ordinarily suited up and got buckled in somewhere towards the end of the course, where there was rarely an audience; then they were pushed up the strip, they fired the engine (by engaging the clutch at about 35 miles per hour), they swung an arc behind the starting line, they pulled up to the line, they waited for the signal to go, and they went. Most dragsters had no bodywork ahead of the engine, so there was little way to tell one from another. During the race itself the drivers were entirely obscured by billowing smoke from the tires. The funny car routine was quite different. The cars emerged from behind the starting line and were fired by means of electric starters. There was a showy ritual: "The car comes to the line, the announcer starts his buildup, the crewmen (and sometimes even the star himself, the driver) spread and sweep the resin, the driver makes at least three blasts off the line through the resin, his opponent doing likewise, treating the crowd to a display of that awesome leap off the line several times before the actual race even begins."[53] From the start there were racers who could cap off all this preliminary "display" with impressive times. But, even when funny cars performed well only in a theatrical sense and poorly otherwise, the fans did not seem to care: "The nature of the 'funny' spectator is dashing the hopes of the dragster devotees," wrote *Drag World*'s Doherty. " 'Funny' fans will pay more, demand less, and show more enthusiasm."[54]

If purpose were defined as a matter of drawing spectators, there was no

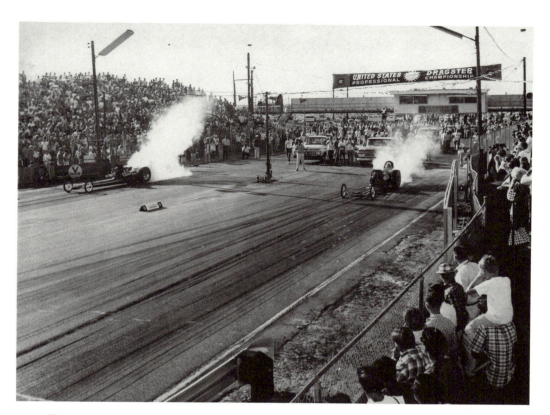

The opening round of competition at Doug Kruse's inaugural Professional Drag-ster Championship in 1967. Although the turnout was extraordinary, this shot is indicative of a major problem from which the fuelers were suffering: The casual spectator could not tell one from another. (Smithsonian Institution collection)

arguing with the gate receipts. When funny car meets were staged in direct competition with dragsters they generally attracted bigger crowds. The UDRA still had some life in it, and at the beginning of 1967 it offered local strips guaranteed dragster fields (which could be advertised in advance) in return for $100 per round. A spokesman for Lions was noncommittal, ad-vising UDRA to "start building personalities." Irwindale's Steve Gibbs was dismissive, suggesting that UDRA find "a way to please the crowd with something different." *Drag News* noted that on February 17 Irwindale "put on an all funny car show," with no dragsters at all; in addition to the eight floppers in competition there was Chuck Pool's "Chuckwagon," a wheel-stander, and John Smyser's twin-engine "Terrifying Toronado," which gave the fans their ultimate thrill when it veered sharply and jumped clear over the guardrail.[55]

In 1965 Gil Kohn, an eastern promoter, had taken over the U.S. Fuel and Gas Championship in Famoso. Kohn's associate was Ed Eaton, the former NHRA kingpin. Two years later, when they instigated a funny car program,

Kohn announced that this made Bakersfield "one of the first of the major national races to accept the highly controversial pseudo stockers as a fully qualified element of competition, rather than a sideshow." As it turned out, the funny car competition was mediocre compared to dragsters. All the same, the dragster payout was exactly as advertised, a total of $7,450, only $1,800 shy of the winner's share alone the year before and, in sum, $6,500 less than in 1966. Pete Robinson's response to what he perceived as an insult was to pay only half his entry fee.[56]

There was no doubt that dragsters *could* draw crowds if vigorously promoted. Ever since leaving the UDRA Doug Kruse had been planning an all-dragster meet: "I'm gonna run a professional race with professional racers with more money up than anyone has ever put up and they [the spectators] will pay to see it," he promised. "What's more, they'll get their money's worth." In the summer of 1967, at Lions, Kruse put on the first United States Professional Dragster Championship. It was a box office success, and he did it again in 1968. *Newsweek*—even *Newsweek* covered this event—reported that the gate was the largest in Lions's history; 18,400 people paid a total of $80,000. Sixty-four fuel dragsters shared a $38,000 purse, nearly 50 percent of the gross, and yet Kruse still made money. In the flush of success he predicted that "dragsters can become just as professional as Indianapolis cars."[57]

In fact, sheer enthusiasm would sustain top-fuel competition in the face of the funny car invasion. Eighty-two top-fuelers showed up at Pomona for the 1969 NHRA Winternationals, and in June fifty of them came to New York National Speedway on Long Island, where Kohn and Eaton had moved the U.S. Fuel and Gas Championship. Fifty was a good turnout, but not even 10 percent of all the top-fuelers still active nationwide: In his spare time a GI in Vietnam compiled a card file of five hundred that had run better than a 7.79, one hundred of which had run better than 7.05.[58] With slider clutches fuelers needed to do burnouts (starting in a shallow trough of water) to heat up the tires. The crew would then push them back to the starting line by hand. This increased their time onstage by as much as a minute, and that helped. But only a handful of fuelers could get bookings, whereas there were dozens upon dozens of funny cars on the match race circuit by 1970. Although the floppers had narrowed the gap significantly, they were still a half-second off on elapsed times (6.90s to 6.40s) and 15 miles per hour slower (215 to 230). Yet there were continual forecasts of dragster doom: "It looks like fuelers are headed down the drag strip toward a permanent resting place in an obscure museum," concluded *Super Stock* magazine. "The dragster?" asked Woody Gilmore in 1970. "She's a dead duck, financially speaking."[59]

A lot of people with funny cars seemed to be in drag racing for what they could get out of it quickly. A lot were on the scene for a year or two and then gone forever. Highly touted match races were often dismal flops. Racers seemed disinclined to "thrash" when they had trouble, something for which dragster people were famous. A lot of the show seemed to be ritualistic. At worst there were scenes reminiscent of professional wrestling: shouting matches on the starting line, even pie throwing.[60] Some fans seemed to like the burnouts better than the races, because these often resulted in towering wheelies, but races were often full of close calls, too. As early as 1966, Don Nicholson began to suggest that "this 'run what you brung' policy is getting out of hand," and, ultimately, he was instrumental in the creation of another set of rules closer to those that had once governed super-stockers.[61] But the funny car's momentum was slowed scarcely at all, nor the performers' excesses. When C. J. Hart began booking funny cars he would provide accompanying music over the public-address system— circus music. In fact, the whole thing was a lot like a circus. People like Art Chrisman and Ed Donovan, who believed drag racing was about "technological progress," were not pleased.

8

HUSTLING

*You know how that is; that's part of
drag racing.*
    KEITH BLACK, 1989

For drag racers who thought of performance in traditional (i.e., technological) terms, funny cars, wheelstanders, and all "exhibition" machines were sheer travesty. "Why don't the strips just hire the Joey Chitwood Thrill Show to come . . . and roll some stocker every week, if that's what the spectators want?" UDRA activist Dawes Wafer demanded to know.[1] Eventually, some promoters did exactly that, but, admittedly, not all exhibition machines were the same; recall, first, that this is essentially how the National Hot Rod Association had treated fuel dragsters for many years. Moreover, by the time the gasoline-only rule was rescinded there was a type of exhibition machine on the scene which was capable of running away from even the fastest fueler.

If the purpose had *only* been to maximize acceleration from A to B, there was a whole new breed in the early 1960s which would have been the ultimate, a breed pioneered by Walt Arfons and Romeo Palamides. Arfons debuted the first jet-powered Green Monster at Lions on August 6, 1960. It was equipped with a Westinghouse J-46 engine designed for the navy's F-70 Cutlass. (Other "wienie roasters" would use the General Electric J-47 designed for U.S. Air Force F-86 Sabrejets, the General Motors J35-A designed for F-89 Scorpions, and even more powerful engines.) Times were mediocre at first, but after Arfons added a thunderous afterburner, 7-second clockings came up regularly. Romeo finished the first of his "Untouchable" jets in 1962, his driver Archie Liederbrand turning 7.73 on April 14 at Vaca Valley and 7.48 the next day at San Gabriel. That same day, in Kansas City, Walt Arfons was reported to have recorded elapsed times in the 6s. Meantime, Art Arfons was leaving crowds awestruck with his own jet dragster and similar numbers.[2]

Soon promoters hit on the idea of matching jets against fuelers. Because igniting an afterburner was problematic, and because holding still on the starting line was impossible once it was lit, the jets behaved erratically, sometimes fouled, and lost as often as they won. But Liederbrand regularly clocked speeds well above 200 miles an hour, as did Doug Rose, a former navy J-46 mechanic whom Walt Arfons hired to drive for him in 1962, and Gary Gabelich, who drove Bill Fredrick's "Valkyrie." Without any doubt the jets put on a compelling show, literally lighting up the sky and shaking the earth. And they put a smile on the face of many an impresario; from coast to coast, at one track after another, the initial appearance of a jet dragster resulted in the best gate ever, and subsequent performances drew even better.

Promoters loved jets, and so did fortunate fuel racers such as Ivo, who could garner bookings to match race them. But the rank and file felt threatened. Don Moody, driver of Dave Zeuschel and Kent Fuller's new slingshot, was emphatic about jets: "I'm flatly against them," he snapped. "It takes away from the whole idea of drag racing." But jets were quicker than any fueler, and wasn't the "whole idea" to get from A to B quickest? Not exactly, Moody indicated. There was something else on his mind: "For every $1,000 they give for a jet appearance, that in turn takes $1,000 from you." "I would rather spend $5 to see good racing than 2¢ to see one of these freaks," said another southern Californian.[3]

The resentment of racers was pretty transparent. Wally Parks had different sorts of reasons for looking askance at jets, but, of course, NHRA did not necessarily see the objective as getting from point A to point B as quickly as possible either. NHRA had long since indicated that it would not permit technology to be the sole driving force in drag racing. While Parks was no longer associated with Petersen Publishing, it was no coincidence that NHRA proscribed jets at exactly the same time as an article appeared in *Hot Rod* elaborating their dangers.[4] Subsequently, NHRA revoked sanctions at Ramona Raceway and elsewhere for violating that mandate, and track owners such as Gil Kohn severed their ties with NHRA rather than comply.

Why NHRA's ban on jets? Did it have to do with maintaining cordial relations with Detroit, which naturally wanted to keep attention focused on names like Chevrolet, Ford, and Chrysler? No, NHRA insisted, it was because jets were unsafe. To begin with, there were those speeding impellers and the chance of an engine disintegrating after ingesting a foreign object. But that was not the primary safety consideration. At least three times heavier than a fuel dragster, a jet was far harder to stop, even with an enormous ring-slot parachute. Indeed, there were numerous accidents precipitated by parachutes that tore loose upon deployment. Bob Smith of San Jose, who took over the controls for Palamides after Liederbrand went to work for Garlits, sustained only cuts and a few broken bones when Untouchable 2 crashed off the end of a strip at Union Grove, Wisconsin, but later he suffered extensive injuries when Untouchable 4's parachutes failed at Milan, Michigan. It was not an edifying story. Palamides, sure Smith would die, "grabbed all the appearance money and took off for California," then filed for bankruptcy. Smith spent three years recuperating, but he did survive. Later there were fatalities involving jet machines built not only by Palamides but also by Fredrick and both the Arfons brothers as well.[5]

Yet fuel dragsters, which NHRA readmitted to the fold around the same time it shut out jets, were hardly a safe proposition either. In the mid-1960s there was a rash of parachute failures that took the lives of, among others, Lou Cangelose, one of the pioneers on the tour, and Lefty Mudersbach, a member of the team that won the 1957 NHRA Nationals. There were also fatal fires and crashes in which drivers died as a result of structural failure. In truth, protective standards were inadequate. With a chassis of .047-inch tubing a top-fueler could be built which weighed only a few hundred pounds more than the engine alone; drivers usually survived flips, but sometimes they did not. Throughout the 1960s not a year went by without several fatalities, often at NHRA strips and despite safety inspections regarded as rigorous at the time.

To be sure, jets seemed to have a more fearsome potential for getting "into the crowd" (as the expression went). But spectator fatalities—and they were never commonplace—typically involved flying parts from fuel engines, or out-of-control stock-bodied automobiles. One of the most shocking incidents involved Richard Petty, who left the NASCAR circuit briefly in 1965 and went drag racing; an eight-year-old boy died when Petty's exhibition stocker crashed into spectators at an unsanctioned (and uninsured) strip in Dallas, Georgia.[6] If the layout of a strip was inadequate for crowd control, any machine could pose a danger to spectators, not just a jet. As it had done previously with fuel dragsters when NHRA dropped them, the AHRA welcomed jets. Ultimately, NHRA did so too, in 1975, after design and construction standards had been much improved. By then it was also clear to all concerned that they were "a valuable marketing tool," too valuable to be ignored.[7]

By the 1970s conventional fuelers were turning 5s, and jets could not match their elapsed times. On the other hand, exhibition machines powered by hydrogen-peroxide rockets were far faster and quicker than anything else—clockings of well over 300 miles per hour in the 4s were

Roger "Lucky" Harris is seen here slowing down with the assistance of a ring-slot parachute. "U.S. 1" was a jet that Art Malone purchased from Mickey Thompson, who had in turn bought it from Bill Fredrick. Harris lost his life in this machine. (Courtesy Museum of Drag Racing)

reported—but their vogue was brief. The rockets had an atrocious safety record, worse than anything else in the whole weird menagerie of exhibition vehicles—everything from motorized skateboards to jet-propelled Mack trucks. The rockets were too much. For some promoters, however, theatrics were everything, and their excesses were truly wretched.[8] Some of them turned to acts that did not even involve driving from one end of the strip to the other, particularly daredevils whose archetype was Robert Craig Knievel, "Evel" Knievel, the bad boy from Butte, Montana. A variety of radio spots later dubbed "screamers" was born at the hand of Ben Christ and proved ideal for publicizing such acts among youthful devotees of Top 40 stations. "We made it sound like the whole world was going to come to an end at the drag strip," remarked Bill Holz, who ran York Drag-O-Way in Pennsylvania and dreamed up the idea of having "fire breathing jets" incinerate junk automobiles with their afterburners.[9]

While ink for the top-fuelers seemed chronically sparse in daily newspapers, the local press would often feature the appearance of an outrageous daredevil act. Even the trade papers could be guilty of relegating the regular racing to secondary status: "The famed Evel Knievel brought the fans out in droves as he wheeled his Harley Davidson over 13 cars at Lions this past weekend. A reported 14,000 folks witnessed Evel's dare-devil antics and also saw Supernationals champion Rick Ramsey outdistance an invitational Top Fuel field."[10] In the mid-1970s *Drag News*'s Dave Wallace, Jr., cautioned that promoters had "apparently forgotten that most of their fans still come out to a drag strip to watch a drag race, not a carnival. Too often the

In August 1974 Russell Mendez makes a run in Ramon Alvarez's "Free Spirit" rocket, in which he would later be killed. (Smithsonian Institution collection)

On the way to what Brock Yates called "fantasyland." Doug Rose's "Green Mamba," chained in place, torches a station wagon. (Courtesy Museum of Drag Racing)

Exhibition machines like Bill Golden's "Little Red Wagon" were designed to pull the front wheels up as soon as they left the starting line and to keep them up the full quarter-mile, like this. (Courtesy Bill Golden)

show becomes a sideshow, with race cars inserted as 'filler.' "[11] But that was more than a decade after promoters started booking exhibition acts regularly. As far back as 1966, *Car and Driver*'s Brock Yates had issued a trenchant warning. Drag racing had a lot of appeal, Yates wrote, because of "the unearthly excitement it provides when giants like Don Garlits, Don Prud-homme et al., blast down the strip in their fuel dragsters." Nevertheless, he added:

> Unless its mentors change course in the very near future, drag racing is an odds-on favorite to replace roller derbies and professional wrestling as the Great American Non-Sport. This means that dragging is on the threshold of losing whatever value and prestige it has as legitimate auto-motive competition and is about to slide into limbo with all the other

hoked-up hippodrome acts . . . a complete fantasyland of bogus race cars, driven by bogus race drivers in bogus non-races.[12]

At bottom it was all a matter of divergent opinions about purpose. For many spectators the essence of the drags was theater, not progress in technology. Blackie Gedjian, the promoter at Fresno Dragway, was fond of bringing in four fuelers to race side by side—a "tag team match."[13] At Fontana four *jets* raced all at once. But spectators seemed easily bored by a steady diet of anything. And, even while decrying the influx of "cheap sensational 'circus acts,'" Terry Cook acknowledged "the strip owners' problem of drawing a crowd every week and making the money they deserve because of their large investment."[14]

For their part the exhibitionists typically maintained that they had just "decided to pocket some money themselves rather than give all their financial support to the already huge drag racing industry's parts business."[15] Hayden Proffitt, who performed in both jets and rockets, had driven most types of traditional machinery from slingshots to super-stockers. Bill "Maverick" Golden, whose "Little Red Wagon" became the most popular wheelstander, had been a super-stock star.[16] Palamides had, of course, built fuel dragsters before turning to jets. Some racers regarded exhibition vehicles as the only way they could possibly continue to pursue their enthusiasm.

Because of the financial success of people like Garlits and Ivo, and because of the conservative influence of Wally Parks, Brock Yates's doomsday specter never fully materialized—which is not to say, however, that people cast in the mold of Christ and Holz did not take drag racing a long way towards "fantasyland" in the 1970s. There were pathetic stories of young men like "Leapin' Larry" McMenamy, who was featured in screamers produced by Bill Doner and Steve Evans as a means of shoring up the sagging fortunes of Orange County International Raceway (OCIR), which they took over in May 1975. Oklahoman McMenamy, twenty-seven, had devised an act that consisted of leaping sixty feet through the air in a go-cart. An early appearance at Irwindale (also owned by Doner and managed by Evans) sent him to the hospital; his next appearance, at OCIR, sent him to his death.[17]

Promoters like Evans and Doner, and eager pawns like Larry McMenamy, had one conception of putting on a performance at a drag strip. Another conception focused on devising superior technology for getting quickly from A to B—or, if not single-mindedly fixed on that objective, at least to keep an eye on drag racing's image, as was always the case with Parks. Although it seemed touch and go for a decade or more, Parks and those who shared his sense of purpose would ultimately prevail; in 1990 some of drag racing was still carnival, but what one saw at NHRA's national events was usually closer to legitimate theater.

It could all have been different—all carnival, like the "monster trucks" that would eventually command nearly as much TV time as drag racing—had not a significant number of enthusiasts continued to define performance as essentially a matter of technological progress. While some of them might have had a keen instinct for commercial reward, they also had strong feelings about the bounds of legitimacy. Besides Garlits and Parks, two other men central to defining those bounds were Keith Black and Don Prudhomme. Black built one of the first manufacturing establishments whose clientele was largely involved with dragsters (in contrast to firms like Edelbrock's or Iskenderian's, whose bread-and-butter market remained street machines). Prudhomme was among the first racers to look beyond purses and appearance money, to perceive the crucial import of hustling commercial backing: "The sponsorships on race cars are everything," was how he would sum it up years later.[18] Both men had a deep concern about minimizing personal financial risk; while they could provide requisite skills, somebody else should pay the bills. Yet, no less than Garlits and Parks, both were enthusiasts at heart, and both would continue to be driven, in part, by the sheer excitement of getting a machine from A to B very quickly.

Keith Black was born in Huntington Park in 1927 and first probed the inner workings of an automobile engine when he was fourteen, in 1941, the year Don Prudhomme was born a few miles away. In 1943 Black joined the air corps cadet program. He wanted to fly; after all, this was southern California, with "all the aircraft plants, and AT-6s and P-51s and P-38s zipping over the housetops all the time." He finished high school in June 1944 and enrolled at Compton Junior College but was called up in December. When V-J Day came he was stationed at Minter Field in Kern County, still waiting for preflight. In November 1945 he was separated at Amarillo. "They asked me to re-enlist," Black recalled, "but they couldn't guarantee that we'd go to flight school . . . so I came back to Southern California and started peddling parts."[19]

Black was an itinerant wholesaler to service stations and repair shops, a wagon peddler. His own garage was his warehouse. He also did some mechanical work there, and in the late 1940s he got interested in boat racing. With a hull purchased from Rich Hallet, who would become famous in hydroplane circles, and a flathead Ford he had modified himself, he set a world record the second time out on the Salton Sea, in Imperial County. Already Black was learning a lot of "little secrets and tricks"—he managed, for example, to machine 23 pounds of excess weight off his destroked flathead crankshaft—and in the early 1950s he began reworking flatheads for other boat racers. Later he worked on Chevrolet, Cadillac, Buick, and Oldsmobile engines, but mostly he concentrated on the Chrysler Firepower, whose hemi head design he regarded as far superior. "Reverence for engineering" had always been Chrysler's talisman.

In 1959 he bought a piece of property on Atlantic Boulevard in South Gate and opened Keith Black Racing Engines. Don Garlits's exploits were already storied, TV Tommy Ivo was about to go on the road, and Don Prudhomme would soon be gaining a bit of a name locally. Black-powered racing boats held dozens of national and international records, but he had only been to the drags a couple of times—once to Santa Ana and once to Pomona, when Art Chrisman invited him and Hallet to "come out and see us run. . . . Keith maybe thinks he does good with boats," Chrisman needled, "but he ought to get into something difficult like drag racing."[20]

"Aren't challenges what we're after?" asked Black. Soon *Drag News* was reporting occasional wins by the Keith Black Racing Team at Colton and San Fernando, in gas coupes and street roadsters. Then his friend Cliff Collins asked Black for a favor. Collins, co-owner with Kenny Harman of a firm that manufactured camshafts and magnetos, was involved with Chuck Gireth. After a stint driving for Ted Cyr, who had been injured in an accident, Gireth had built a new chassis and gone racing with a new partner, Dave Carpenter, who owned a 354 Chrysler. Collins told Black: "Keith, these guys need some help. I think it will run pretty good, but they can't keep it together."[21]

Gireth and Carpenter brought the engine down to South Gate. Black recalled that he "took it all apart and went through it, just sanitized it." He left the combination alone but did his best to make certain everything would be reliable, mostly a matter of fastidious attention to tolerances, something many drag racers did not yet understand. Black had come from a different school, from boat racing, and there, as he said, "you can't afford to break." "You break, you get hurt—the prop locks up. . . . I had to learn how to make something stay together, and I worked very hard at it."[22] The car did quite well afterwards, and it provided Harman and Collins with some nice ad copy when Carpenter set several track records, speeds in the 180s. Yet Black knew that, except for the cam, pistons, magneto, and blower, the engine was "a stocker practically." It would "just run and run and run and run—never hurt a thing."[23]

Next Holly Hedrich, who worked for Paul Schiefer and had been around the drags since the dawn of competition at Santa Ana and Paradise Mesa, brought Black the Chrysler he was running in his roadster and asked him to go through it. "*He* never had any more trouble," Black recalls. "So my boat education became a very good thing. All of a sudden here's two drag deals we did, and they not only run very well, they don't have trouble, they don't break. They work."[24]

In late 1961 Black was approached by Tom Greer, who proposed that they campaign a fueler together. Black responded that he had only dabbled in "asphalt" competition. Besides, he told Greer: "I can't afford to do any type of racing. My business takes everything I've got." But Greer, who owned a prosperous company in nearby Cudahy, had plenty of money. He said he would pay all the bills. At that point Gireth, in Black's words, "started hustling Tommy."[25] Next thing Carpenter was out, Greer owned Gireth's chassis, and Greer and Black were about to go racing with Gireth as their driver. Black did not foresee any profit: "I don't lose anything, I don't make anything, I just spend a lot of my time on it. But it's a challenge."[26]

Black put together the best engine he could, but after that the car began having traction problems. Gireth did not seem to know the reason. Black did not pretend to know; as he reiterated time and again, he was an engine man, not a chassis man. All he knew was that here was a car "that was such a killer with no horsepower in it [i.e., with Carpenter's engine]. We put horsepower in it, now it's junk." Gireth had a reputation for building chassis that handled well, but his design was flawed in terms of transferring power effectively to the ground. Black knew there were those who held that "the harder you smoke [the tires] the quicker it goes," but that did not make sense to him. Chuck Gireth could not hustle himself out of this fix, and, with Greer's blessing, Black bid him good-bye.[27]

Black's plan was to find a proven chassis, have Greer buy it, then "go out and work that turkey until we get it to do what it has to do."[28] They bought a machine that Kent Fuller had built in January 1961 with Rod Stuckey. It had a 112-inch wheelbase, the engine sat nose down, as low as possible (the oil pan cleared the ground by only an inch), there was a torsion-bar front end patterned on that of a Volkswagen, Fuller's most distinctive trademark—and, in sum, it embodied Fuller's perception that, rather than merely welding tubes together and then installing the components, the chassis should be built *around* the components. It worked extremely well for Stuckey; he recalled that "everywhere I ran I was usually 5 to 10 mph faster and ½-second quicker than all my competition."[29] After only two months, however, Stuckey was badly burned by flaming oil and had to sell the car to pay medical expenses. The buyer was Lou Senter, proprietor of a speed shop called Ansen Automotive and Lou Baney's original partner at Saugus.

Early in 1962 Stuckey had arranged with Senter to resume driving the car, and just before Greer bought it Stuckey had won the AHRA's Winter Nationals in Arizona. Greer and Black retained him as their driver. As soon as they installed Black's engine, however, they were once again plagued by what Black regarded as excessive wheelspin. Furthermore, Stuckey reminded Black of Gireth; he seemed to be "looking for a pigeon." "So many of the guys in drag racing are that way," Black remarked.[30] He preferred to think of his friend Tommy Greer as a patron.

So Stuckey got cashiered, too. Meantime, Black had taken the car back to Fuller in Van Nuys and asked him to "clean it up"; Stuckey had detailed the bare chassis, fabricated the various brackets and mounts, and Black did not care for his workmanship. While it was there, Fuller's neighbor Wayne Ewing—who worked mainly on champ cars but occasionally on dragsters such as Chuck Jones's Magwinder and Ernie Alvarado's Shudderbug—made a set of aluminum body panels. Greer and Black agreed that there were "a lot of terrible-looking dragsters. . . . There's nothing wrong with making it a nice looking car; drag racing needs to look better." But what they needed in particular was a driver who was talented yet not too flagrant a hustler. Fuller recommended the young man who had been victorious for him the year before at the U.S. Fuel and Gas Championship, Don Prudhomme.[31]

Prudhomme's father was a Texan, his mother from Louisiana. Together they had moved to Los Angeles in the late 1930s, then shortly after Don was born in 1941 the family settled in Van Nuys. Although he was younger than Black, Prudhomme told many of the same stories about growing up in southern California. They both did some street racing. Black helped his uncles in their garage; Prudhomme hung around his dad's body shop. At fourteen Black began tinkering with engines, and Prudhomme bought a Mustang, a motorcycle of the sort that would later be termed a minibike. He delivered newspapers, mowed lawns, fed the chickens at a nearby ranch— there were still plenty of ranches in the San Fernando Valley in the 1950s. He attended Van Nuys High School but never graduated. "I was only interested in cars," he recalled. "Nobody could tell me anything back when I was that age."[32]

Prudhomme earned money painting cars. He bought a '48 Merc, a '50 Olds Rocket 88. He hung around nearby drag strips, San Fernando and Saugus. Then, like his friend Tommy Ivo, he bought a 1927 Model-T Ford roadster, along with a Buick V-8 engine. Soon he put that engine in a dragster owned by Rod Pepmuller, a fellow member of a club called the Road Kings of Burbank. In 1960, besides going back East with Ivo, he campaigned locally in the dragster he bought from Ivo after Fuller had built his twin. Later Prudhomme replaced the Buick with a supercharged fuel-burning 392 Chrysler assembled by Dave Zeuschel, whom Prudhomme had met through Fuller. (Fuller had been a flame cutter for Don Clark and Clem TeBow at C-T Automotive in Van Nuys and taught that art to Zeuschel.) In the fall of 1961 he teamed up with Zeuschel and Fuller, and in 1962 they won the world's premier race for fuelers; nearly ninety of them were entered, and the reported attendance was fifty thousand.[33]

Prudhomme did not have the mechanical savvy of Black or Zeuschel. Nor did he have Ivo's flair for showmanship, or even Garlits's, though there was in him something of both men. But he was an exceptional driver, no doubt about it, and occasionally he allowed himself to dream about becoming "a professional racer." After all, following their Bakersfield win, Zeuschel had quit his job at C-T and gone into business for himself, while Fuller was swamped with orders for chassis. It seemed that drag racing was rapidly becoming specialized; why not specialist drivers? "I'm interested in just driving," Prudhomme told an interviewer. "I want to spend full-time thinking about driving, not worrying about the engine."[34]

When Black started calling around for Prudhomme, he located him at the Sherman Oaks upholstery shop of Tony Nancy, another drag racing enthusiast. Prudhomme went down to have lunch with Black and Greer. They offered him half of the car's earnings, and he agreed on the spot. To be sure, they wanted him to do a lot more than just drive. He had to run errands, to take care of the routine maintenance. He painted the car, and the nosepiece was lettered "Greer-Black-Prudhomme." In its own way this was a fueler every bit as significant as Garlits's first Swamp Rat.[35] Skilled and resourceful racers like Garlits could do well on a limited budget; someone with plenty of money like Ivo could do well with relatively modest abilities himself. But to take three talented specialists—Black, the engine man; Fuller, the chassis man; and Prudhomme, the driver—and provide them with the sort of financial resources which Tom Greer could provide, that was something new.

The team debuted in the summer of 1962. They seldom raced out of state and never won a California meet the likes of Bakersfield or the Winternationals, but in week-in, week-out competition they were almost unbeatable. There were also two-out-of-three match races, with Garlits, with Ivo, with the Greek, with jets. Their going rate for such performances was one thousand dollars, and Black recalled that they grossed twenty-two thousand dollars the first year. Nine times out of ten, when Prudhomme came to the line, he got to the other end first—something like 270 individual races in two and a half years the team was together. "It was unbelievable how smart this Keith Black was," Prudhomme marveled.[36]

Tom Greer got into drag racing because he was a technological enthusiast, and with a first-rate team he stood a decent chance of staying out of a deep hole financially. Prudhomme envisioned a career as a professional driver. Black, whose business burgeoned because of the "G-B-P" car, regarded that machine as "a four-wheeled dynamometer, a means of experimenting with various combinations of compression, fuel mixtures [and] component reliability." He reiterated time and again that chassis were a distinct specialty. The engine was his key variable, but the central challenge was "getting hold of the race track," and that required both the right engine and the right chassis. Sooner than just about anyone else Black understood that billowing clouds of tire smoke might look dramatic, but they were undesirable in terms of getting from A to B quickly. Smoke signified excessive wheelspin. He "went to work *not* smokin' the tires."[37]

Partly it was a matter of matching horsepower to "what the track would take." While a savvy engine builder could overpower any surface, backing off on power entailed its own problems; when a car "bogged" coming off the starting line, it often did a wheelstand. "All this stuff was gradual learning," Black recalled. "We got lost lots of times." What he worked on most was "making the car repeat and repeat, because the driver could not pedal and play games [i.e., vary the throttle] and be consistent. . . . It had to be hammered every time. Everything automatic."[38] By the very nature of drag racing, a sprint from a standing start, the key mechanical component was the clutch, the clamping device that transmitted power from the engine to the driveline and tires. Some people still thought otherwise, but Black realized that it should be designed for controlled slippage. Development of the slider clutch, begun in concert with Schiefer, was a long and perilous road, but it would be one of the crucial factors in cutting two full seconds off elapsed times in the next ten years.

Greer provided Black with his R&D budget until a wave of setbacks drove him into bankruptcy court. Black "ended up owning the dragster" for the money Greer owed him, but he was not keen about racing "on his own dollar" and brought the car out only a few more times before retiring it towards the end of 1964. Fortunately for Black, there was someone waiting in the wings to take Greer's place, a twenty-year-old Hawaiian named Roland Leong.

Leong was what Don Garlits would have called "a rich guy" and Prudhomme regarded as "bucks up." Someone else might have regarded him as a patsy. For Black he was a godsend. Roland's parents had prospered in the insurance business in Honolulu as well as in other ventures; his mother Teddy was part-owner of a speed shop. When he was only eighteen he got his own Dragmaster-chassied gas dragster. Although he had spent time around the Dragmaster shop in Oceanside, raced at various southern California strips, and worked with the savvy Danny Ongais (also a Hawaiian), Prudhomme's first impression was that he "didn't know how to unscrew spark plugs yet."[39] They met when the G-B-P car was flown to the islands for an appearance in conjunction with the opening of a new strip. Leong fell in love with it. He had to have a fueler just as nice—no, even nicer.

The chassis and engine were almost identical, but the detailing was something else: "It was the most beautiful car you had ever seen," Prudhomme recalled, "blue, sparkling, with the name 'Hawaiian' on the side."[40] It was assembled at Black's shop with the best of everything. One autumn afternoon in 1964 Black convened a crew and headed for Lions to debut the Hawaiian. Roland had never driven a fueler, but he had gone as fast as 180 miles per hour on gas, and there had been no question about who would drive. The car had 40 percent nitro in the tank, and Prudhomme came over at the last minute to suggest that Leong put a rag under his helmet because of the noise. Black pushed Leong up to the strip and signaled him to fire and then swing around and stage. Lots of people could describe what happened next, including Gary Cagle, who debuted a new dragster himself that same Saturday and was watching from the return road. But Roland himself tells it best:

> I came off the line OK, the tires were lit, and then got out of shape right away. Went about another 1,000 feet, and got out of shape [again]. I put

Keith Black peers down the Lions strip, as Prudhomme, hand on brake, awaits the signal to stage. It is early afternoon (hence, the sparse crowd), and this is one of the car's last appearances before retirement. (Photograph by Roy Robinson, courtesy Museum of Drag Racing)

my foot back in it, not really hard, but enough to get going pretty well again, and the car went through the end at 191 MPH. I went to reach for the 'chute, and I was used to the Dragmaster cars that all had levers instead of pull rings. For a long time I couldn't find the ring, and when I finally did I put my right hand over my left shoulder, and bumped the steering wheel, and ran off the edge of the strip. The car went down the left edge of the strip, hit a sign, and finally stopped, thanks to the railroad tracks at the end.[41]

Only Leong's pride was hurt (and his mother's pocketbook), but this incident had significant repercussions. Both UDRA and NHRA were in the process of instituting licensing programs, and Leong's crash reinforced the perception that nobody should ever just climb into a new fueler and "leg it" all the way through. A checkout procedure, with incrementally faster runs under the eye of official observers, was required ever afterward, and drivers would spend time familiarizing themselves with how cockpit controls were arranged while just sitting by themselves in the garage. Leong lost his license, but, quite independently of this, he had "decided that [he] didn't want to drive the car anymore": "I needed somebody with a lot more experience than I had."[42]

From the Lions starting line the mishap looked more serious than it was. Fuller, who had moved his shop to northern California by then, was able to make repairs in a couple of weeks. With the Greer-Black deal already on the rocks Prudhomme was available. What Leong liked most about Prudhomme was that there were few better drivers anywhere; by this time he had been dubbed "the Snake" by a crew member awed by his reflexes. What Prudhomme liked most about Leong was that he "wasn't afraid to spend money. Whatever it took."[43] The team of Leong, Black, and Prudhomme debuted in December 1964, and the Hawaiian was a winner all winter. After the Snake won the Pomona Winternationals he and Leong began preparing for an eastern tour and a series of booked-in match races.

Both were badly undereducated in mechanical matters—as they pulled their rig out of Black's shop, he told them, "You two together are like the blind leading the blind"—but that did not seem to make any difference.[44] "It was like a dream come true," Prudhomme recalled. "The car, thanks to Keith, was set up perfectly. We were killers. We toured four months and got beat only twice." Prudhomme topped his season with a win at the NHRA Nationals (Black himself flew back to Indiana to make sure everything was all right), defeating Tommy Ivo, who had introduced him to touring five years earlier. Not a good loser, Ivo complained that the Snake owed his successes mostly to the quality of his machinery. Prudhomme admitted that this could be so, but he added, "Ivo's comment kept eating me, so I started learning more about engines, because some day I planned to have my own car in the winner's circle."[45] Aren't challenges what we're after? Black understood.

In 1966 the Snake did go it alone, acquiring a Woody Gilmore chassis from Bob Spar of B&M Automotive, returning to Dave Zeuschel for his engine work, and setting out on tour as the B&M Torkmaster Special. Black continued to work with Leong, and their new driver, Mike Snively, pulled off the same trick as Prudhomme, winning both of NHRA's premier events. Indeed, Leong, Black, and Snively enjoyed a more successful 1966 season than Prudhomme did on his own, and they kept right on rolling in 1967, winning at Famoso in March.

Meantime, things had been breaking fast at Keith Black Racing Engines. At the beginning of 1964 Chrysler had come out with its updated hemi, and

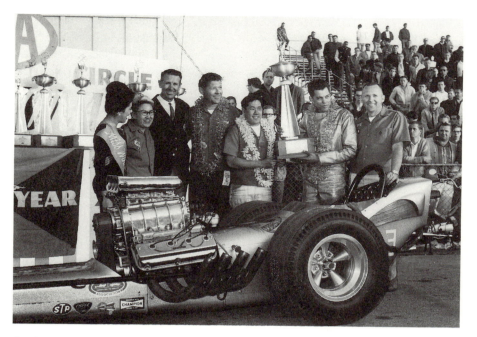

Posing victoriously after the 1965 Winternationals, Don Prudhomme holds trophy with Roland Leong. On Leong's right are Keith Black, Wally Parks, and Leong's mother Teddy, whose money had financed the "Hawaiian" — "the most beautiful car you had ever seen," said Prudhomme. (Photograph by Eric Rickman, courtesy Petersen Publishing Co.)

it soon became a mainstay in the factory experimental ranks. Fuel racers had tried the 426, the so-called Elephant Motor, but a lot of problems needed solving if it were going to work with a supercharger and nitro. The Ramchargers made some progress, as did Garlits. Black had been offered 426s gratis, but his response was altogether typical: "What good's the engine? I want the money you develop it with. The money's in the development, not the engine."[46] Black had been spoiled; he wanted someone else to pay the R&D bills.

And, again, somebody was there—this time the Chrysler Corporation itself. In late 1965 Black was approached by Chrysler executives who wanted him to develop a marine racing program around the 426, and, at about the same time, he was asked to look at a 426 that had overrevved in the Summers brothers' Bonneville streamliner. After taking that engine apart and replacing what was damaged, he was simply astounded when he put it on a dynamometer; on gasoline, with carburetors rather than injectors, "it made more power," he said, "than my super-duper double-whammy 392 SK [racing boat] engine." The potential of the 426 as a fuel engine was enormous. After a few trials in a hydroplane Black persuaded Chrysler to let him use a dragster for testing. "I could do everything there in a year that would take me five years in a boat," he remarked. By 1967 Leong owned two identical dragsters, state-of-the-art chassis from Don Long, a new rival to Fuller, Gilmore, and Huszar. One of these became the test vehicle for Black's 426 program; Surfer Mike Sorokin, out of a ride, was hired to drive.[47]

Black was a masterful empiric. "We really flogged it," he recalled, "make a change, and run; we kept floggin' it, floggin' it." After they had made substantial progress Sorokin was released, and the 426 went into the car Mike Snively drove. Then they started having traction troubles. Leong was distraught, understandably so, as he was providing everything but the engines. Black pondered. He had been building 392 hemis for SK boats and 426s for super-stockers. On gasoline both had the same horsepower, but the smaller engine had to turn 1,000 RPM faster, and it did not have as much torque. "I'm going to kill the torque," Black thought, recalling how he had once shortened the stroke of a flathead crankshaft. That did the trick with the 426, and, when Black started selling engines designed this way, Don Prudhomme was one of his first customers.[48] By the early 1970s the 426 hemi was almost standard in top-fuelers and fuel funny cars as well. Thinking back about his R&D program in 1989, Black recalled: "Sure I put a lot of my time into it, but we got paid to do it, and they [i.e., Chrysler] have never regretted it. It's been a good deal for them for a long, long time."[49]

All the while that he was sorting out the 426, Black was pursuing his fixation with "getting hold of the track." The ideal was not to spin the tires at all, to make "no-smoke" runs. Killing torque helped, but the real key was "playin' with the clutch." Others were certainly onto this besides Black. One of them was Sid Waterman, who had graduated from MGs and gymkhanas to a Kent Fuller dragster in 1962. His friend Bob Smith drove, but after funds ran low Smith went to work for Palamides, and Waterman moved from San Jose to the San Fernando Valley, where he got a job at C-T Automotive, which had once employed both Fuller and Zeuschel. Later he worked several years for Mickey Thompson at M/T Enterprises in Long Beach. After his Fuller car was demolished in an accident at Pomona, Waterman ordered a new chassis from Frank Huszar, which debuted in October 1964. Three months later Waterman's driver set track ET records at both Fontana and Lions. Waterman attributed these marks largely to experimentation with a slipping clutch.[50]

But Keith Black had been a step ahead of everyone. People recalled a memorable pass that Prudhomme had made at San Gabriel two years before, on January 20, 1963. Holly Hedrich, who later became Black's second-in-command, described it:

Getting the green flag, Don put the throttle right on the bellhousing and unloaded the clutch. Almost immediately it appeared to those competitors in a position to see the start of the run that they were not going to witness a new record on this pass. The car moved a full length before there was any trace of smoke from the tires, and when it did appear it was just a haze that lay straight back of the top of each tire. Further, at about the 800-foot mark, all smoke disappeared completely. At a time when almost all good runs were accompanied by clouds of tire smoke, Prudhomme's pass just had to be a loser.

In fact, this was Prudhomme's stunning 7.77, and he backed it up with a 7.84. Hedrich recalled that "until that time seven-second runs were almost 'science fiction.'" Six weeks before Ivo had clocked a 7.99, and on New Year's Day Prudhomme ran a 7.90, "but most of the better-running fuelers were still in the 8.05 to 8.20 bracket."[51]

The secret, Hedrich indicated, lay in "endless hours" spent in experimenting with the clutch. On his 7.99 run Ivo had slipped the clutch himself, with the pedal (a trick at which Leonard Harris had likewise been adept in the Albertson Olds Special), but this was impossible to do consistently, and Black treasured consistency. His idea was to design a clutch with automatic slippage. Since the 1950s various manufacturers had marketed dual-disc clutches for fuelers, with the two discs separated by a steel floater plate. Schiefer's "Velvatouch" linings were considered best because they slipped least. The aim of Black's R&D program, carried out in concert with Schiefer and Hedrich, was to come up with linings and pressure plates that would permit controlled slippage for the initial part of a run and then lock up. Prudhomme's input was essential because of his ability to report things "that we weren't able to see."[52]

Unfortunately, we get results we don't want along with those we do. A slipping clutch behind a 1,500-HP engine built up tremendous heat, sometimes to the point of total disintegration. Both sanctioning bodies mandated bellhousings designed to contain flying pieces, but these were generally made of aluminum, and, as late as 1966, Hedrich could warn that no bellhousing could "contain a pressure plate, flywheel, or even a clutch-disc explosion at anything over 5000 rpm." NHRA reported that exploding clutches and flywheels were "the major cause of personal injury or death and drag strip accidents in general." Black and Hedrich were still working with sliders in 1967. So was Tony Waters, for whom Mike Sorokin was driving. One December night at Orange County, Black recalled, "we pitched a unit out . . . and practically cut the car in two." Mike Snively was not hurt, but that same night the clutch in Waters's slingshot disintegrated, just as Sorokin won his first-round race at more than 200 miles per hour. The chassis was ripped apart, and Sorokin died instantly.[53]

"The thing had become so dangerous that you might as well [have handed] the driver a hand grenade with the pin pulled," Black observed. In the face of an NHRA order to "make all clutches safe, or go back to smoking the tires," Black and Hedrich ascertained that the discs served as heat dams, while the floaters and flywheel were heat sinks; they absorbed heat. Their solution was to add another disc and floater, to increase the amount of heat sink area. "If everyone uses his head," Black thought, "the sport will be safer for everyone concerned."[54] Not everyone did use his head, and things got even more dangerous before they got safer—1969 was the worst year ever—but the situation did improve after that, with the development of centrifugal multiple-disc clutches and particularly with the refinement of steel and, later, titanium bellhousings that would contain disintegrating components.

At the end of the 1960s Black curtailed his active participation in racing in order to concentrate on manufacturing. Eventually he produced nearly all major components for fuel engines, even blocks. Chrysler had made its first version of the hemi for only eight years, and racers had just about exhausted the supply. Even the 426 was phased out in 1971, unable to meet new emission-control regulations. The 426 was all steel, but once a block was damaged by a broken rod or similar mishap it either required costly machine work for sleeving or else had to be discarded. The idea of casting blocks from aluminum with replaceable steel liners was in the air, and Howard Johansen, among others, had actually made a prototype. But the first to get into actual production was Ed Donovan, in late 1971. Black clearly recalled his feelings about this. Even though Donovan's blocks were cast on patterns almost identical to a 392 Chrysler, they were called Donovans, not Chryslers: "Well, I was in the engine business. The first thing that came to my mind was 'if I go buy Donovans they're not Keith Blacks anymore.' So you either get out of the engine business, or you better have your own, because that's what they're gonna call it. I thought, 'Well, why not?'"[55]

Unlike Donovan, Black patterned his block on the 426. Although he received no direct assistance from Chrysler, much of the design work was done by Bob Tarozzi, a former Chrysler engineer. The "KB" debuted in 1974 and pretty well dominated fuel racing for more than a decade, though Black had some competition from Donovan and from Don Alderson's "Milodon" and strong competition after the mid-1980s from Joe Pisano and his "JP-1." For a time there were indications that the McGee "Quad Cam," developed in Australia, might supersede all the Chrysler clones. Black was unconcerned. "I'm not going to fold this place up," he said in 1989. "If it's not drag racing it will be something else. You walk out there [on the shop floor] and you'll see aerospace work going through here. We're doing that also. I'll adjust."[56]

Like Don Garlits, Keith Black became a wealthy man because of drag racing. Unlike Garlits, he attained his wealth with only a limited direct involvement with race cars. He knew the risks of technological enthusiasm. He was convinced that "whatever you win off a race car you'll eventually put back anyway."[57] Aside from a period when he and John Mazmanian fielded a funny car, with Mazmanian paying the bills, Black never competed personally after 1970. The young man with whom he teamed in 1962, on the other hand, was hardly ever absent from the racing circuit in the next quarter of a century. But the Snake, too, had a keen instinct for reducing his own financial risk to a minimum.

Prudhomme had been aware of that risk from the beginning. Racing an unblown gas dragster close to home, he could break even at best, even though expenses were minimal. He broke up with Zeuschel and Fuller when "everyone got low on money."[58] While driving for Greer and Black, he made more than he earned as a car painter (a job he held simultaneously), but that was not saying a great deal. The Hawaiian grossed something like sixty-five thousand dollars in 1965, but the lion's share of profits were Leong's. Prudhomme's 1966 season in the Torkmaster was a disappointment at least partly because he did not have the all-around skills to excel on his own. By his own admission he "didn't really know a lot about engines"; Black or Zeuschel had always been there to take care of that end of things. Indeed, he later confessed that he was afraid to remove a magneto because, "I wasn't sure I could get [it] back in right." At the beginning of 1967 he hired on to tour a fueler owned by Lou Baney, powered by a Ford single-overhead-cam engine from the Van Nuys shop of Ed Pink and sponsored by a mainstay of Ford Motorsports, Carroll Shelby.[59] This time he paid attention to everything, and after two years with Shelby's "Supersnake" Prudhomme felt ready to strike out on his own once again. He put all his savings into a new car, but he also wanted money for operating expenses, ideally from a commercial sponsor. "To have a sponsor back in '69, that was something else," Prudhomme recalled.[60]

Not that it was unheard of. There had been modest sponsorship deals since well back into the 1950s. Neither Ollie Morris, Calvin Rice, nor the Bean Bandits paid all their own expenses. As early as 1956, Maurice Richer campaigned a fueler as the Nesbitt's Orange Special, and in the 1960s various racers provided advertising for purveyors of beverages ranging from A&W root beer to Smirnoff's vodka. In 1967 Bill Back, PR man for a fast-food chain called Der Wienerschnitzel, began sponsoring the "Top Dog," a fueler owned by Jim Nicoll and Don Cook. And, of course, there had been arrangements with parts manufacturers, tire makers, and refiners of specialty lubricants. Garlits struck a deal with Wynn's Friction Proofing to call his dragster "Wynn's Charger." In 1969 the Snake made a similar deal to campaign his new machine as "Wynn's Winder." Most drag racers were driven by enthusiasm, but that was not enough for Prudhomme: for him "money was the key."[61]

It was not the Snake, however, who put together drag racing's first bigtime sponsorship package. Rather, it was his old friend and rival, Tom McEwen. Ed Donovan, for whom McEwen had once driven, knew his Kipling, and so he knew that there was only one animal that could be counted on to best a snake. McEwen had been the "Mongoose" ever since, and at some point he began making the *s* in Mongoose into a dollar sign. Although their backgrounds were similar, the Snake and the Mongoose had, as Prudhomme's wife, Lynn, put it, "different needs from the sport." McEwen relished the celebrity. Prudhomme liked it too, but he liked winning even more; "I don't do this because I think I'm going to turn the crowd on," he told a journalist in 1988. "I race because it's something I need *real* bad."[62] Both of them started out as specialists. Prudhomme eventually mastered all aspects of his chosen pursuit. McEwen never did learn much about what made the cars go, but he was a hustler par excellence. He always managed to find what Keith Black would call a pigeon.

In 1969, with backing from the maker of Tirend Activity Booster and Gold Spot Breath Freshener, McEwen campaigned both a dragster and a funny car on national tour. By then some racers had begun to perceive that "nonautomotive" sponsorships were where the real money could be found. It happened that McEwen's mother worked for the Mattel Toy Company as a secretary, and his stepfather worked for the same firm as an attorney. This gave McEwen an opportunity to make a pitch to Art Spears, a Mattel vice president. He did so on behalf of "Wildlife Racing Enterprises," the Snake and the Mongoose. Next thing the two of them were going racing with two dragsters and two funny cars "on the biggest budget ever attained by quarter-milers." In addition to Mattel's "Hot Wheels" toys the package also included Coca-Cola, Plymouth, Goodyear, and Wynn's. People from a firm that represented the likes of Mario Andretti, A. J. Foyt, and O. J. Simpson told McEwen and Prudhomme that there were lots of other money trees, but they could not manage to shake them, so Wildlife went back to negotiating its own deals.[63]

Wildlife did just fine. There were two or three match races every week from March through September. There were endorsements, interviews, a TV special called "Once upon a Wheel"; McEwen and Prudhomme had to join AFTRA, the television and radio entertainers' union. A writer for *Drag Racing* magazine exclaimed that they were assured of an income "no other drag racers have ever approached."[64] In one sense this was not saying much, since, as Garlits often reiterated, it was unlikely that more than a handful of "professional" racers actually showed a profit year in and year out. But there was no doubt that the Wildlife deal involved a healthy sum. McEwen, not one to downplay numbers, claimed that it came to $250,000. Others said the total was about $160,000, $75,000 of which came from Mattel. Whatever the amount, Steve Evans, another pretty good hustler, was probably not exaggerating when he said that the Mattel sponsorship "revolutionized the business of drag racing."[65]

Yet McEwen and Prudhomme still approached the challenge differently. For Prudhomme "the business" was only a necessary evil; for

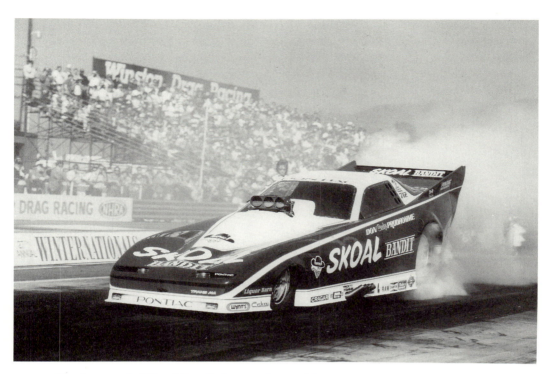

Prudhomme's "Skoal Bandit," an archetypal funny car of the late 1980s, representing one of the more lucrative sponsorships in drag racing. (Courtesy Don Prudhomme)

In May 1971 Drag Racing USA featured the Snake's funny car and the Mongoose's dragster, two of the four cars on the Mattel "Hot Wheels" team. Slingshot's squarish lines imitated lines of Mattel's model kits.

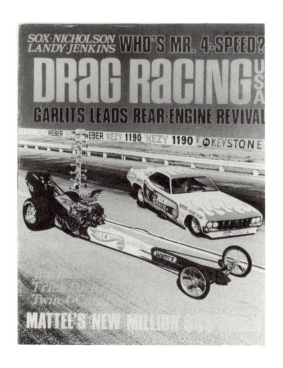

McEwen it was central. Prudhomme always raced for keeps; McEwen never worked as hard. "I was the bullshitter," McEwen said. From his perspective the whole Wildlife deal was something, yes, "like wrestling."[66] Their match races often drew record gates, and the crowds would boo McEwen roundly. He loved it. "I was the villain and the villain has always made more money than the good guy, right?"[67] The Mattel deal lasted three years. In 1972 McEwen grossed somewhere between $360,000 and $480,000 (depending upon whose figures one believed). Part of this came from big wins in open competition, $23,000 at the first NHRA Supernationals at Ontario Motor Speedway, and another healthy sum at the Professional Racers Association meet in Tulsa, an event promoted by Don Garlits. Even so, McEwen was not known as someone who won often. "He kind of nurses his car through the races," remarked promoter Bill Doner, while hastening to add that the Mongoose was "definitely a very strong PR guy."[68]

After the demise of Wildlife Racing McEwen negotiated sponsorships with Revell model kits and English Leather cologne, and for many years he was backed by the Adolph Coors brewery. The Snake was no slouch in the sponsorship game—he had long-term contracts with the U.S. Army Recruiting Command, Pepsi-Cola, Plymouth, and Wendy's—but he was also driven to win. He won the NHRA Nationals twice running, something only Garlits had done, then again did it twice. After NHRA instituted a season-long points competition he captured that four times in a row, 1975–78. He did this in a funny car, not a dragster. Funny cars were no longer a circus act. They could not perform (technologically) as well as a dragster, but it was certainly every bit as difficult to be a consistent winner.

Next to Garlits the Snake was the toughest competitor of all. He was Garlits's equal at stirring up controversy. Garlits had his most memorable clashes with Wally Parks; Prudhomme had his with Larry Carrier, head of the International Hot Rod Association. Carrier could not abide the Snake's army sponsorship, the idea of the government "spending $100,000 for any race car to compete against the less fortunate ones." Prudhomme, said Carrier's house organ, was "the highest-paid soldier in Uncle's modern Army. . . . The Army pays Snake better than double the salary of a four-star general." Prudhomme countered that he was simply "advertising for the Army," and that, if advertising was objectionable, he himself objected to Carrier "promoting cigarettes at his race track[s]": "I have to bring my daughter out to the track, and that's enticing her to smoke cigarettes, you know what I mean?"[69]

Prudhomme would later change his tune about *that* matter: "Winston came into our sport and I think they saved it," he said.[70] In 1987 he would sign a long-term deal with U.S. Tobacco to race as the "Skoal Bandit," and Skoal representatives would distribute free samples of "smokeless tobacco" to folks walking through the pits at NHRA events.

With tobacco came television, and with television came the beer companies and cutthroat competition for sponsorships that ranged into seven figures. Even those racers who helped to create this big-time milieu feigned second thoughts: "You know, sometimes I think it was more fun in the old days," Prudhomme reflected in 1977, upon taking delivery of a new funny car worth fifty thousand dollars. "I kinda liked poking around in junkyards for pieces." "Money has ruined it," said Tom McEwen a few years later. "They all try to steal your sponsors, and no one helps nobody anymore." Not long afterwards Coors diverted its backing to Dan Pastorini, the former National Football League quarterback, and McEwen was reduced to touring an exhibition machine. But he was still a hustler, a trouper, still a star. And that was always what he needed most: "I had to read about myself in *National Dragster,*" he said. "I couldn't do anything else."[71]

The Snake confessed too that he "couldn't imagine [him]self without drag racing," but he was never motivated simply by celebrity. He also loved getting a technological edge. In 1975 he was the first to run in the 5s in a funny car and seven years later he was first to top 250 miles per hour. For a long time there had been talk that he would supersede Garlits as drag racing's Big Daddy. Actually, the two were cut from different cloth. Like Garlits, Prudhomme was always thinking, and he was never as comfortable within an established paradigm as many racers were. But neither did he dream incessantly about possibilities for radical departures. At about the same time that Prudhomme began campaigning a funny car—not for any technological reason but only because it had greater commercial appeal— Garlits was thinking about a fundamental reconfiguration of dragster design, a reconfiguration that may well have saved it from extinction in the face of the funny car's superior theatrics.

9

REVOLUTION

Every time I make a pass in the car I feel
that I am endangering my life.
   MIKE SOROKIN, 1965

Damn *those slingshot dragsters.*
   DON GARLITS, 1970

have been trying to show that drag racing provides a stage for
different kinds of innovation—technological, theatrical, entrepreneurial; some innovators, of course, ply all three realms. The
maestro of the American Hot Rod Association, Jim Tice, liked to revamp
his competitive format, and in 1970 he came up with a variant on what
NASCAR called "the plan." During the off-season Tice signed a select group
to contracts. Each racer agreed to appear at ten AHRA events, the Grand
American Series of Professional Drag Racing. While everyone else who
entered these events would have to vie for one of the "open" positions, the
"seeded" racers would not risk coming away as a DNQ (did not qualify),
with nothing. And Tice's organization could headline the booked-in participants without undue concern about whether they would actually show up
at any given event. In addition, partly to encourage others to tour the entire
circuit, there would be a system for accruing points at each event, with a
jackpot to be divided among various classes. Twenty thousand dollars was
earmarked for the winning fuel dragster.

Tice scheduled the debut of the Grand American Series for the AHRA's
Winter Nationals in Phoenix. Five of the seeded racers had moved into the

spotlight only recently: John Wiebe of Newton, Kansas; Jim King of Providence; Bob Murray of St. Louis; Texans Don Cook and Jim Nicoll, each on his own now. There was also Ed Donovan, teamed with Oklahoman Bob Creitz, both of whom had been racing dragsters as long as Garlits; there was Chris Karamesines; and there was Big Daddy himself. Despite inroads made by funny cars and exhibitionists, Garlits and the Greek remained top draws.

Garlits won the inaugural event handily. He had equipped Wynn's Charger with something novel which gave him an edge. In the early days most dragsters had transmissions, and drivers shifted gears a few hundred feet out (normally, the shift was from second to high with an automobile "tranny" from which low and reverse had been removed). Then, in 1957, Emery Cook and Garlits broke into the 160s and the 170s, and into the 8-second range, with direct drive, no tranny. This became the conventional setup and remained so for well over a decade. By the late 1960s, however, with elapsed times dropping into the 6.60s and speeds pushing 230, blown Chrysler engines were sometimes being twisted to 10,000 RPM. This was a major factor in the rash of clutch explosions. Some racers were pondering the desirability of a setup that would permit upshifting on the top end, and in the spring of 1969 Bruce Crower told Garlits about a San Diego firm named Lenco. The owner, Leonard Abbott, had designed a planetary transmission, an "overdrive." Garlits placed an order at once.

Using a Lenco unit, Garlits won a July 4 meet at Union Grove, Wisconsin. With a blown fueler a gear change is not readily audible, and nobody seemed to realize what was going on. In midweek he went to Indianapolis Raceway Park, where he was scheduled to test tires for Goodyear. Don Prudhomme was there, as well. "It didn't take long for a sharp guy like Don to discover that I was shifting gears," Garlits recalled, and before long the Snake had a Lenco, too.[1]

Sometimes when a racer's "edge" became common knowledge others emulated it en masse; this was the case, for example, in the mid-1960s, when Garlits's rivals discovered that he had nozzles squirting fuel directly into the ports of his hemi as well as through the injector atop the blower. It was the case in the mid-1980s when Joe Amato showed up with an airfoil on seven-foot struts. It was not so with the two-speed. For one thing, the first Lenco units had internal weaknesses and were unreliable. Most racers elected to stay with direct drive. But Garlits liked the concept so much that he contracted with a machinist named Ed Stoeffels, proprietor of Quartermaster Industries in Skokie, Illinois, to build an improved version. Garlits had recently formed a partnership to produce and market racing components, and by late summer of 1969 he was advertising this unit as the "Garlitsdrive": "Tired of . . . winding your engine to death? Or three runs out of your clutch? Install the Garlitsdrive two-speed transmission and solve these problems. Tested and proven to be safe and reliable by Big

Daddy Don Garlits."[2] The price was $495, which might have deterred low-budget racers. Weight was 45 pounds, which might have been a deterrent at a time when people were going to extremes in paring down. Wynn's Charger weighed a little more than 1,200 pounds, but on the West Coast Jim Nelson had used a lot of .035 tubing to build a slingshot that weighed less than 800 pounds.[3] In any event Garlitsdrives were still not generally available, so John Mulligan and his partner Tim Beebe could not have had one when they arrived at Indianapolis Raceway Park for the 1969 NHRA Nationals.

It had been a good year for the "Fighting Irish," who had a reputation as racing one of the strongest fuelers in the land but also as perpetual "bridesmaids." Beebe and Mulligan broke that jinx with a win at Pomona to begin the season, then put in a crowded slate of match races. At IRP they qualified with a 6.43, the quickest credible elapsed time ever recorded. Then, in the first race of the opening round of competition, tens of thousands watched in horror as their kelly green machine "burst off the starting line, then swallowed itself in flames half-[way] down the strip and tumbled end-over-end like a bouncing stone." The clutch had disintegrated, the engine had over-revved, and everything had exploded. Fire burned through the safety harnesses, and Mulligan was thrown free. He hung on at Methodist Hospital for more than three weeks, but finally he succumbed to internal injuries that physicians could not properly diagnose or treat due to the severity of his burns.[4] If there had been a two-speed and he had been able to drop the revs partway through the run, perhaps it all might not have happened. True, the two-speeds were problematic in themselves, frequently breaking. So far, however, breakage had resulted only in lost races.

Stoeffels went to Florida to help sort things out, and, when Garlits won the first AHRA Grand American in January 1970, everything seemed to be under control. But a few weeks later, at Gainesville, Florida—the inaugural of a new event on the NHRA's championship schedule called the Gatornationals—Garlits's tranny again gave out on the starting line, leaving him "as helpless as a beached flounder."[5] He would soon know that he could be left a lot worse off than that.

After the Gatornationals Garlits and his helper, Tommy "T.C." Lemons, headed west again for the second race in the Grand American Series at Lions, an event that had drawn an exceptional level of advance interest.[6] After that they planned to go to the venerable Bakersfield meet at Famoso, but the Grand American was postponed on account of rain, and the contract with Tice stipulated that missing an event meant losing eligibility for points. Garlits had to stay over. The next weekend the weather was sunny, and everything started out on a high note. Garlits installed a new two-speed, which seemed to work perfectly. He turned in a qualification round of 6.61, which would remain the low ET of the event. On race day, Sunday, March 8, he eliminated an unseeded competitor in the first round, Don

Cook in the second, and John Wiebe in the third. That put him in the final against Creitz and Donovan's brash and talented driver, Richard Tharp. Tharp recalled that both of them completed their burnouts, then he staged, then Garlits: "I saw the lights go on the tree and I stomped on it. Oil started spewing all over me and when I got to the end, my goggles were full of oil. I couldn't see. I couldn't figure out where Don was. I kept asking, 'Did we win?' It was a $10,000 payoff and I was anxious. Someone finally told me Garlits didn't get off the line."[7] After Tharp got his eyes rinsed out he returned to the starting line. There he saw Mickey Thompson carrying Garlits in his arms. Blood was gushing from his right foot, or what was left of it.

As soon as Garlits had engaged the clutch, with the engine at about 6,000 RPM, the car lurched slightly, something broke inside the transmission, directly between his feet, and then it grenaded. Pieces came right through the quarter-inch seamless steel casing. One large chunk had just missed Tharp's head. Another hit a young spectator in the arm; others had broken Don's left leg and foot and severed his right foot at the arch; still others had cut the entire car in half and sent him tumbling, still strapped in the roll cage. Garlits recalled that it came to rest upside down and that Thompson had unbuckled him, lifted him out, and helped him remove his helmet. He woke up in the intensive-care unit of Pacific Coast Hospital in Long Beach, with his wife, Pat, at his bedside; Brock Yates had driven down from Famoso to pick her up at LAX and take her to the hospital. What remained of Garlits's foot had been stapled together with seventeen stainless steel clips.

During his six weeks in the hospital Garlits had lots of company—Tom McEwen came by often to bring reading matter and to sit with him and watch "Star Trek"—but he also had plenty of time all to himself, time to think. After his fire at Chester in 1959 and six years later, after a high-speed collision at Cecil County, Maryland, he announced that he would retire. This time he was certain he would drive again, but he was also certain that he would try "to design a car that wouldn't put the driver in such a dangerous driving position."[8]

The necessity to shield clutches and flywheels had been obvious for years, and protective devices had certainly improved. A one-piece 365-degree "can" was now required. Even so, if inertial forces were powerful enough, even the best that money could buy was not enough; disintegrating components would simply shear the bolts holding the can to the back of the engine.[9] There had been some spectacular failures involving fluid couplings, used in most funny cars, but nobody had considered the consequences of a planetary transmission disintegrating. A Garlitsdrive had a heat-treated casing, which was itself wrapped in a nylon ballistics blanket; in racers' parlance it was regarded as "bullet proof." Garlits at first believed that there was "an inherent destruct deal built into these things that we

This photograph was taken a split second after Garlits's transmission exploded, just off the Lions starting line, on Sunday, March 8, 1970; shrapnel has cut the roll cage completely loose from the rest of the chassis. (Photograph by Jim Kelly, courtesy Museum of Drag Racing)

don't truly understand."[10] But the Specialty Equipment Manufacturers Association—SEMA, the trade association that set safety standards—quickly mandated different specs that took care of the problem. Yet clutches and flywheels, not to say blowers and entire engines, remained extremely volatile, and in Garlits's mind there was no doubt that "to be safer, you had to put all that noise and machinery someplace else," behind the driver.

Conventional wisdom held this configuration to be dangerous in itself, for a different reason—poor directional stability. Maybe. But did it have to be? As it turned out, Garlits simply did not have time to address the problem at once. He had contractual obligations.

T.C. Lemons and Connie Swingle put Wynn's Charger back together, and Garlits arranged with his old pal Marvin Schwartz to fill in as driver. By the time of the AHRA Spring Nationals in Bristol, Tennessee, the fifth race in the Grand American Series, Garlits was getting around well enough to go along. Schwartz had taken delivery on a new car of his own, so Swingle was now slated to drive Wynn's Charger. Because Tice had already agreed that Schwartz could take over Garlits's points from the first two races, his was the seeded car; Swingle would need to qualify for one of the open spots. Having driven only exhibition machines such as wheelstanders since 1966, he made a poor run. When he came back Garlits hobbled over and said, "Swingle, I'm going to make a pass."[11] He turned both low ET and top time of the event.

He drew Tharp in the first round. Said Tharp, "So you're going to try again, eh, Big Daddy?" And Garlits answered, "Yeah, I didn't like the start in that last one."[12]

Garlits defeated Tharp but lost two rounds later to Wiebe, who went on to win the $20,000 AHRA bonus. While fulfilling his remaining commitments for the season, Garlits recalled that he was beginning to get over his apprehensions about slingshots. At the NHRA Nationals he made it to the final four, losing to Jim Nicoll, who faced Don Prudhomme for the money. The Snake had signed his "Hot Wheels" deal and was lavishly financed; Nicoll was not. The Snake had a two-speed and a top-of-the-line Black Elephant, the best of everything. Nicoll was still using direct drive, a 1950s-style 392 hemi, and a scattershield that was not the best available. In order to keep the RPM up for maximum horsepower he had his clutch set up very loose.

The two left the starting line evenly and were side by side in the lights. Then Nicoll's clutch disintegrated and sawed through the can and the chassis. It was like a replay of a series of tragedies that had begun with Mike Sorokin at Orange County in 1967, only more surreal than most. The front part of Nicoll's car, with the engine still running, veered toward the Snake, who narrowly averted a collision. It skidded to a stop alongside Prudhomme, essentially undamaged. The roll cage, with Nicoll inside, tumbled over and over, finally coming to rest far beyond the edge of the shutoff area. Prudhomme climbed out of his dragster close to hysterics, certain that he had seen Nicoll die.

Actually, heaven could wait. Nicoll climbed out by himself and was able to walk to the ambulance. He was only bruised. Even so, a lot of racers who witnessed this episode—Nicoll was dubbed "Superman" until the end of his racing days—were thinking the same thing Garlits was: "I knew that the only place for a driver was in the front section, not the rear." As he and T.C. headed home from a final 1970 match race with the Snake, Garlits recalled that he could not wait "to commence work on the rear engine dragster I had built a thousand times over in my mind."[13]

Typically, racers had become concerned about danger only after maximum provocation. At first it was almost impossible to get them concerned under any circumstances. Don Jensen recalled that "there was not a lot of pressure about safety" and that this was linked to the prevailing attitude of the 1950s:

> You have to realize that most of the guys out there racing . . . were like
> Old West gunfighters, if you will. They just did not worry about the

A clutch explosion at high speed often precipitated a conflagration like this, with the driver sitting helplessly in the middle of an inferno fed by oil and nitro. Many drivers were badly burned, despite their fire-protective outfits, but here, at Irwindale Raceway in May 1972, Tom Toler rode out this calamity uninjured. (Photograph by Randy Dacus)

AHRA champ John Wiebe (center) with two other Grand American stalwarts, Chris Karamesines (on his left) and Jim Nicoll, wearing the lower half of a driver's fire-protective outfit. (Smithsonian Institution collection)

possible consequences. There was a lot of macho. Some drivers almost took pride in the fact that they were running cars with no safety equipment like roll bars. One of the attitudes that was very dominant was that you *did not* abort a run, a bad run. You stayed with it, you got it straightened out, you stabbed it and steered. . . . It was a foolish attitude, but it was very, very dominant.[14]

Safety rules were characteristically lax. In 1954 at Santa Ana, for instance, a gutted '34 Ford coupe could pass technical inspection even though it had nothing to retain the wheel if a rear axle should break, even though it was ballasted with 50-pound sacks of sugar stacked over the spring mount, even though the fuel tank consisted of a paint thinner can held down with plumber's tape (which was used a lot of other places, too), even though the kill switch on the magneto consisted of a loose wire that the driver had to grab and ground on the steering column, even though the scattershield consisted of a sheet of boiler plate bolted to the bottom of the firewall, and even though the roll bar consisted of arc-welded water pipe.[15] Rarely was anything magnafluxed. In a flip, welds sometimes came apart completely.

Even if some cars were fairly safe, the tracks themselves were often perilous. Jensen recalled seeing Red Case killed in late 1958 at Vaca Valley:

> He was driving a pretty standard and fairly sanitary slingshot with a double tube frame—a large tube with a truss underneath—and a reasonably good, well-constructed roll bar. . . . I'm sure Red would have survived had he just gone off the end of the track and rolled in the dirt, or run through the fences, or whatever—it would have been no problem. But . . . there was a hayfield around the outside of the track and, way, way down there, there was a side-delivery hay rake. The car hit it and was pretty well destroyed. There would not have been a lot of dragsters that could have survived a wreck like that.[16]

Referring to the "recent rash of drag strip fatalities," cam grinder Phil Weber asked the readers of *Drag News:* "Is drag racing on the way out?" It was not, but the toll would mount; Doris Herbert estimated that there were fifteen fatalities in 1959.[17] A good proportion of these could have been prevented if the places where people raced had been safer. Some had dips and bumps that sent cars flying, all four wheels off the ground. Some had shutoffs that were way too short. Some lacked guardrails, and, even where there were barriers, they were usually placed some distance beyond the racing surface—the idea being that when drivers lost control they would have leeway.

This was perhaps arguable in theory, but it did not take into account daredevil tendencies; drivers often treated the open space beside the pavement as part of the racing surface. *Drag News* described how Robert Lee Lace died at LaPlace, Louisiana, in September 1963: "Lace . . . appeared to lose control about 180 yards out of the starting chute during a solo time trial. The machine veered across the . . . track and onto the grass bordering the strip. Lace skidded through the grass and was apparently trying to gun the dragster back onto the track when his wheels caught the edge of the strip and he started to flip." The same thing happened to John Wenderski, whose Zeuschel-powered "Black Beauty" had been an overnight sensation on the southern California scene in the fall of 1963. At Ramona Raceway in February 1964 Wenderski was killed "when he failed to shut off after getting out in the loose stuff" and flipped several times in a rocky field.[18]

Wenderski's chassis had been fabricated in Tommy Ivo's shop using sturdy .130-inch tubing, and he might have been safe if he had simply aborted the run when he got off the track. Indeed, as one *Drag News* correspondent pointed out, drag racing was "by nature, the least risky of all auto racing": "There are no turns and never more than two cars competing at any one time and even then they are more widely separated than in any other type of racing."[19] But it took a lot of time to minimize the risks. While it was clear by the mid-1960s that the optimum location for guardrails was right at the edge of the racing surface, such barriers had been installed at only a few places, and there were still dozens of strips at which an out-of-control vehicle was in real danger of ending up someplace like a rocky field.

The situation improved as measures were taken to curb "macho" instincts. With disqualification automatic for crossing either the centerline or the outside edge of the strip, some of the incentive was removed for trying to "drive out of trouble." Both the UDRA and NHRA licensing programs mandated suspensions for reckless driving.[20] Licensing entailed a "Federal Aviation Administration–type" physical checkup beginning in 1966, and 20 percent of the initial examinees failed because of inadequate eyesight!

Progress towards safer competition was painfully slow throughout the 1960s. At least fifteen drivers died in slingshot dragsters between 1963 and 1966, including two the same weekend in late June 1965—Lou Cangelose at Springfield, Missouri, and Tex Randall at Aquasco, Maryland. That same weekend, at another Maryland track, Don Garlits, unable to see because of flaming oil, crashed into the dragster he was racing against and promptly announced that he would "never drive again."[21]

But one of Garlits's best quips was: "Retiring is easy. I've done it lots of times." In 1965 he had dozens of commitments; that March he had dominated the Bakersfield meet and had greatly enhanced his drawing power. He recanted within a few weeks. Just in that interim Gary Taylor died when his slingshot hit a bridge abutment at San Fernando, and Joseph "Q-Ball" Wale died when he hit a barrier at Bob Harmon Memorial Speedway in Monroe, Louisiana. Dead racers included both tyros (Taylor, twenty-one, was driving a fueler for almost the first time) and veterans (Randall and Wale were in their thirties, Cangelose in his fifties).[22] In spite of all the perceived dangers of the Indianapolis 500, more drivers died in slingshots

A bellhousing from Joe Schubeck's Lakewood Industries, of the sort prevalent in the late 1960s. Photograph also shows a Lenco two-speed transmission, steering box, and a hydraulic throttle control. (Photograph by Leslie Lovett, courtesy National Hot Rod Assn.)

at Lions Drag Strip during the 1960s than at the Brickyard. Across the nation, but mostly in California, the toll continued to mount. Bruce Johnston. Bruce Woodcock. Boyd Pennington. Dub Irwin. Denny Milani. Gene Goleman. Harrell Amyx. Pete Petrie. Ray Marsh. John Crews. John Wilson. John Martin. Chuck Suba.

For a long time details of accidents were deliberately cloaked in secrecy, and only gradually did the view prevail that specifics should be aired to avoid repeating past mistakes.[23] After a horrible 1960 episode in which a youthful driver was incinerated it became the rule that drivers of closed cars could not be fastened in from the outside. Funny cars were equipped with escape hatches in the roof. Requirements for fire-protective apparel progressed from motorcyclist's leathers to treated cotton and rayon to aluminized glass and asbestos layered with Nomex to even more fire-resistant synthetics like Carborundum's Kynol. In 1970 Tom McEwen installed a system of Halon extinguishers in his funny car similar to what NASCAR required, and others followed suit. Yet a complement of fire bottles and a state-of-the-art firesuit cost well over five hundred dollars, and by one estimate as few as 5 percent of drivers were as well protected as they might be. Terry Cook asked how much adequate protection would have been worth to Gaspar "Gas" Ronda, Ford's leading funny car star. Ronda suffered third-degree burns over 30 percent of his body in a high-speed, oil-fueled fire that roared into the cockpit "like a huge blowtorch." He said that he had intended to install NASCAR-type automatic fire extinguishers in "about a week."[24]

Disintegrating clutches often set off a chain reaction, as with John Mulligan. In January 1972 SEMA mandated that current containment devices were inadequate. Aluminum cans with steel liners could generally withstand the explosion of a triple-disc slider, but not the newer units with more discs, which could weigh as much as 60 pounds. After that all fuel-burning cars had to have an all-steel can that was attached to a steel motor plate with a full circle of aircraft-grade bolts, and even this was not always adequate. Superchargers were as disintegrative as clutches; typically, a rocker arm would break, an intake valve would hang open, and the combustive force would roar back up through the manifold into the blower, sometimes shattering the housing, sometimes launching the whole thing skyward. Eventually, SEMA specified and NHRA mandated nylon retaining straps on blowers, full Kevlar ballistics blankets (the design identical to that of a military flak jacket), and "burst plates" on blower manifolds, but that was not until the 1980s.[25]

From the beginning burned pistons had been commonplace when using nitro, and, indeed, replacing pistons eventually came to be regarded as routine maintenance. With a hole through the "backside" of a piston the combustive force would go directly into the crankcase, spraying oil out through the breathers. Occasionally, the oil would be ignited by the exhaust headers, which happened to Garlits at Cecil County. Even if not ignited, an "oil bath" often made it impossible for drivers to see where they were going; this happened to Richard Tharp in his race with Garlits at Lions. Eventually, the rules stipulated closed blowby systems with catch tanks located behind the driver, but this and similar precautions would have been way too late to help any driver sitting in a slingshot fueler in 1970, with the engine just in front of his face and nothing at all in between. At least in a

Bill Burke ran this machine at El Mirage from 1947 to 1949, and it was one of the featured attractions at the first Los Angeles Hot Rod Exposition. Lakesters configured like Burke's "Suite Sixteen," the bodies adapted from aircraft drop tanks, were thoroughly familiar to the first drag racers. (Photograph by Walt Woron, courtesy Museum of Drag Racing)

funny car everything was a little further away, and the body, windshield, and interior "tinwork" (actually, sheet aluminum) could afford pretty good protection from oil baths—provided there was no fire—and even from blower explosions. While the number of fatalities in slingshots diminished in the latter 1960s, it was obvious that the configuration was inherently dangerous.

The sociologist John Law has called attention to the growth of a body of literature which posits that "creative genius is best seen as a process in which the end results of a system of collaborative production are gathered up and attributed to a particular individual."[26] The invention of the mid-engine dragster was unquestionably a collaborative process. As an "innovation"—in the sense of rendering the mid-engine design "commercial"—the credit largely belongs to Don Garlits.[27] But there had been such machines right at the beginning, one archetype being the "lakesters" that first appeared in the late 1940s, with bodies adapted from aircraft drop tanks.

Ollie Morris fielded a winning mid-engine dragster in 1954. In 1955 the poster advertising the NHRA's first national championships at Great Bend depicted a mid-engine machine edging out a slingshot. Emery Cook broke records in a mid-engine roadster in 1956, and Red Greth kept doing so into the late 1950s. Jack Chrisman almost won the NHRA Nationals. Many others tried the mid-engine configuration. James Warren and Roger Coburn did so in 1959, when they first teamed up. Ken Droesbeke of Long Beach, Jack Friend of Memphis, and the Coleman brothers, Marylanders, raced similar machines in the early 1960s. So did Dave Marr and Lynn Carruthers of Sacramento and Warren Welsh of Reno, who used blown engines. Welsh's "Shoehorn" was turning in 170-MPH, 9-second clockings by 1962. Paul Shapiro of Miami debuted a mid-engine dragster called the "Israeli Rocket" in 1963 and won several times at Masters Field in 1964 and 1965. The best-financed effort of the mid-1960s was mounted by Tony Nancy, the custom upholsterer from the San Fernando Valley, who had Frank Huszar build him a mid-engine "Wedge." *Hot Rod* asked: "Has the 'slingshot' dragster come to the end of its string?" But Nancy's Wedge had a tendency to "twitch," and it was demolished in a crash at Sandusky, Ohio. Nancy thought he understood how to solve the handling problems, but his second version was likewise problematic, and soon he was back in a slingshot.[28]

Nearly all mid-engine machines from the early and mid-1960s were gas burners, although there were exceptions such as Speed Sport II, Holly Hedrick's roadster, and "Nasty I," run out of Modifications Unlimited, a shop in Kensington, Maryland. "Nasty II," which debuted in 1965, featured streamlined after-parts like Nancy's Wedge, and was quite a nice piece of work overall. But it was aptly named. "Don't you know every rear engine [i.e., mid-engine] car that was ever built turned right or left in the lights?" asked Jerry Tiffin of Goodyear.[29]

Some mid-engine machines were conceived mostly with the idea of attracting attention on the show circuit; some were "amateur efforts" conceived essentially "for the notoriety of something different." Dave Scott speculated that amateur involvement was actually desirable: "No true professional racer can halt his schedule to begin the lengthy and expensive process of building a better 'mousetrap.'" He thought that breaking the bounds of convention would require "the full-time effort of some well-heeled and technically knowledgeable amateur who can afford the growing pains of the new machine." People like Scott were not thinking so much about driver safety as about a chassis design that could more efficiently utilize the "brutal amount of horsepower" a fuel engine could produce, something lighter and more compact, and also a configuration more suitable for streamlining.[30]

In drag racing the mid-engine machine was considered (in Scott's term) "a bad actor." Yet this configuration had likewise garnered a negative reputation among road racers and oval trackers in both Europe and America. All that had changed after John Cooper reintroduced the concept to Formula 1 in the late 1950s; Cooper's Climax was followed by Colin Chapman's Lotus, and, in Australia and New Zealand, Jack Brabham and Bruce McLaren brought forth their own mid-engine designs. A Cooper finished ninth at the

Tony Nancy spared no expense on either of his "Wedge" machines, the second of which is seen here in the pits at Indianapolis Raceway Park in 1965. Watching Nancy balance a wheel is the designer of the Wayne Ewing body, Steve Swaja, a graduate of the Los Angeles Art Center School. (Photograph by Eric Rickman, courtesy Petersen Publishing Co.)

Brickyard in 1961, and A. J. Foyt won the Indy 500 in a Lotus in 1964. By the latter 1960s mid-engine designs dominated most forms of sports car racing as well as the United States Auto Club (USAC) and Grand Prix circuits, but drag racers possessed what most considered to be a "workable general configuration," and mostly they spent their time "polishing and refining . . . a pattern that is seldom broken."[31]

With a new configuration dominant in other racing constellations, however, there had to be questions about this pattern. In 1969 STP godfather Andy Granatelli ordered a mid-engine machine from the Logghe brothers' shop. Granatelli was on the right track. His machine had a negative airfoil over the engine to apply aerodynamic down-force and an exceptionally long wheelbase to enhance the driver's directional awareness. Later that same year Denver fabricator Mark Williams built a mid-engine machine

This mid-engine fueler, built in Denver in 1969, survived its early gremlins to race throughout the 1970s under the aegis of the Kaiser brothers. Here crew members wipe the tires—an enduring ritual—before a match against a slingshot at a Las Vegas strip. (Photograph by Leslie Lovett, courtesy National Hot Rod Assn.)

At Orange County International Raceway Pat Foster does a burnout in Woody Gilmore's first mid-engine machine. A week later this car was history. (Photograph by Gary Densford, courtesy National Hot Rod Assn.)

which ran 200 miles an hour right off and was ultimately campaigned for at least a decade. At the opposite extreme was the machine that Woody Gilmore built in the fall of 1969, which was checked out by Pat Foster at Orange County on December 6 and demolished at Lions a week later, when Foster lost control in the lights.[32]

There were some modest successes in 1970. In June Bernie Schacker of Lindenhurst, New Jersey, ran a 6.98 at New York National Speedway on Long Island. Not long afterwards Dwaine Ong's "Pawnbroker"—built by Gilmore and Foster with Foster's hard-earned lessons in mind—won an AHRA Grand American with 6.80s. These were the best elapsed times that had ever been clocked by mid-engine machines. They were a half-second slower than the best slingshots and even shy of the marks being posted by certain funny cars, but more and more people were becoming interested in the concept, and, in a loose sense, a collaborative process of invention was clearly underway.

Having taken this process into account, and having conceded that invention can certainly "be treated as something other than the expression of personal creativity," I still suggest that one is not mistaken in attributing the particular device under consideration here to a specific individual.[33] It was not as if Garlits had not "made" inventions before, not least of which was

At Garlits's Tampa shop in the late 1950s Connie Swingle helps the boss assemble an engine. (Courtesy Museum of Drag Racing)

the very notion of drag racing as a professional activity. Dave Scott was wrong about the key being a "well-heeled and knowledgeable amateur." Such a description would have fit Granatelli, but he was not the one to solve the puzzle. Woody Gilmore had a lot of skill and experience, and he was coming close to the solution, but Woody had never actually driven one of these things. Perhaps what was essential was a *general* level of empirical knowledge nobody else could match; by 1970 Garlits stood alone among drag racers in his versatility—and definitely alone in his capacity to maintain a winning edge year in and year out.

Garlits also knew how to enlist the right helpers, Lemons and Swingle. Although he had had his ups and downs with Swingle, they had worked together for years, and he regarded him as the best fabricator in the business. Swingle had been a driver, too, and ultimately it was Swingle who provided the final piece in the puzzle. All of them knew enough to leave other things alone (trying to implement several novelties at the same time was a common mode of failure), and what they built was simplicity itself: a chrome moly frame with tubing ranging from .045 to .095 inches thick, a 215-inch wheelbase, a 426 hemi out 20 inches, abbreviated body panels— "mostly pipe and a big Dodge mill," said Garlits. Though Garlits had made sketches while hospitalized in Long Beach, there were never any "complicated blueprints."[34] It has been suggested that "the greatest single difference between a craftsman and an engineer is that the craftsman nearly always designs 'in the material,' transferring his conception directly from mental image, through his hands, straight into the trial or finished object."[35] In this light, consider how Garlits achieved the 82:18 static balance he wanted: "With me in the car, Swingle using scales, and moving a lot of components."[36]

Garlits called his friend Billy Herndon, owner of Tampa Dragway, and asked if they could come down for a test. Making a short checkout pass, Garlits realized something was wrong, so they went back to the shop to fine-tune the chassis. Next they tested at Sunshine Drag Strip in St. Petersburg. They were pleasantly surprised by how well the car hooked up, but it still did not handle well; it "seemed to be floating the front end in the lights." Mounting an airfoil above the axle helped, but Garlits knew that the car "was still a far cry from the beautifully handling slingshots I was used to." At a public debut in December 1970 he turned a best of 6.81, 220. On at least one run he got sideways on the top end; speaking to reporters, Garlits attributed this to "a gust of wind," but privately he knew better.[37]

"Pat Foster at Lions! Tony Nancy in Ohio!" People had been telling Garlits that it would never work. He believed from the start that an unlocked differential was essential; a positraction (limited slip) might be satisfactory in a slingshot because, when it "throws you around, you have enough time to catch it."[38] Yet the car still would "dart" in the lights, and he invariably overcorrected. Driving a dragster requires very subtle corrections "to *prevent* car movement, not cause movement,"[39] and for years there had been a consensus that he was a master at steering a straight course by anticipating where a car was about to head. But *this* machine he could not steer. He finally admitted it to Swingle.

"Why the hell didn't you say so, Gar?" Swingle answered. "I'll slow it down to 10:1" (nearly all slingshots had a steering ratio of 6:1). Swingle machined some new parts that changed the geometry. Then they packed up and went to another of Herndon's tracks, in Orlando, where everything worked flawlessly and Garlits promptly broke both ends of Prudhomme's strip record. The mid-engine dragster had, finally, been "perfected." Was Garlits responsible? One might as well say that Swingle was, because he provided the last answer.[40] But attributions to "a particular individual" are not always capricious, and certainly not in this instance. Garlits did not make a revolution single-handedly; he was not a sufficient cause, but he was necessary.

"**W**e drove back to Tampa from Orlando that afternoon," Garlits wrote, "and if anybody had been with us they would have thought we were drunk! In a sense we were. After . . . having been almost to the point of giving up, the success of that last run had put all of us on a natural high." Truly, Garlits thought, all his troubles were behind him. "It goes just like it has eyes," he told *Hot Rod*'s Ralph Guldahl, Jr. "Wait till them Californians see what the 'Okies' built," T.C. added.[41]

It was time for the winter meets in the West, the first stop being the opener for the AHRA Grand American Series at Lions. Pappy Hart put Garlits right next to the pit gate. Nobody could miss seeing him there. Everyone gawked, and the Snake himself came over and said, "Garlits, you've gotta be kidding." Garlits liked Prudhomme, but he loved giving the "California loud mouths" a comeuppance. He stormed through the first three rounds, and, even though he fell short in the final, there was no doubt that he had built something that "worked." "For anyone who's had any second thoughts about rear-engine dragsters," ran the lead in *Drag News* on January 16, "Don Garlits ended them this week at Lions. . . . The car hooks-up beautifully and goes super straight." While they realized that clutch and tire combinations that enabled "hooking up" consistently were a crucial factor, longtime partisans of mid-engines like Holly Hedrich still felt the thrill of vindication.

A week later Garlits raced at Orange County, and again he lost in the final round, but a week after that he won the Pomona Winternationals. Then he won Bakersfield and just kept right on winning. By the end of the 1971 season he had banked the AHRA's $20,000 points bonus, he held both the NHRA and AHRA elapsed-time records, and he was knocking on the door of the five-second zone.

Garlits's first thoughts about the mid-engine configuration had been in

In the final round at Famoso in 1971 (the event now called the "March Meet," as the Smokers no longer had rights to the original name), Don Garlits gets out on Rick Ramsey. Garlits's was the only mid-engine fueler in 1971; a year later only a handful of entries would be slingshots. (Photograph by Bob McClurg, courtesy Museum of Drag Racing)

Bob Bofysil's slingshot was not unusual in the amount of weight it carried on the axle, although most racers were a little neater about how they attached it. (Photograph by Tim Marshall)

terms of safety, and certainly it was safer. "That's the most pleasurable ride I've ever had," he told Shav Glick of the *Los Angeles Times*: "It's just great out in front of all that noise. There's no fear of flying parts. The water, clutch dust, and oil don't blur your goggles."[42] Bantering with T.C., he said, "Saw a big ol' 3/8-inch bolt layin' down there in the lights." "Oh yeah?" T.C. answered. "Yeah," Garlits said, "*fine thread*."[43] But he had not really anticipated progress in other realms; having less weight on the rear end than before, he initially cut back on power. As it turned out, however, the configuration was more efficient in getting from A to B quickly.

With the tire compounds that Goodyear and M&H had developed by the early 1970s the racers, for once, had about as much "bite" as they could use. To keep the front wheels down, slingshots were often carrying a hundred pounds or more of lead clamped to the axle; Garlits needed hardly any such ballast. Furthermore, a mid-engine car was inherently simpler; it required less tubing and, hence, could be lighter. Beyond that the design was better aerodynamically because the center of pressure was behind the center of gravity. The best speeds ever attained by slingshots were in the 230s, and these did not improve appreciably between 1968 and 1972. By 1973 Garlits had topped 247.

And even beyond all that there was another consideration. As I have reiterated from the first, the word *performance* can be defined in technological terms, but the primary dictionary definition has to do with theatrics, with entertainment. Funny car racers had been stealing a lot of the applause, and, in addition, the best of them—Gene Snow, Mickey Thompson and Danny Ongais, Ed McCulloch, John Mazmanian and Keith Black— were fast closing the performance gap in terms of getting quickly from point A to point B. The floppers already commanded the lion's share of match race money, and payoffs in open competition were at least equal. (Tice's Grand American Series, for example, offered the same deal to a group of seeded funny cars as to the dragsters.) Some dragster stalwarts such as Roland Leong had switched camps altogether; others like the Snake and the Mongoose, even Kalitta and the Greek, were hedging their bets by campaigning both diggers and floppers. Garlits understood the situation perfectly: "The [fuel dragsters] have made mistakes in recent years. We were playing a pat hand. We weren't making any changes. It was the same show, the same basic cars. . . . We weren't getting with the show business side of things. We weren't offering something new. Now we can."[44]

There were those who suggested that Garlits's motive had been *primarily* to get away from a "pat hand." "Garlits probably won't admit this," said Don Moody, "but the main reason he switched . . . was to create some interest." Certainly, the mid-engine revolution rejuvenated top-fuel racing. But it was by no means just a matter of "tricking up the show." First, there is a possibility that slingshots might soon have been banished outright as too dangerous. Even if not, they could well have gone by the boards anyway as savvy fuel racers climbed on the funny car bandwagon and closed the gap. The head of the International Hot Rod Association, Larry Carrier, predicted that there would be only one class for fuelers ultimately.

One thing Garlits unquestionably precipitated was an era of intense experimentation. The matter of configuration might be settled, but there was still the matter of seeking (in Walter Vincenti's phrase) a "particular instance of it." The mid-engine revolution coincided with major changes in other auto racing constellations, especially Jim Hall's breakthrough with the Chaparral, as Formula 1, Can Am, and Indy cars bristled with wings, air dams, and other aerodynamic accoutrements. Everything seemed to open up. There were experiments with negative airfoils of every conceivable size, attitude, profile, and placement—above the cockpit and below, over the engine and out behind. There were wings forward, wings aft. There were vertical stabilizers. There were dragsters and even funny cars with canard wings to the sides.[45]

After all this a wing above the rear tires was the only thing that became part of the conventional combination. But there were those who remained enthusiastic about further potentials for "air management." Robert Johnson, to whom we will get a proper introduction in the next chapter, had long been an advocate of full streamlining, and any number of racers had experimented with partial fairings. Pete Robinson became particularly interested in utilizing air-management devices as a means of minimizing weight—with Pete, an enthusiasm that verged on fanaticism and stirred controversy time and again. He miniaturized all sorts of parts and ran a parachute the size of a toy.[46]

Then there were his so-called jumping jacks, designed to help him regain an edge when his dragster, with its wedge-head Ford engine (which he liked because it was so light), was getting overpowered by Chryslers. The jacks were extremely simple: a cam at the back of his frame connected to a lever, which enabled him to lift his rear wheels slightly off the pavement after he had staged. He would rev up the engine, then release the lever when the starter gave him a green flag—or, more and more frequently around this time, 1963, when he got a green light. The aim, he said, was twofold. His first dragster, the one in which he won the 1961 NHRA Nationals, had operative differential gears, but his second dispensed with the differential to save weight. The trade-off was reduced directional stability, and Pete claimed that "pre-rotation" of the rear wheels had the effect of a gyroscope in stabilizing the car.

NHRA rules made no mention of such a device, and when Pete showed up at Indianapolis Raceway Park for the 1963 Nationals he cleared tech with no particular problem (though he never found this preliminary hurdle easy). After his first qualifying pass, however—it turned out to be the low ET of the entire event—Ed Eaton ordered him not to activate the jacks anymore. But NHRA did not control all of drag racing, and Pete kept them for other times and other places, one being an autumn match race with Garlits in Atlanta. Pete won easily. As Garlits put it, as soon as he came down off the jacks, "he literally shot by me 4 or 5 car lengths." Certain that "a new era in drag racing" was at hand, Garlits began constructing a chassis with jacks, too. Yet he claimed that any such setup was potentially hazardous, especially if one wheel were to hit the strip before the other. If jacks were going to be permitted, Garlits said, he could not afford to be without. Better they should be banned.[47]

Both Robinson and Garlits knew that the idea was scarcely new. In the 1950s jacks had appealed to certain enthusiasts of the sidewinder configuration because it was difficult to rig a clutch. An Ohio racer, Lee Pendleton, had been using jacks with an airplane-engined dragster for some time. Pete contradicted Garlits on every point. Jacks were superior because the driveline was not shocked as violently as when a clutch was engaged. Nor could one wheel contacting the pavement before the other result in a loss of control. "The installation of our jacks," he said, "has provided us with a car far safer than any now running in the country due to its ability to drive straight, [and] reduce drive line damage and clutch blow up."[48]

A debate ensued on the pages of the trade papers. Pendleton agreed with everything Pete said and also argued that jacks helped counter uncontrolled wheelstands, which were becoming a commonplace occurrence. An Illinois racer, Ron Colson, admitted that Pete deserved credit "for working out an unusual idea" but argued that cars with jacks should be classed with jets "and other weirdies." Another correspondent insisted that jacks could have no effect on wheelstands.[49] There were comments on the irony of Garlits holding a brief for safety when he himself had once seemed willing to try anything to gain an edge. Robinson and others reminded everyone that the NHRA's motto was Ingenuity in Action. In 1964 experiments with jacks continued at West Coast AHRA strips. But with Garlits and Wally in accord the issue was essentially closed. Did jacks really improve technological efficiency? Was this a revolution thwarted? Perhaps not, yet the episode certainly added to the evidence that claims about efficiency were simply a matter of opinion.

Robinson never used a Chrysler, preferring lighter engines, even if he had to give up horsepower. In 1966, however, he got a chance to go the Chryslers one better, signing an agreement to run a powerful single-overhead-cam Ford and enjoying considerable success at first, much to the pleasure of the people in Dearborn. Ford kept a high profile in racing throughout the 1960s (in 1966 Ford claimed victory at Le Mans, the Indy 500, and Daytona as well as Pete's win at the NHRA World Championships at Tulsa), but in 1970 all the factories backed off, Ford especially. "Cammers" consumed a lot of parts, and everyone who had been running them switched to other power plants, everyone except Robinson, who had spent a lot of his own money refining the SOHC design and was reluctant to give up on it.

He had plenty of other tricks up his sleeve. Terry Cook recalled sitting around a Detroit motel room with Pete and Larry Shinoda discussing streamlining, but somebody so weight conscious could never have been truly enthusiastic about streamlining. Rather, he became fascinated with the phenomenon of ground effects. In 1968 he anticipated Jim Hall's Chevrolet Tech Center with the idea of inducting air from a chamber underneath the engine. He said that this made "the tires 'think' the car weighs about a thousand pounds more than it does in reality, thus . . . providing more traction." Pete reluctantly concluded that this particular setup worked *too* well, bogging the engine off the line.[50] He kept working on variants, however. In 1971, when he showed up at Lions for the opener of the AHRA Grand American Series, his car attracted almost as much attention as Garlits's "middie" because of a strange device with a rubber flange hanging underneath. It was worth about "two-tenths," Pete said, and indeed he backed up this claim by qualifying with low ET, 6.50.

He still had the device when he arrived for the Winternationals three weeks afterwards. Late Friday, after waiting around in a gloomy overcast

Pete Robinson pauses for a photographer as he works on his single-overhead-cam Ford engine. Robinson manufactured various components for use when these engines were converted to superchargers and nitro, including blower drives. (Smithsonian Institution collection)

most of the afternoon, he came out of the staging lanes for his first qualifying run. He left the line strongly and clocked the day's quickest ET, 6.77, then, at top speed, his car turned directly into the righthand guardrail and disintegrated. In utter silence otherwise the roll cage could be heard tumbling for hundreds of feet. Robinson had survived previous California crashes at Lions and Irwindale, but the third time his luck ran out. He died at Pomona Valley Community Hospital a few hours later. Terry Cook wrote: "The drag dilettante named him 'Sneaky Pete' . . . but he wasn't sneaky at all. He was warm and honest, intelligent and trustworthy. Above all, he was an innovator."[51] For a drag racer there could be no fonder eulogy.

There was talk that Robinson had "tricked himself out of the park" with his ground-effects device. A photo taken a millisecond before he crashed showed some extraordinary force pulling him to the right and showed both front tires leaving the rims. Whatever happened, Pete was gone, killed in a slingshot at thirty-seven, ironically at almost the very moment Garlits showed that this design was passé. Pete probably could not have survived

The ground-effects device that Robinson attached to his car for the 1971 winter meets in southern California. (Photograph by Gray Baskerville, courtesy Petersen Publishing Co.)

such a crash by any means, but with a mid-engine machine he might not even have been using such a device.

Two weeks after the NHRA Winternationals, at the AHRA's Arizona counterpart, Texan Paul Prichette died when his slingshot blew an engine and flipped beyond the lights. But Pete Robinson and Paul Prichette were the last drivers ever to lose their lives in a front-engine fueler. Within a couple of years there were virtually no such machines still in competition. Prudhomme had a middie completed by the time of the 1971 NHRA Spring-nationals, its body showing much influence from Can Am canons. Woody Gilmore, who had been just a step behind Garlits in figuring out the puzzle, started turning out mid-engine machines one after another; Bob Murray of St. Louis ordered one, as did Herm Peterson of Poulsbo, Washington, and Californians Mike Kuhl and Leland Kolb. Kolb's "Polish Lotus" imitated the Snake's machine. Kuhl's driver, Carl Olson, won the 1972 Winternationals, a race at which only a handful of the thirty-two top-fuel qualifiers were slingshots. Tony Nancy and one or two others stuck with the old configuration longer than most, but by 1973 the slingshot fueler was gone. In his own words, Garlits had made "about a $1 million junk-pile."[52]

Apprehensions that the mid-engine design harbored a whole new set of dangers proved baseless. When a slingshot rolled over, the tires afforded the driver some protection, but in a crash it was better to be separated from the engine's kinetic energy, a precaution more readily attainable in fabricating a middie. In collisions with guardrails engines broke away and flew up and over. A driver's lower extremities were at greater risk: One driver had a leg amputated as a result of a crash at Epping, New Hampshire, and another lost both legs at Orange County, but in subsequent litigation the design of the guardrail was cited as a contributory factor. Drivers would die in mid-engine dragsters—among them, Marvin Schwartz in 1980—but the number was infinitesimal compared to the number who had died in slingshots. Time and again there were spectacular explosions behind drivers who scarcely knew it. Time and again drivers survived incredibly violent top-end spills.[53]

Drag racing remained a dangerous enthusiasm, some competitors less than fully attentive to safety precautions. "If racers were regimented, un-enthusiastic, and without daring they would not be racers," explained one journalist.[54] With dragsters aerodynamic devices remained a perilous frontier. Just as Robinson did not quite know what forces he was dealing with in the realm of ground-effects, in their experiments with airfoils others were flirting with structural matters they did not truly understand. When something gave way at high speed the results could be calamitous, yet the drivers survived. They also survived funny car mishaps—fires, to be specific. Aerodynamic instability was rarely problematic anymore.

In addressing the revolution in dragster design, I have perhaps not made

At Orange County an inadequately braced wing has given way, but this time the driver has the car under control and the chute safely out. (Photograph by John Shanks)

something altogether clear: Through it all the funny car remained the same as always—driver in front of the axle, engine in front of the driver. There were experiments with mid-engine funny cars, but the old paradigm proved invulnerable. On-board fire extinguishers, as well as the fiberglass, acrylic, and aluminum that were interposed between the engine and cockpit, forestalled many of the dangers that had become overwhelming in an open dragster. As important a revolution as the mid-engine dragster was, it is arguable that it was less significant than the way in which funny cars were transformed while leaving the basic configuration the same as the Comets and Chargers of the 1960s—or, for that matter, the same as the fuel-burning deuce coupes of the 1950s.

At some point people began to realize that the funny car could be something other than an exhibition machine, that it might be capable of getting from A to B nearly as quickly as a dragster, even quicker maybe. The static weight distribution might be disadvantageous, but there was a lot to be gained from a low coefficient of drag. As concepts of purpose began to shift and technological designs were refined, funny cars began to emerge as "real" race cars. Specific improvements are attributable, but nobody has yet suggested that any one person deserves credit for the transformation as a whole. Someday, however, the "end results of a system of collaborative production" may be gathered up. What happened is so intriguing that people may be tempted to try to wrap a neat package.

What happened is this: While remaining essentially linked to the chassis configuration of dragsters in the days before slingshots, by 1990 a funny car could perform (in a technological sense) nearly as well as a dragster. The NHRA top-fuel records stood at 294.88 with a 4.909 ET, the funny car records at 284.18 with a 5.140. Remarkably, the marks could be that close, even though funny cars were required to weigh 250 pounds more. If, as the rule of thumb held, "100 lbs. is worth a tenth," their capability of getting from A to B was identical to that of a machine whose design bore little resemblance to its precursors. The evidence for there being no "one best way" to maximize acceleration was compelling.

One of the funny car's clear-cut technological advantages lay in the realm of aerodynamic drag, a factor that had been taken into account with dragsters only haphazardly. The wheels, around which turbulence was always greatest, were almost invariably out in the open, and the slicks presented a tremendous expanse of unfaired frontal area. The mid-engine configuration proved superior aerodynamically, but that was considered only an incidental benefit. What, however, if aerodynamic finesse was considered a matter of top priority with a top-fueler? For a long time a few enthusiasts had dreamed about what might result.

10

FINESSE

These open wheels are bad business.
ROGER HUNTINGTON, 1956

There appears to be a lot more horse-power available in these motors.
GENE SNOW, 1987

n the summer of 1972 Art Marshall took top-fuel eliminator at the second running of the NHRA's Grandnational at Sanair Drag Strip in St.-Pie, Quebec. Marshall was driving one of the original Mattel "Hot Wheels" machines, purchased secondhand from Tom McEwen by Pete Van Iderstine, proprietor of a New Jersey speed shop. Nobody knew at the time that this would be the last major event won with a slingshot, but Marshall rode to victory on a streak of luck, and there was reason to suspect it.

Fuel dragsters were star performers once again. Competition between the old and the new, slingshots and middies, stirred excitement at first. Then, as the mid-engine design took complete command, there was a spectacular improvement in elapsed times. Skeptics shrugged when Tommy Ivo was credited with a 5.97 at a small Pennsylvania strip in October 1972. But nearly everyone believed when Mike Snively received a 5.97 time slip at Ontario Motor Speedway on November 19, at the NHRA Supernationals. Although another veteran of the slingshot wars, Don Moody, turned 5.91 just an hour later, secure credit for "first in the 5s" belonged to Snively, driving a Keith Black–powered Woody Gilmore car owned by "Diamond

Jim" Annin of La Cañada. Because the track had been prepped with a sticky substance, there were suggestions that this invalidated the quick times—but afterwards traction would be enhanced the same way at *all* major events.

Gilmore had predicted that 1972 would be a banner year for chassis fabricators, and he was right on. His own Race Car Engineering had customers waiting in line, as did other southern California shops, such as Frank Huszar's, Don Long's, Don Tuttle's, and Roy Fjastad's. Up north Kent Fuller built mid-engine machines for Don Prudhomme, Chris Karamesines, even Tony Nancy. And new talent was coming into the trade in other parts of the country, such as Ed Mabry in Texas, Lester Guillory in Louisiana, and Bill Stebbins in Kentucky. Don Garlits had been selling chassis to other racers for nearly a decade. People said that Garlits's chassis worked much better for him than they worked for anyone else, and his volume was never large. Even Californians bought his middies, however; Tom McEwen won the 1972 Bakersfield meet with a Garlits chassis.

One day Garlits got a phone call from Robert Johnson, "Jocko," as he was known. Jocko had been involved in drag racing from the start, earning his living by contouring and polishing the ports of cast-iron cylinder heads (and, in flathead days, cylinder blocks), an art that required consummate finesse. In the latter 1960s his interest had shifted to wood sculpture, using the same portable tools, and he sold his porting business and moved to the desert. But Jocko's love of smooth contours had yielded one enduring automotive passion, streamlining.

In 1956 Jocko had conceived a dragster with a full-envelope body, the form somewhat reminiscent of hot rodders' efforts with Bonneville streamliners. Over a two-year period he laminated a fiberglass shell, which totally enclosed the engine, the running gear, and all four wheels, except where they contacted the pavement. The cockpit was in front of the engine and had a canopy like an airplane's. Jocko's streamliner debuted in July 1958. The driver was Jazzy Jim Nelson from Venice Beach—like Jocko, a free spirit who reflected some of the contemporary values of beatdom. Jazzy and Jocko. For nearly a year times were nothing special. But on Memorial Day weekend of 1959 they showed up at Riverside with a new 450-cubic-inch Chrysler that Jazzy had built; looming above everything was a GMC blower with Stromberg carburetors adapted to nitro, pretty much the same combination as Garlits had adopted after Bakersfield. They received a time slip for an 8.35 ET at 178 miles per hour.[1]

Very few fuelers were going faster, and none had run any quicker. People recalled the furor over Jazzy's reported 9.10 in his Fiat coupe, and they wondered about that 8.35. Jazzy could have done it again, perhaps, but a month later Jocko's lovingly crafted shell cracked and disintegrated at high speed.

Jocko began forming another streamlined body, using aluminum this

Seen on the Bonneville Salt Flats in the early 1950s, this streamliner had body-work fabricated by George Barris, later the most famous of the southern California "customizers." Chet Herbert, the owner, was also involved with Jocko Johnson's streamlined dragster. (Photograph by Fred Wohlfarth)

Jocko's mid-engine machine of 1958, the first dragster with a full-envelope body. (Courtesy Museum of Drag Racing)

time. Aluminum was lighter than fiberglass, but he found it almost as time-consuming to work. At last his new machine debuted at Lions in 1964, with Allison power, of all things. (Having switched sanction to the AHRA, Lions still permitted airplane engines.) Elapsed times were noncompetitive; if the streamlining helped any, it did not offset the weight penalty. Emery Cook took this machine on tour for a couple of years as the "Thundercar," capitalizing on the growing vogue for "exhibition" machinery, but it never ran in heads-up competition.

Meantime, Jocko had made full fiberglass bodies for two of Gilmore's standard slingshots, one owned by Woody himself, the other by Gene Mooneyham. Although he managed to keep the weight within reason, there were other problems. Mooneyham's driver, Larry Faust, reported that the car could not be controlled above 190 miles per hour. After Pete Ogden made a couple of passes in Woody's machine, with inconclusive results, Woody and Jocko clashed, and development work ceased. Now Jocko felt certain that successful streamlining entailed something more than simply shrouding a conventional dragster in a smooth shell, be it aluminum or fiberglass. A mid-engine configuration was essential.

That is why he was so excited about what Garlits had done at the beginning of 1971 and why he got in touch: "Jocko filled the old man's newly opened mind with ideas. Phone calls flowed across the country, models were built, and an agreement was reached. 'If you build the body,' Garlits said, 'I'll put a frame and motor under it and run it.'"[2] The "old man" (he was now thirty-eight) had just scored an innovative triumph—set the world on its ear, as the journalists were fond of saying. People warned him that Jocko's bodies were heavy, that they induced handling problems, that they "wouldn't work." But Garlits kept thinking that this was just what they had said to him the last time.

Jocko began carving a "plug" out of Styrofoam. This became the male mold, from which he made a two-piece female mold. In January 1972 he strapped this to a trailer and, with his son Benny and two helpers, headed for Garlits's shop in Florida. By April they had finished the fiberglass shell—76 inches wide, 44 inches high, and 240 inches long—and were starting to fabricate a matrix of bulkheads and supports for reinforcement and attachment. Connie Swingle had already begun the chassis, which was novel only in that it was truncated about six feet in front. This was Jocko's other sine qua non; not only did the vehicle have to be mid-engine, it also had to be short enough to put the center of aerodynamic pressure and center of gravity on the same axis. Nearly every new mid-engine fueler had 220 inches of wheelbase, and some had even more than that; the streamliner's was to be 145.

As work stretched out much longer than anticipated, Garlits's ambivalence was palpable. Sometimes he would confess to having second thoughts about the entire project; sometimes he would predict speeds of

275. Usually, he would reiterate that Jocko thought it would work, and, if Jocko thought so, so did he. One well-regarded writer reminded his readers that "history has shown that the Swamp Rat usually has the last laugh."[3] Still, the debut of the liner kept getting set back.

*F*rom the very start hot rodders had been aware of the effect of frontal area on speed though less attentive to the phenomena of form drag and turbulence. Mid-engine modified roadsters, with sharp prows and both the engine and driver pretty much out of the airstream, were known in the 1950s as good top-end machines. There were experiments with cockpit canopies on slingshots, as on Ed Cortopassi's "Glass Slipper" and Hank Vincent's "Top Banana," but, as early as 1956, Roger Huntington had told readers of *Drag News* that "minor niceties in body streamlining" such as canopies and pointed noses would have little effect on total air resistance. Turbulence around open wheels and axles was the main problem: "The first step towards really effective streamlining on a competition car would be to *get those wheels cowled in*."[4]

While the idea of cowled-in wheels appealed to other enthusiasts besides Jocko, the conceptual technique was typically what has been called "eyeball aero." Such a rule-of-thumb procedure, as one expert indicated, "usually leads to designs that look clean at a glance, but actually embody serious air flow errors."[5] It took a long time for the nature of this fallacy to sink in, but another facet of the problem was readily comprehensible. As Huntington said, "Streamlined body-work weighs up." It apparently took "only a few pounds of additional weight to offset any practical benefits of streamlining."[6]

The fate of Jocko's first effort stood as a continual reminder of the necessity for rugged internal bracing: His shell had come apart when air pressure pushed it into the front wheels. The way to get off most lightly, it seemed, would be to enclose just the rear wheels. Mickey Thompson had done that with his first slingshot in 1954. In 1960 he drove "Assault I," a Dragmaster chassis with a similar fairing, to standing mile and kilometer records at March Air Force Base in Riverside County. For going after additional international records in 1961 he fabricated a shell that enclosed all four wheels, like Jocko's, but he abandoned the quest after encountering a "gusty sidewind."[7] Later Larry Faust and others also mentioned "sidewinds" that caused handling problems, but what was often at fault were designs based on an inadequate understanding of lateral forces operating at right angles to drag forces. Configurations sensitive to lateral forces often left drivers with the impression that they were being buffeted by winds.[8] Even if this was not the problem with Mickey's full fairing, it obviously took a lot of fiberglass to make. In Roger Huntington's phrase, it weighed up, and it would have posed serious inertial resistance in a quarter-mile sprint.

Most attempts to streamline dragsters entailed enclosing after-parts

A mid-engine fueler outfitted with pants on the front wheels; most racers flirted with these devices only briefly, and NHRA later prohibited them. In this photograph driver James Warren has just hit the parachute release while still under power, and the spring-loaded pilot chute is pulling the main chute out. (Photograph by John Shanks)

only, usually using aluminum. Craig Breedlove did an attractive job of it in 1963, with the help of designer William Moore and fabricator Quincy Epperly. And with its truncated after-section, the "K-form" (devised in the 1930s by Wunibald Kamm of the Motor Vehicle Research Institute in Stuttgart), Breedlove's "Spirit II" appeared to be configured with some attention to theoretical considerations. Otherwise, it seemed like a classic example of eyeball aero. Breedlove had merely selected one of Moore's renderings, then added "aerodynamic styling features" that he found appealing. Intuitively, he thought the nose was "right" because it imitated "mother nature," specifically, the duckbill platypus. The front "pants" may have looked right too, and Breedlove believed they would assist him in steering "with wheels on or off the ground."[9] Actually, they were a menace, making for dangerous oversteer at high speed, as many racers discovered when such pants enjoyed a vogue a few years later.

With a blown-fuel Chrysler built by Dave Carpenter, Spirit II had plenty of power. Yet its marks of 8.50, 185, in 1964 were far off the times of the best conventional slingshots. The whole spirit behind Spirit II had been "build it and try it."[10] Among hot rodders such unabashed pragmatism had yielded tremendous dividends: Had not "calculations by formula" shown that the maximum feasible speed for a dragster powered through the wheels was 167 miles per hour? But aerodynamics did not lend itself so readily to cut-and-try. "Aerodynamics is an empirical discipline," writes James Flink, but it is "dependent upon wind tunnel testing to determine air flow patterns."[11] A little wind tunnel time should have indicated right off what was wrong with those front pants. There was an irony, for Breedlove came to the idea of a streamlined slingshot directly from his quest for the land speed record; the design of the pants was copied from the rear-wheel fairings on his three-wheeled "Spirit of America," a vehicle conceived with plenty of input from aerodynamicists.

Born March 23, 1937, a graduate of Venice High School in West Los Angeles, Craig Norman Breedlove had a classic hot rodding background. He first competed at Bonneville at the age of sixteen. He turned nearly 150 at El Mirage in a '34 Ford. He was a good craftsman, and a very good hustler. Even in 1960 he was telling *Hot Rod*'s Eric Rickman about plans he and Epperly had for a streamliner "that would carry him faster on earth than any man had ever before travelled."[12] But in the rarified climes of the quest for the land speed record (LSR) eyeball aero would rarely suffice; wind tunnel work was almost essential.

For many years the LSR had been largely an enthusiasm of English playboys driving costly leviathans that seemed the very antithesis of the hot rodder's ideals of spare simplicity. Since the 1930s the prime venue for LSR attempts had been the salt flats one hundred miles west of Salt Lake City. With two Napier Lion airplane engines, a London furrier, John Cobb, had set a record of 368 in 1939 then returned in 1947 and gone 394.

Spurred by Wally Parks, the Southern California Timing Association had begun running annual meets at Bonneville two years later. Top time during that initial "Speed Week" was less than half as fast as Cobb's mark.[13] Within a decade, however, hot rodders were getting familiar with the idea of 300 miles per hour. The fastest of all was none other than Mickey Thompson, who had raced across the salt even before making a name at the drags. In 1959, in his "Challenger I," Thompson clocked 360, with a two-way average (a stipulation of the FIA's rules aimed at compensating for tailwinds) of 330. After that there was a flurry of American interest in the LSR. Along with Athol Graham of Salt Lake City Mickey headed the "purist" camp, which accepted the FIA's stricture that the LSR could be contested only by machines with internal-combustion power. The other camp included Nathan Ostich, a Los Angeles physician, Breedlove, and several names familiar to drag racers, Walt Arfons, Art Arfons, Romeo Palamides. This was the jet set, and their machines resembled aircraft without wings.

In 1963 Breedlove exceeded 400, as did both Art and Walt Arfons. Two years later, with plenty of outside technical counsel and with financing from Goodyear (which was just beginning to reestablish a foothold in auto racing), Breedlove hit 600. For the purists Mickey's Challenger I topped 400 one-way in 1960 but never managed to complete the run "back up" the course that would have permitted breaking Cobb's FIA mark. Five years later, however, another four-engine machine owned by Bill and Bob Summers established an official 409 average, a record that stood into the 1990s.[14]

The quest for the LSR involves a discrete technological idiom and a distinctive theater of machines. Suffice it to say here that attaining speeds over 300, 400, 500 miles per hour has generally necessitated some pretty keen attention to theoretical matters. People tried to do it with eyeball aero, but this could be extremely dangerous. It was said that Athol Graham epitomized a "pure hot rod philosophy." Rather than surrounding himself with swarms of technicians like all those rich Englishmen did, and like Breedlove did, he believed in doing everything "with his own two hands." The body of his "City of Salt Lake" was made from a surplus B-29 bellytank split down the middle, the two halves joined with sheet aluminum; the cockpit was surplus P-51. Graham had previously attained a two-way average of 344 and felt confident of hitting 400. But on the morning of August 1, 1960, he got sideways four miles down-course, and the sheet aluminum peeled away. His car leaped high into the air, crashed upside down, then bounced and slid for nearly a mile. Graham was killed. Slingshot veteran Glen Leasher, the driver of Romeo's "Infinity," died in a similar crash two years later.[15] Like Graham and Palamides, Art Arfons seemed rather indifferent to aerodynamic formalities, instead relying primarily on huge amounts of power. In 1966, pushing 600, he too crashed. He survived, but Breedlove's vehicles had more finesse, and the jet record was ultimately his.

For a generation of purists, the heroes were the Summers brothers, hot rodders from Arcadia, California. In 1965 Bob was twenty-seven, Bill a year older. They were not hustlers on the order of Breedlove or Mickey Thompson, but they managed to garner six-figure support for their LSR venture from Firestone, Hurst, and Chrysler, which furnished four of its brand-new 426 hemis. Mobil Oil, which helped finance a number of Bonneville efforts over the years, also became involved. Previous four-engine machines had been configured with two side-by-side pairs. The Summers brothers sought to minimize frontal area by putting the engines in tandem. There were exhaustive wind tunnel tests with models at Cal Tech in Pasadena. When they finished their "Goldenrod" it had only nine square feet of frontal area and a coefficient of drag (Cd) of .117, less than any previous LSR machine.[16] Its configuration was the shape of all things to come in "purist" Bonneville realms, from Thompson's "Autolite Special" of 1968 (his last LSR venture, abandoned when one of his primary sponsors, Ford, backed away from such involvements), to the downscaled machines of Al Teague and Nolan White, both of which were finally closing in on the Summers brothers' record in 1990.[17]

Bob Summers had been a machinist in the aerospace industry, and he

On the salt flats in 1964, Walt Arfons peers in at Tom Green, driver of Arfons's "Wingfoot Express," powered by a Westinghouse J-46 jet engine. Green briefly held the land speed record at 413 miles per hour. (Photograph by Fred Wohlfarth)

The Summers brothers' four-engine "Goldenrod," posed at Bonneville for a publicity photo in 1965. This machine was configured like a slingshot dragster but with a far superior Cd. It would hold the land speed record for piston engine power, 409.277, into the 1990s. (Photograph by Fred Wohlfarth)

and Bill went into business together manufacturing drivetrain components, gears and axles. They had plenty of drag racers as customers, and over the years they must have winced at most of the efforts at streamlining. The popular magazines were full of articles with titles such as "New Concepts for Fuelers," and the concepts were often unsound on the face of it. William Moore, who had designed Breedlove's dragster, was enthusiastic about "streamlining separate components in individual canoe-like fashion so as not to create a flat horizontal wing."[18] Eyeball aero suggested that this "pod concept" might be viable, but in terms of creating turbulence it could hardly have been worse.

Most of the "new concepts" remained on sketch pads. Between 1964 and 1966, however, a number of dragsters were outfitted with aluminum fairings aft. Nye Frank's twin Chevy-powered "Pulsator" vied with the Chrysler-powered "Scrimaliner" of Ron Scrima, Don Alderson, and George Bacilek for various "Best Appearing Car" awards. Both were visually striking, and both ran quicker and faster when stripped of their sleek bodywork. Tommy Ivo gave up on his streamlined "Videoliner" after encountering problems of high-speed instability; he thought the slicks were "floating" on smoke and bits of rubber that would ordinarily have been blown away by zoom headers. Pete Ogden built a similar car for a customer in Sacramento, and it had similar problems. Indeed, the problems were much the same as Tony Nancy had reported in 1964. "There are . . . a lot of guys who can sketch up a pretty body shell," said Jack Chrisman a few years later, "but few who have the aerodynamic savvy to make them efficient."[19]

The Logghe brothers of Detroit, Ron and Gene, were involved in two well-publicized failures. They overcame the problem of excessive weight in a machine they built for Roy Steffey and Maynard Rupp, which came in at under 1,000 pounds with an unblown Chevy. But driver Rupp reported that, like others of its ilk, it handled poorly on the top end. Steffey had an idea of the cause, "a low-pressure area present along each side of the body, producing an unbalanced force."[20] Yes, the body could have been redesigned, but the readier fix was simply to remove it. Later, enthusiastic Ford engineers conceived something similar. The Logghe brothers built the "Super Mustang" in 1966, Connie Kalitta tested it, and Tom McEwen was signed to drive. But performance was inadequate in all senses of the term, and it was reportedly consigned "to a stall in Dearborn next to the inventory of unsold Edsels."[21]

After concluding a series of interviews in the summer of 1965, *Hot Rod*'s Bob Greene came away with the impression that streamlining was a touchy topic. "Those who have tried it," he said, "have not met with overwhelming success and those who have ignored it are reluctant to go the consequences and surely sacrifice not only the cost of an experimental car but their regular cut of the weekly win booty in the bargain."[22] Aside from the immediate drawbacks there was confusion about fundamental principles.

The Steffey-Rupp streamliner from Michigan, like Breedlove's on the West Coast, had a K-form design aft, and it also had ample provisions for venting tire smoke. It is seen here being prepped for a run at the 1964 NHRA Nationals. (Photograph by Bob D'Olivo, courtesy Petersen Publishing Co.)

Jocko thought that the center of gravity and the center of aerodynamic pressure should be as close to the same point as possible. But the Super Mustang had the engine located more than four feet out, which put the center of pressure eighteen inches behind the center of gravity. That was actually an advantage, said Ford's designers, because it would increase high-speed directional stability "much as the heavy weight of a throwing dart . . . allows it to fly straight and true." Racers were fond of such allusions, but the Super Mustang was afflicted with handling problems, too, even though it had purportedly come "slipping out of the wind tunnel." And who knew whether there was any viable analogy between a dart and a dragster?[23]

*J*ocko was a hero to Terry Cook, for the same reason that Pete Robinson was: Jocko was a thinker, a doer, not a mindless follower. When Cook arrived in

"Wynns Liner" is seen here prior to a pass by Don Cook at Lakeland, Florida; T. C. Lemons has just set down a bottle of bleach, used to enhance traction. The machine had raised hopes for the emergence of a new dragster paradigm but in the end proved to be an abject failure. (Photograph by Jeff Tinsley, courtesy Museum of Drag Racing)

California from the East his first stop was Jocko's Porting Service in North Long Beach; of all the hundreds of specialty shops in the Los Angeles area he felt sure this would be the most exciting.[24] Naturally, he was disappointed when Jocko's collaboration with Woody unraveled, but when he next addressed the subject of streamlining it was with renewed faith. It was 1971. Garlits's mid-engine revolution was a fait accompli. Larry Shinoda had concluded years of painstaking design and unveiled a model of a monocoque (i.e., unitized) mid-engine streamliner. Doug Kruse had a full-scale streamliner nearing completion. Cook noted how well the new paradigm seemed to lend itself to streamlining.[25] And Cook had not even heard what, to him, would have been the best news of all: Jocko had made a deal with another thinker and doer. Together they were going to create a fully streamlined dragster.

While Jocko was still laminating fiberglass, Garlits had decided to build a mid-engine car that was much shorter than normal in order to get the feel of such a machine. The streamliner was to have a 145-inch wheelbase, the interim chassis was 175 inches. Garlits first tested it at Bristol International Speedway in Tennessee, flagship track of the fledgling International Hot Rod Association. Handling was poor and its 6.80 ET was far off the pace set by an up-and-coming driver from Mississippi, Clayton Harris. After making some changes back at his shop, Garlits towed cross-country to Orange County. Again he encountered handling problems, and the shorty went into early retirement; with the streamliner Garlits hoped he could count on aerodynamics for better stability. Jocko wound up his work in late 1972 and headed home. Finally, the next year, Garlits and Swingle finished the new machine, dubbed "Wynns Liner."[26]

The debut was slated for an AHRA Grand American at Orange County. Garlits elected not to drive, which should have sent some sort of a signal. Instead, he hired a local journeyman named Butch Maas. Maas qualified "on the bump" (i.e., at the tail end) of a thirty-two-car field with an aborted run. In the first round Garlits claimed the Liner would not fire. Maybe it would have fired; what it would not do was go straight. There was one more try. While Garlits was competing at a Grand American event in northern California, T.C. Lemons took the Liner to an IHRA meet in Lakeland, Florida. This time Don Cook drove. Cook had steel nerves and a lot of skill, but he could not get it to go straight either. Garlits later revealed that he had made a couple of midweek passes in the Liner and "vowed to never drive it again."[27]

Spooky handling was of course endemic to short mid-engine machines. Yet some of the racers who were experimenting with mid-engine funny cars could at least get down the track, and Jazzy Nelson had been able to do so in Jocko's original streamliner, even with a ninety-nine-inch wheelbase, in the 1950s. Garlits's mid-engine prototype of 1970 had handling problems at first, yet he doggedly worked them out. Why, then, did not he try harder this time? The answer seems clear: The streamliner had one problem that he knew was beyond him. Jocko had estimated that the shell would weigh 250 pounds. Actually it weighed more than 400, and, by the rule of thumb, this would have meant giving away four-tenths of a second. Nobody could afford that. Garlits simply dropped the project cold, selling the chassis to Russell Mendez, who envisioned running it with rocket power. He wrote off something like fifteen thousand dollars in R&D costs.

Meantime, history had been repeating itself. While Jocko was doing things his own way, others were trying partial streamlining. Within a few weeks after Garlits first appeared on the West Coast with his middie, metalsmith Tom Hanna was advertising the advantages of a design faired up to the front of the slicks, with everything else pretty much in the open, a design termed the "Super Wedge." Hanna, who had done work for Garlits, Ivo, Prudhomme, Don Cook, Roland Leong, Connie Kalitta, and dozens more, had learned much of his craft from George Bosckoff, who made the aluminum skin for Breedlove's LSR machine. Now his expressed aim was to take the mid-engine concept "one step further." He had devised, he said, "an advanced body-chassis design which offers fuel dragster racers a needed breakthrough for lower ET's and higher speeds."[28]

In concert with various chassis makers Hanna ultimately fabricated several wedge-type machines. Customers included the team of Vince Rossi, Tom Lisa, and Bob Corbett, all of whom had been drag racing for twenty years or more, initially as part of a team called the "Spaghetti Benders," which included Lou Baney (Corbett had worked for Baney at Saugus). There were no handling problems, and driver Bill Tidwell turned 240 the fourth time out; at the 1972 Supernationals—the event at which Mike Snively and Don Moody broke into the fives—Danny Ongais drove this machine to the fastest speed ever, 243.34. But elapsed times were only in the 6.20s, and Garlits had turned 240 with something completely conventional.

Among several other machines with similar Hanna bodies there was one other of special note. Byron Blair, who had once campaigned a top-fueler in partnership with Hanna, fabricated the chassis. The owner was Harry Lehman, twenty-nine, an engineer from Alexandria, Virginia, who had a good theoretical understanding of aerodynamics. The cockpit was fully enclosed, as was the engine, save for the fuel injector. The front wheels were discs rather than the all but universal spokes. Dubbed the "American Way," this was a stunning piece of work; one *Drag News* correspondent pronounced it "the most beautiful digger to ever come down the pike."[29] Lehman debuted it at Columbus in 1972, and later it ran 230s at Englishtown, New Jersey. The car was not overweight, directional stability was good, and the potential seemed excellent. In April 1974, however, the American Way was destroyed in a top-end collision at Maple Grove Raceway in Pennsylvania. Sometimes, a racer who lost a car in a crash would immediately field another, even if it meant going in a deep hole financially. But Lehman was not prepared to do that, and the potential of a car designed like his would remain an enigma.

The enigma was not resolved, even though others played variations on the same wedge theme. One was John Buttera, a chassis man who had started in business in Kenosha, Wisconsin, and then moved to southern California, working for Mickey Thompson at first, then opening his own shop in Cerritos. He built his streamliner in concert with Louie Teckenoff, a metalsmith who also had worked for Thompson, and sold it to Barry Setzer, a double-knit mogul from North Carolina. Pat Foster had been running a Buttera funny car for Setzer with considerable success, and everyone anticipated similar success with the dragster. But it was a total disappointment. Although Buttera was a marvelous craftsman, his creation had a basic flaw. With its monocoque design—the entire structure was .050-gauge magnesium sheet riveted together—it was too stiff to get hold of the track and never made a competitive outing.[30]

Beyond the wedge bodies there was speculation about how dragsters might borrow aerodynamic tricks from the sophisticated world of formula racing.[31] And there were several machines whose design showed a lot of Can Am influence, notably one built for Prudhomme by Buttera in 1971 and

In full streamlined dress Harry Lehman's "American Way" is seen here perform-ing at the last drag race ever held at Lions. (Photograph by Steve Reyes)

Topless funny cars, a short-lived fad, would perhaps have appealed most to an old-timer like Bob Sullivan, accustomed to feeling the wind in his face in the cockpit of a dragster. Whether dragster or funny car, Sullivan called his machines by the same name. (Smithsonian Institution collection)

another built in 1974 by Woody Gilmore for Herm Peterson. But the extent of the bodywork conjured up an old bugaboo: too much weight. Peterson shelved his machine after a few outings. As if on signal, all the excitement over streamlining once again subsided as racers returned to a world in which they were much more comfortable, a world centered on devising means of making more and more horsepower. Between the two experimental epochs of the 1960s and 1970s speeds increased about 40 miles per hour, and elapsed times dropped from the 7.40s to the 5.60s. Virtually none of this progress had anything to do with aerodynamic finesse.

Within a conventional frame of reference, however, the limits were at hand. In 1975, at the Supernationals, Garlits set both ends of the NHRA record, 5.63 at 250 miles per hour. The ET mark would stand for nearly seven years. Racers could indeed make more and more power, but it did not make any difference. The key to further progress lay somewhere else. Somebody remarked that dragsters "had the aerodynamic finesse of a pile driver." But hadn't finesse been tried?

By the mid-1970s the funny car had transcended its disreputable heritage, even though it could never shed its original appellation, despite the efforts of partisans of "fuel coupe." Prudhomme broke into the 5s in 1976, and at major events the best floppers would sometimes exceed the speeds of the dragsters. The key had been making them stable, an arduous process, since it was mostly cut and try. At first the bodies were pretty much stock. As racers turned up the horsepower, they opened a veritable can of worms. Funny cars betrayed a tendency to take off; this was the fate of Tim Woods's initial venture and also Tom McEwen's and Roland Leong's. McEwen came out all right, as did Leong's driver, Larry Reyes, but Doug Cook's driving career was ended (as was Reyes's, in a later accident). Learning the hard way about going airborne, funny car racers ascertained how to employ air dams and spoilers effectively.[32]

Funny cars had a better Cd than dragsters but more frontal area. To decrease it Detroit configurations were stretched and sliced, and, as usual, there were those who thought that, if a lot was good, too much was better. There was a "topless" phase. But that was counterproductive in terms of spectator interest, for much of the funny car's appeal lay in its resemblance to the family car—the phenomenon called product identity, which was central to the popularity of the NASCAR circuit. Commercial considerations dictated that the bodies had to look "right," sort of right, anyway.

Yet stock dimensions could be fudged a great deal, particularly if the "product" had a slippery shape to begin with. The Plymouth Arrow of the mid-1970s was excellent in that regard, and so was the Chevy Monza; it was a Monza that Prudhomme first put into the 5s. But where funny cars really excelled was on the top end, and that was only partly because they had less drag. Their bodies also generated stronger aerodynamic down-force. Tires for fuel cars were now designed to "grow" centrifugally, and this growth, tantamount to a continuously shifting gear ratio, was partly responsible for the speeds both dragsters and funny cars were attaining. But a narrow "footprint" made traction elusive at high speed. Dragster racers would seek to alleviate the problem by means of airfoils. The rules prohibited airfoils on funny cars—too detrimental to product identity—but the funnies were already ahead of the game. Doug Kruse had foreseen this when he designed his full-envelope dragster: The crucial advantage would lie not in diminished drag but, rather, in being able to apply more power down-course without losing traction.

Nearly all aerodynamic considerations in drag racing remained empirical until the 1980s. True, Kalitta had put his dragster in a wind tunnel (for the cover of *Hot Rod*) in the mid-1960s, and a few funny car racers had done so in the 1970s. But wind tunnel time was extremely expensive. Only with the lucrative sponsorships of the 1980s were any racers in a position to buy enough time to do some serious testing. With financial backing from Hawaiian Punch and the expertise of engineers from Dodge, Roland Leong put in long sessions at Lockheed's facility in Marietta, Georgia. Kenny Bernstein and his crew chief, Dale Armstrong, along with a team of Ford engineers spent even more time there. Bernstein was a relative newcomer to the NHRA circuit, but, with sponsorship from Budweiser and nearly a dozen others, he budgeted more money for R&D than any other racer. What emerged from wind tunnel testing were configurations with closer attention to the fine points of aerodynamics throughout, particularly around the wheel wells, the hood scoop, and the skirts.[33]

Some of the reconfiguration was not so subtle. Bernstein appeared at the 1985 Winternationals with a Ford Tempo body that severely tested NHRA's rules, which were vague except with respect to width (the impetus for that being a Duster only thirty inches wide with which Jim King had surprised everyone in 1972). In 1987 Bernstein showed up with a body that had slipped completely away from any ready factory identification. It was purportedly a Buick LeSabre, but the fans dubbed it the "Batcar," and NHRA nixed it. Yet 1988 templates were only a little more conservative. Drivers familiar with both dragsters and funny cars remarked that, despite their short wheelbase, the latter felt more stable in the lights. Though probably mystifying to anybody who still believed that form followed function, there were two wholly distinct forms, each of which entailed trade-offs yet "worked" for the intended function of getting from A to B. The funny car's chief advantage lay in the down-force generated by its full body and spoiler and came in to play in the latter half of a run. The dragster's lay in the long wheelbase, engine location, and static weight distribution, which enabled it to get off the line quicker.

Some people wondered whether the two forms could be merged. There was a flurry of interest in mid-engine funny cars in the early 1970s, but most

In 1985 Roland Leong (left) *observes a session in the Lockheed-Georgia wind tunnel. (Courtesy National Hot Rod Assn.)*

of these machines did not handle well. Ed Shaver, who raced a mid-engine flopper in England with (of all things) a Vauxhall body, remarked that "the car used to enjoy side trips." The same was true of a machine that Roy Fjastad built for Robert Contorelli of Long Beach. Yet Gilmore built a similar car for Jim Dunn which seemed to "work" fine, took the award for best engineering at the 1972 NHRA Winternationals, and capped the 1972 season with a win at the Supernationals.[34] Perhaps there was more to the human element. Contorelli was only a novice, whereas Dunn had been drag racing for two decades, driving slingshots in the latter 1960s and before that a short-wheelbase fuel altered. The day Garlits debuted his ill-fated Wynns Liner at Orange County, Dunn qualified in top-fuel using his funny car chassis fitted with Doug Kruse's body and ran well enough to win the first round. Pulling quick body swaps between rounds, he took the same chassis to the third round of funny car competition. In the hands of a capable driver, at least, a mid-engine funny car could be competitive. Yet the hybrid design offered no obvious technological advantages. The rules did not stipulate it, but the two paradigms remained quite distinct.

In the 1980s various observers noted how the overall profile of a state-of-the-art funny car was reminiscent of a Jocko-style streamlined dragster.[35] Why, then, were not contemporary dragsters streamlined? Gary Wheeler, an aerodynamicist who worked for Bernstein, explained it this way:

> I'm a proponent of using aerodynamics for downforce but not for streamlining if you have to add a pound of weight to do it. In other words, if you

The richly sponsored Kenny Bernstein won the NHRA's 1987 funny car championship, his third in a row, in this machine. The aerodynamics were outstanding, but nobody could see much resemblance to a Buick or, indeed, to any known automobile. (Courtesy Susan Arnold)

Woody Gilmore (left) poses with Jim Dunn and the mid-engine funny car he built for Dunn. The body is on the ground beyond them; Dunn drove with his face almost up against the windshield. (Smithsonian Institution collection)

In 1964 Joe Schubeck powers off the line at a strip in San Diego County. At this point he is counting on the airfoil to prevent the front end from coming up any higher. (Smithsonian Institution collection)

At Pomona in 1963 Emery Cook adjusts the wing on Don Garlits's "Swamp Rat V." (Photograph by E. Pat Brollier, courtesy Petersen Publishing Co.)

When this car was switched from gas to nitro it would skate to and fro. "You should have a wing on there," said Al Barnes, a machinist who worked for Howard Johansen. The wing, not unlike the style later featured by "World-of-Outlaws" sprint cars, remained a fixture throughout the latter 1960s. "Wild Willie" Borsch was one of drag racing's premier performers; note his nonchalant one-handed driving style. (Photograph by Jere Alhadeff, courtesy National Hot Rod Assn.)

> have to have a body, like in a Funny Car, it naturally makes more sense to make it run with low drag; but if you have to add a body to a dragster (to run with low drag), that adds weight to it. I think you are barking up the wrong tree. . . . Drag racing aerodynamics should be thought of in terms of increasing downforce and not in reducing drag.[36]

Increasing down-force and not reducing drag. As it turned out, there would be yet another wave of experimentation with reduced drag for dragsters, but the more important discovery was that, by means of airfoils, down-force could be multiplied almost infinitely.

Airfoils on dragsters had a long history, of course, and even a prehistory involving inclined plates and panels; the Howard Cam Bear ran a stark sheet of plywood in front of the engines. About this same time racers started exploring the possibilities of exploiting the Bernoulli effect, that is, inducing a low-pressure area by means of an inverted wing. When a slingshot was in optimum dynamic balance the front wheels were just on the verge of

Leland Kolb was one of several enthusiasts who followed Don Prudhomme's lead in trying a body configured like this, both for streamlining and to gain down-force. The scene here is Orange County International Raceway in the summer of 1971; note the slingshot, by then a threatened species, in the opposite lane. (Photograph by Alan Earman, courtesy National Hot Rod Assn.)

By the time of the NHRA Supernationals in the fall of 1972 Kolb had pared back to this setup, with wings called canards (following aircraft terminology) in front of the tires; the aims were the same, the weight considerably less. Buster Couch, longtime chief starter for the NHRA, watches, Christmas tree control in hand. (Photograph by Steve Green, courtesy Petersen Publishing Co.)

dancing. On the top end, anyway, an airfoil was just as effective at keeping the front wheels down as weights fastened to the axle and certainly more effective than any flat surface. A firm in Michigan tooled up to produce aluminum airfoils commercially, and Cleveland fabricator Joe Schubeck made these available on all his new chassis.[37]

Meantime, Garlits had showed up at the 1963 Winternationals (the first major NHRA event to permit fuel since 1956) with a fiberglass wing set up on struts above his engine. The idea was Bruce Crower's; the aim was to exert down-force on all four wheels and enhance stability. Tony Nancy equipped his mid-engine Wedge with a similar wing, and so did Jim Harrell and Willie Borsch, who campaigned a short-wheelbase altered roadster. Contemplated in retrospect, it was reminiscent of Douglas Davis testing an airplane wing by cantilevering a section above a large Packard and racing through the southern California outback.[38]

Garlits said that his airfoil helped him at a slick spot on the track, where others were freewheeling. He won the Winternationals and even set a national record. When the setup later began showing signs of structural fatigue, however, he simply removed it. With a few notable exceptions—Borsch campaigned his "Winged Express" for many years—a device of this sort was rarely seen in the latter 1960s, largely because of the discovery of "free" down-force available from zoom headers.

Wings aft occasionally appeared on latter-day slingshots, and they soon became part of the conventional mid-engine repertoire, two or three feet above the slicks. After giving up on a Can-Am body configuration, Leland Kolb tried mounting four small wings, two behind each tire and two canards in front. Canards had first been tried by John Wiebe on a slingshot, Garlits had them when he turned his 5.63, and they later showed up even on funny cars. NHRA wrote them out of the funny car rules, along with airfoils in general. Funny cars were not permitted to have any aerodynamic device that would allow air to pass underneath (and thus create a vacuum), and, ultimately, that was why dragsters would keep their edge.

Following the experimental epoch of the early 1970s, dragsters settled down to a combination that was pretty routine, though not entirely. In front there was sometimes an airfoil, sometimes a flat panel over the axle, sometimes both, sometimes neither; weights could always be substituted, and that was as good a place as any to put weight if a car was shy of NHRA's minimum. Aft there was always a wing. Though variable, the angle of attack was often quite steep. Profiles varied as well, and occasionally somebody tried something "trick" like dual elements or vortex generators. Once in a while someone ran a wing way up high, but the advantage of doing that did not become obvious until later.

In the early 1980s Garlits was absent from the NHRA circuit, spending a lot of time preparing to open his Museum of Drag Racing in Ocala. The strongest top-fuel competitor was Shirley Muldowney, followed closely by

Gary Beck. Beck had won the first major event he entered as a driver, the 1972 NHRA Nationals, then he repeated in 1973, and he had been a major force ever since. Muldowney won the NHRA championship in 1977, in 1980, and again in 1982. That was the year Joe Amato stepped up to top fuel after competing in the alcohol ranks for many years. He brought along a capable crew chief (the expression was becoming commonplace), Tim Richards. In 1983 Amato and Richards finished second to Beck for the NHRA championship (Muldowney slipped to fourth), then in 1984 they defeated Beck. Along the way they captured the NHRA speed record. During qualifying at the Gatornationals in Florida Amato turned identical 257.87s four times and 259 twice. On race day he turned 260 and 262 in the final two rounds. Later that year Amato went almost 265; the record a year before had been 257, and the fastest Amato had ever previously run was 254.

The only thing noticeably different about his car when Amato showed up in Gainesville was the wing, designed by Eldon Rasmussen, who regularly occupied his time with Indy cars. It was out behind the plane of the rear tires and seven feet above the ground. The height put it beyond turbulence created by the engine and tires. The distance behind made it function like a lever, capable of exerting tremendous down-force even with a shallow angle of attack and thus minimal drag. "All these were modifications aerodynamicists would have suggested years ago," wrote John Brasseaux in *National Dragster,* "if only someone had asked."[39]

Most of the other racers installed a "Kareem Abdul" wing at once. When Garlits showed up at the NHRA Nationals in September, his first time at IRP in several years, he brought a conventional setup not much higher than the injectors, and he brought a tall wing. The former was put away in his trailer after one qualifying pass. Garlits closed out the 1984 season strongly, then built a new car for 1985, Swamp Rat XXIX. On his way to Pomona for the Winternationals he stopped in Arizona for a match race with Beck. In the third round he clocked a track record 5.45, 259 miles per hour, then one of the wing struts collapsed. He lost control and crashed in a muddy field. Garlits was safe, and the car could be fixed, but there would be many other failures of this sort as, without much finesse, racers tapped what was essentially a newfound source of power. Indeed, there had been similar failures long before, when down-force had been measured in the hundreds of pounds rather than thousands. As far back as 1972, one commentator suggested that SEMA and "some people who specialize in aerodynamics" should "look into the wings, material, position, etc. and curtail the rash of breakage of wings and struts."[40] Some improvement was made, but there were still accidents precipitated by failures resulting from all that down-force, particularly after 1984.

*T*he tall wing opened the door to performances that would keep the dragsters three-tenths of a second and about 10 miles per hour ahead of the funny cars. Funny cars were restricted to a spoiler, a device that did not permit the airflow to pass underneath. True, a spoiler could help a lot; on the very day in 1984 on which Amato broke 260 miles per hour Kenny Bernstein did so, too. But an airfoil had tremendous potential for multiplying down-force, whereas even a huge spoiler, a so-called whale tail, had limits. Excited by the realization that a major performance breakthrough was still attainable, dragster racers again began chasing off after streamlining. This time most of them realized that, as Gary Ormsby put it, "you need help from someone." Ormsby, a northern California car dealer, had his crew chief, Lee Beard, solicit help from Nigel Bennett, the designer of the Lola. Bennett ran a model of Ormsby's 1984 machine through Lola's wind tunnel and suggested a variety of little tricks that showed up on his 1985 model, such as panels flaring outward in front of the slicks. But Beard and Ormsby had something more radical in mind.[41]

In 1986 they outfitted their machine with a full bellypan and engine enclosure as well as much more sweeping flares in front of the rear tires. Eyeball aero? Not this time. Everything had been designed through computer simulation by an aerodynamicist, Pete Swingler. The fairings had been fabricated in a lightweight composite material (carbon fiber and Kevlar) by Eloisa Garza, and the mounting had been done by Jackie Howerton, both of whom worked mostly on Indy and Formula 1 cars.[42] After a disappointing debut at Pomona Ormsby went on to Gainesville and turned 263, two miles per hour better than the car had ever run before. At that same event Garlits debuted Swamp Rat XXX, which also had a couple of aerodynamic tricks, albeit less dramatic than Ormsby's. Garlits had simply enclosed the cockpit with a Lexan canopy and faired the *front* wheels, something Roger Huntington had suggested thirty years before: "It would be much better if we could somehow cowl in the two front wheels. This is because the air flow is quite turbulent by the time it gets ⅓ of the way back on the body, and rear end streamlining is thus less effective than cleaning up the nose of the car."[43] All things considered, Ormsby's performance was nothing for him to get too excited about (the computer simulations had predicted he would gain 6 or 7 miles per hour), but Garlits topped 272, ten miles per hour faster than Ormsby, and he also won the event in convincing fashion. Was a new epoch finally at hand? People had thought so twice before.

Ormsby went to only one final round in 1986 (losing to Garlits), and he finished sixth in the NHRA standings, having slipped from fifth in 1985 and fourth in 1984. Nevertheless, he debuted a new streamliner at Atlanta in 1987 which, in addition to everything else, had a front end something like Garlits's. After losing in the first round, he beat a quick retreat: "We went back home and had our . . . car rebuilt to a totally conventional dragster," he said. "We cut off the front end, took off all the aerodynamic body panels and started fresh."[44] Despite the advent of composites, Ormsby's machines

Joe Amato, who pioneered the "Kareem Abdul" wing in 1984, is seen here doing a burnout at Pomona with his 1988 version. Virtually every other top-fueler had an identical setup. (Courtesy Joe Amato)

were too heavy. Others tried setups similar to Garlits's, including Darrell Gwynn, a popular twenty-five-year-old who had hit the top-fuel scene in 1985, finishing sixth in the points, then finishing second to Garlits in his second year. Gwynn quickly reverted to a more conventional combination, as did Amato and Gene Snow (who had switched from funny cars to dragsters in 1980). Still others who had announced plans for streamliners quietly dropped them. The firm that produced Garlits's Lexan canopy took orders for two or three dozen more, but most of these had been removed by the end of the 1987 season.

A canopy weighed 30 pounds, and the racers were finding that they needed to keep all such components to a minimum in order to accommodate a variety of new engine setups like solid billet cylinder heads (no water passages), dual magnetos, dual fuel pumps, and four- and five-disc clutches. It was the same old story. Wings enabled turning up the horsepower. "If you are making a lot of horsepower, you'll go fast," said Amato. "All aerodynamics are worth something, but they also cost something."[45] "We can never select the one result we want to the exclusion of all others," is how David Pye would have put it.

But what about Garlits? Despite a mid-season accident, Swamp Rat XXX had won the NHRA championship in 1986. Unquestionably, it was a suc-

Gary Ormsby debuted his 1986 machine in conjunction with sponsorship from Castrol GTX. Besides the streamlining around the cockpit and aft, note how far back the wing has been located.

Cockpit canopies had been seen on dragsters in the 1950s; fairing the front wheels rather than the slicks was a concept that had been suggested but never tried. Before Don Garlits retired this machine in 1987 it ran faster than 275 miles per hour. (Photograph by Francis C. Butler)

An inside joke. After having flipped over backwards, Rhode Islander Jim King rebuilt his slingshot and equipped it with a caster atop the roll bar, a precaution in the event of a recurrence. (Photograph by Ed Sarkisian)

cessful design. Garlits declared that "the way to go fast is to punch a nice, clean hole in the atmosphere and drive through it."[46] There were, however, other ways to explain its success. Patrick Hale, an Arizona aerospace engineer who had devised a computer program for analyzing dragster performance, suggested that Garlits's aerodynamic devices actually "contributed very little to his dramatic performance gains." Rather, the key was his relocation of the fuel tank ahead of the front axle, thus altering the weight distribution without any basic reconfiguration. This permitted him to run a deeper low-gear ratio and tighten up the clutch, thereby transferring more power to the ground. "All the aerodynamic hoopla is just a cover the clever 'Old Man' has contrived to keep the competition off what he is really doing," Hale concluded. "He has done this type of thing repeatedly in the past."[47]

With Swamp Rat XXX essentially unchanged, Garlits kicked off the 1987 season auspiciously with a win at the Winternationals. But then his fortunes went into a steep decline. There were races at which he could not even qualify. Looking around, he was the first to concede that Gwynn and Amato were running better than he was "with the front wheels just hanging out there in the air." How much horsepower could be wrung out of a KB or JP-1? 3,000? 4,000? "A lot more," said Gene Snow. "I've got so much work to do on the engine," Garlits snapped to a reporter in May, "who's got time to mess with the body?"[48] Actually, Garlits was just being cagey. Even then he was working on Swamp Rat XXXI, a subtly refined version of its predecessor. He debuted it in Minnesota in August, then went on to an appearance in Spokane. There it flipped over in the lights, Garlits's third major accident in three years. Again he came through safely. But at the age of fifty-five, and more than thirty years after turning pro, he retired.[49]

By the end of the 1980s dragster speeds were in the 290s, and more than a dozen cars had turned elapsed times in the 4.90s. Indeed, fans who went to races on the NHRA's championship circuit or watched them on television (they were all on television) *expected* such performances. There were no streamliners that were even close to being competitive. But a new combination was standard: Chassis were close to 300 inches long, and they were very flexible. And there were those cantilevered wings that multiplied the down-force as a car gathered speed and kept the rear tires glued to the track throughout the entire run.

But not always the front tires. The wings were essential to stability, Nye Frank had observed not long after the mid-engine paradigm shift; they were essential so "the cars won't fly."[50] It did not always work out that way. Over the years there had been occasional incidents in which dragsters did wheelstands coming off the starting line then went all the way over backwards. Now, however, there was a new phenomenon: a backflip far downcourse, at terrific speed. This happened to Garlits in each of his final two Swamp Rats, and there were those who said it was because his frontal fairings were aerodynamically incorrect. "Curved on top and flat on the bottom," observed Harry Lehman. "On an airplane that would be a lifting surface."[51] But a number of others had the same problem with "blowovers" (the term was borrowed from hydroplane racing), even without any tricky nosepieces; Prudhomme, who switched back to dragsters in 1990, had it happen to him twice in just a few months. As with the ground-effects experiments of Pete Robinson and others, the racers were again "combining and integrating, in various combinations, elements selected from a group of basic mechanisms and forms."[52] What they were really doing, according to Dave Uyehara, was "playing with fire." Uyehara was one of the two foremost chassis fabricators of the latter 1980s. The other one, Al Swindahl, elaborated: "Garlits took two [dragsters] over backwards because of big wings, in my opinion. . . . Big wings that are too far backward are like a big lever. It pries the front end up, and once it's up, it's gone! There were some wings that were really huge and some crew chiefs who wanted them even further back."[53] When the front end came up, the center of gravity shifted, the wings assumed an entirely different attitude, and, all too often, the whole car was "gone."

Less spectacular than a blowover, but more commonplace and just as dangerous, were structural failures induced by thousands of pounds of down-force. Racers seemed to take this danger all too lightly. A writer doing a feature for *Time* in 1986 was amazed to see Kalitta repairing a "wrinkled" wing strut by throwing his weight against a two-by-four.[54] Four years later Kalitta was seriously injured when a strut buckled and sent him crashing into a guard wall. The same year as Kalitta's first mishap, 1986, young Darrell Gwynn had one too, at high speed, but he kept control and came to a safe stop. Then, again like Kalitta, he had another structural failure in 1990. He was making an appearance in England, at Santa Pod, not far from London. It was not a wing strut per se which gave way; rather, the down-force contributed to his chassis breaking in two just behind the cockpit at a speed well over 200 miles per hour. Gwynn came home to Miami alive, but he would never be the same. Even when they crashed at terrific speed the cars had seemed so safe. For his competitors and millions of drag racing fans in the United States what happened to Darrell Gwynn came as a terrible shock.

11

TELEVISION

The sponsorship of our sport over the past 15 years by the R. J. Reynolds Tobacco Company has been greatly responsible for the growth the sport has enjoyed.

DALLAS GARDNER, 1990

Darrell Gwynn was gravely injured in England on Easter Sunday, 1990. Videotapes of the crash were being shown in the United States the next evening, and in the days following there were frequent TV updates on Gwynn's condition from Stoke Mandeville Hospital in Aylesbury. When Leonard Harris was killed, and, later, Mike Sorokin, John Mulligan, and Pete Robinson, they may have been topflight competitors, but they were not celebrities, and accounts of their fate remained sketchy. It was a little different with four-time pro-stock champion Lee Shepherd, who died in a testing accident at Ardmore, Oklahoma, in 1985. It was different with Herb Parks, who had done two long stints as crew chief for Don Garlits, killed in a starting line mishap at Bradenton, Florida, in 1988. Both of them had been seen and heard many times on television. But nothing like Darrell.

Wally Parks was fond of saying that the real stars in drag racing were the machines, in all their glorious variety. Perhaps this had once been true, but now they had a human context; TV was responsible for that. Darrell Gwynn was a TV star. He had a fan club. His manner was familiar to millions, who saw him do countless interviews on shows produced by Diamond P Sports.

Just before every run Darrell Gwynn ever made, his father, Jerry, wished him luck by taking his hand. This photo precedes Gwynn's beer sponsorships, first Budweiser then Coors. (Courtesy Darrell Gwynn)

Even though the championship had eluded him, he had won twenty-five major NHRA events and was instantly recognizable behind Steve Evans's microphone. Viewers had seen him flushed with joy after victory at the 1990 Gatornationals before a partisan Florida crowd. Perhaps they saw him after he lost the final round of the Winston Invitational in Rockingham, North Carolina, on April 8. "At least I'm safe," he remarked. "At least I didn't crash." The crushing irony there was that Diamond P's production of this event did not air until late in April, and Gwynn had crashed on the fifteenth.

Gwynn was brash but amiable and, for a top-fuel driver, exceptionally youthful. He was twenty-eight; Joe Amato, who narrowly defeated him at Rockingham—and for the 1989 championship—was forty-six, and everybody else among the front-runners was even older than that. Eddie Hill and Gene Snow were close to twice his age, and Don Garlits *was* twice his age. Shirley Muldowney's son John was older than Darrell.

Muldowney had been involved in a devastating crash six years before and went through a long period of recuperation and reconstructive surgery, but she had been back on the NHRA tour since 1986. Garlits, Snow, and Hill, Connie Kalitta, Gary Ormsby, and Jimmy Nix—all of them had spectacular mishaps that were played and replayed on television. Only Kalitta was hospitalized, and his life was not threatened. And Darrell Gwynn did not lose his life. Rather, he lost his left arm at the elbow, and his spine was crushed at the fifth cervical vertebra.

Because of TV, millions of people had a feeling that they knew Darrell and his family, the three of them. They knew that Pat Garlits had been to his mother's, Joan's, baby shower in 1961. That Darrell's dad, Jerry, had been a racer since the 1960s and that they had started taking Darrell along "because it was easier than finding a babysitter." That Darrell himself had begun driving in 1980, alcohol machines first, then a top-fueler. Over and over they had seen Jerry reach into the cockpit and shake his hand just before every run, a set ritual. They had seen Garlits's attitude toward "The Kid" change from avuncular to icy as he became a stronger threat. They had heard Darrell tell how he put his overweight streamliner on a diet to regain an edge "and even lost seven or eight pounds [him]self." They had heard about the ins and outs of his sponsorship deals. They had heard him excitedly describe his arrangements to run exhibitions at "the Pod" in England, to match race Liv Berstad, a Norwegian woman his own age. They had shared Eddie and Ercie Hill's enjoyment at "watching Darrell grow up."[1]

They knew of the ironies: his involvement with Nick Buoniconti's Miami Project to Cure Paralysis; his mother's activism in the Drag Racing Association of Women (DRAW), which had been founded specifically to marshal moral and monetary support for injured drivers and their families. On television they watched him returned home to Miami on a British Airways jet and heard physicians describe the grim realities of a C-5 crush injury. They saw his girlfriend, Lisa Hurst, wheel him into a press conference, and they heard Gwynn himself say that "something like this really changes your priorities." It was all on television.

Of all the technological transformations that had swept through drag racing, the continually evolving theater of machines, none was ultimately as significant as television. Aside from the specialized periodicals and an occasional spot on a weekend TV potpourri such as ABC's "Wide World of Sports," drag racing had once had virtually no exposure beyond the strips themselves, places with a slightly unsavory air. Now the strips were nicer places to go to, some of them anyway. But, more important, people could bring drag racing right into their own living rooms, several times a week if they wished, and a racer like Darrell Gwynn could seem as familiar as the guy next door. The roots of this transformation lay in a federal law passed on March 19, 1970, about two weeks after Garlits's transmission disintegrated on the Lions starting line, dislodging a paradigm. It was HR 6543, PL 91-222, which went into effect on January 2, 1971, shortly after Garlits debuted his mid-engine prototype. Garlits aimed to vanquish a situation he perceived as hazardous. PL 91-222 was likewise aimed at a hazard. It made it unlawful to advertise cigarettes on the air.[2]

Tobacco companies had been the biggest of all TV advertisers. The $194.1 million they spent in 1966 was $65.3 million more than what was spent by automakers, who were second.[3] The cutoff date, a concession to both broadcasters and tobacco companies, permitted one last barrage during the bowl games. Or, rather, that was the cutoff for explicit pitches. The tobacco companies had been hard hit by antismoking announcements on TV, and when the ads went off so did these. But what about all the hundreds of events that might just *happen* to have commercial signage as a backdrop? There was a big loophole, and the tobacco companies were already looking to the likes of the Virginia Slims tennis tour, the Camel GT Series, the Winston Rodeo Series, and the Winston Cup in NASCAR. And Winston drag racing. Dallas Gardner, who succeeded Wally Parks as president of the NHRA in 1984, said it as well as anyone: Drag racing took off into the big time because of RJR, the R. J. Reynolds Tobacco Company.[4]

*D*rag racing was in a state of institutional flux in the early 1970s. Though Parks and NHRA were as powerful as ever, there were incessant accusations on the order of Scotty Fenn's—that NHRA milked exorbitant profits from its big events while scrimping on purses, that it played favorites, that there was a "party line" from which no deviation was ever permitted.[5] NHRA countered that there were real and substantial dangers, which required an iron fist, but its autocratic posture only lent encouragement to would-be rivals. Some came and went with scarcely a ripple. UDRA survived only as a sort of booking agency for strips in the Midwest. Even NASCAR, which had reinvigorated its drag racing division in 1965 and staged its own championship series under the direction of Walt Mentzer, had departed the scene.[6] Much more tenacious, however, was the AHRA, an organization that Mentzer had headed back in the 1950s, back when it began treating drag racing performances as a cash commodity.

In 1964 Parks had finally relaxed his stand against purses, but he remained adamantly opposed to any sort of guarantees such as Jim Tice started posting in 1970, when he seeded competitors for the AHRA's Grand American Series. NHRA's house organ, *National Dragster,* would ridicule AHRA as the "Absurd Hot Rod Association" and even friendly critics had to concede that skimpy staffing deprived its events of "the Teutonic efficiency of NHRA's."[7] One could waste hours in a gridlocked parking lot after an AHRA event, even though none of these drew crowds anything like the NHRA's Nationals and Winternationals, or even its newer Springnationals and World Finals. But the Grand American Series, with the likes of Big Daddy and the Greek as booked-in (and heavily advertised) performers, clearly gave AHRA a powerful new hand.

Proposals that there be a commissioner of drag racing to mediate between factions—suggested candidates had included Lou Baney, Mickey Thompson, and C. J. Hart—no longer seemed superfluous. And conflicting claims to "world championships" would get even more bewildering. In the fall of 1970 Larry Carrier, who had opened Bristol International Speedway in Tennessee under NHRA sanction in 1965, then switched to AHRA, formed the International Hot Rod Association, and within a year the IHRA claimed to have more than fifty other tracks in the fold. (NHRA claimed 163 in 1970, AHRA 95.) Parks dismissed the IHRA as inconsequential, and Tice decried "fragmentation," but Carrier's organization seemed to have a lot of vitality.[8] Nor was he the only one who aimed, by one means or another, to grab power away from the establishment. People like Gil Kohn and Ben Christ could have their own ideas about "the good of the sport," too. Each man controlled several tracks, and both of them established umbrella organizations—Kohn's called the United Hot Rod Association, Christ's the National Association of Motor Sports.

Just as Kohn had the biggest operation in the New York metropolitan area, New York National Speedway in Suffolk County (the manager was Ed Eaton, the former NHRA bigwig, who hailed from this area), Christ owned the major strip in the Chicago area, U.S. 30 in Gary, Indiana. He also operated the Gold Agency, headquartered in Evanston and managed by a man named Ira Litchey. Litchey handled bookings for dozens of performers and had gotten the Coca-Cola Company deeply involved in drag racing through sponsorship of the "Funny Car Cavalcade." Because deals such as this could enhance the bargaining power of individual racers, Mike Doherty declared Christ to be "more of a 'kingmaker' than anyone else in drag racing."[9] There were others of the same ilk, however, such as Ed Rachanski, who ran a Chicago booking agency and dreamed up the "Pro Drag Racer Championship Trail." On the West Coast people like Don Rackemann (an old friend of Lou Baney's) and Carl Schiefer (Paul Schiefer's son) had also set up agencies to offer "professional sports management," and this would inevitably lead to further disputes over power.

One of Schiefer's "Action/America" clients was Don Garlits, who was at odds with Wally Parks more often than not. In 1972 Garlits forced a direct confrontation. His vehicle was the Professional Racers Association (PRA), which he brought to life during a protracted rain delay at an AHRA Grand American. Garlits was voted president, Gene Snow and another funny car racer from Texas, Mart Higgenbotham, became vice president and secretary. The treasurer was Steve Carbone, who had cut his teeth at places like Lions in the 1960s, then won the NHRA Nationals in 1971. That was supposed to be the big time, yet Carbone's purse amounted to only $6,100, less than payouts years back at Famoso or at Doug Kruse's Professional Dragster Championships, or even by NHRA at the Indianapolis Raceway Park. Wally Parks saw the prestige of winning the Nationals as more important than the purse, and many winners had indeed reported that a victory boosted their income from match races and their appeal to sponsors. But Garlits had his own opinion. While touring Vietnam with a group of drivers that included Richard Petty, he had listened closely to talk about payouts on the NASCAR trail. NHRA, Garlits said, simply had the racers running scared: "They're the most powerful organization in drag racing and they can hurt you with their 'National Dragster' and by putting pressure on sponsors. In return, the sponsors put pressure on the men. And so, because of this, the racers are afraid of them and so they run over [to Indianapolis] and run for them because of fear."[10]

Garlits sent inquiries asking whether racers would skip the 1972 NHRA Nationals to participate in a meet offering twenty-five thousand dollars to the winner of each of the pro classes, with about the same amount to be divvied up for round money. The format would be similar to Kruse's (and to what NHRA would try with the Supernationals); there would be none of the stockers and other slower classes that competed at all other events—just top fuel, funny car, and pro stock. "For those of you who have been pressured by the powers [that] be," Garlits advised, "let me say . . . it's your right as citizens of a free country to race wherever you please." That meant the right to race on Labor Day weekend at a strip sanctioned by Garlits's old friend Jim Tice. On the basis of several dozen commitments Garlits got Tice to agree to post $150,000 and supply personnel to run the event, which was to be staged in Tulsa. Tom McEwen thought up the name: "National Challenge '72."[11]

McEwen, Don Moody, and Bill Jenkins emerged the big winners. Though Tice suffered a substantial loss, Garlits felt vindicated: NHRA now knew that a boycott (threatened for years) could actually be sustained, and the top-fuel winner at IRP, Gary Beck, received three times more from NHRA than Carbone had the year before.[12] The PRA also capitulated; there would be a National Challenge '73 but not scheduled to conflict with the NHRA.

Tice suffered another loss the next year. It was not overwhelming, and there was every reason to expect a turnaround in 1974. Some people argued, however, that Oklahoma was not the best conceivable place to stage such an event.

Well, then, how about the Big Apple? Or somewhere close, anyway? Garlits opened negotiations with Kohn and Eaton, and National Challenge '74 was slated for Kohn's track near Center Moriches on Long Island. The metropolitan press ordinarily ignored drag racing. But a big event had never been staged within driving distance of New York City, and enthusiasts like Garlits believed there was at hand an opportunity "to establish drag racing on a par with other major sporting endeavors."[13] PRA changed its name to PRO, Professional Racers Organization. Eaton's publicist arranged to have a city street at Coney Island set off for a preview of racing cars. He lined up the U.S. Navy Rock Band. Press releases went out, and the New York media took some notice. But it was all downhill from there. Tickets for the event were overpriced, the crowd was restive, Eaton's staff was both inadequate and officious, and neither PRO nor track management lived up to its end of the bargain. A squall blew in from the

Atlantic, causing a delay, but actually the entire event was stormy.

Garlits and Eaton argued openly about the interpretation of various rules of competition, particularly the amount of time that would be allowed for maintenance and repairs between rounds. The racers, most of whom had formerly been apathetic, turned combative. In the pits there were confrontations the likes of which had not been seen since the first Bakersfield meet fifteen years before. One hotheaded young driver had to be restrained from physically attacking Eaton. Eaton threatened to send everybody home. In the middle of everything Garlits threw up his hands and quit as PRO president.

During the course of eliminations some of the fuelers had been paired up incorrectly, some had received unwarranted bye runs, and there was utter confusion about which cars were still in the running. Eaton had warned how demanding his audiences were, yet there was a period on Sunday afternoon when not one dragster went down the track for two hours. Finally, as a result of heroic efforts on the part of Ted Cyr's driver and partner, Flip Schofield (a law student), the situation was resolved to some degree. It was not equitable to all, but, as Schofield kept reiterating, it was the only way the event could be completed at all. One of the magazines later declared National Challenge '74 "a strong candidate for the most poorly executed drag race in history."[14] Both the track and PRO did emerge in the black. But, of all those who had last words on the event, Eaton seemed to get most directly to the heart of things: "Actors cannot produce and direct a Broadway show—nor can racers put on a national drag racing event."[15]

Schofield salvaged the organizational records, and PRO was not quite dead. After more disputes with Parks in 1975, Garlits reentered the picture and staged National Challenge '76 at Bob Metzler's Great Lakes Dragaway, outside Milwaukee. Like the 1972 event, it was held at the same time as the NHRA Nationals, but the purse was a mere shadow of Tulsa days. This time a much smaller contingent of racers showed up. Metzler's nickname was "Broadway Bob," but Milwaukee was not New York. Even though Garlits threw all his energies into PRO, he could not make a go of it. It petered out at seedy tracks like Lakeland, Florida, its meets thinly attended, and, compared to NHRA's, woefully deficient "on the esthetic side of things."[16] In the end, there was at least one point upon which Garlits agreed with Eaton, that racers were too fractious ever to produce and direct a show themselves. To be sure, funny car owners staged successful walkouts from time to time—even against the NHRA in 1981—but there was never again as serious a challenge to the NHRA's hegemony as at Tice's Tulsa track in 1972.

Wally Parks and Jim Tice disagreed about a lot of things, but they had some regard for one another. A man who knew them both well said they "were cut from the same cloth . . . at least very similar pieces."[17] Larry Carrier, on the other hand, had no respect for either Parks or Tice, and he kept saying so in public. He did seem to have a lot of horse sense (indeed, he owed his status as a self-made millionaire in some part to horse trading), and he was adept at playing on a "little bit of country" image. Terry Cook described him as "sort of a drag-racing Gary Cooper."[18] But his contentiousness was reminiscent of Scotty Fenn, and there was nobody since Fenn who got so much of Parks's dander. For his part Tice vowed to buck Carrier's events directly, all of them. Yet Parks and Tice were both outmaneuvered. IHRA took control of Dallas Motor Speedway, one of the best-appointed strips in the country and formerly the site of two major NHRA events. It got York, site of one of the AHRA's Grand Americans. People listened when Carrier pledged a fair deal all around: "How the money is divided up in the front office is the important thing. Is it being spent on lush offices and company airplanes? This is where the money in drag racing is going, not back into the tracks or to the racers."[19]

But there was some nasty business. Someone started a rumor that Carrier had Mafia connections. At Dallas Art Arfons crashed his "Super Cyclops" jet, killing a local TV personality who was along for the ride (the machine had a passenger compartment!) and two young track employees. Later two bystanders died, along with the driver, when a rocket crashed. The litigation was interminable. Carrier did garner some nice publicity when Shirley Muldowney scored her first major victory at an IHRA event. Like Tice, Carrier signed racers to annual contracts, and in 1976 he induced Garlits to ink an exclusive. Yet there were always lawsuits, court orders, charges of slander and worse. Carrier seemed to have a flair for antagonizing everyone, racers as well as rival promoters: Muldowney later on and Prudhomme and Beck. And for all Carrier's ambition IHRA remained pretty much a regional (southeastern) outfit, with nowhere near the power of the NHRA—or, in a phrase that a former NHRA publicist named Dave Densmore popularized in the 1980s, the "high sheriffs." When Carrier sold out to Texan Billy Meyer in 1987 NHRA's Gardner remarked: "We have never looked at the IHRA as competition."[20] It was the sort of imperious pronouncement for which the high sheriffs were notorious, but it was essentially accurate.

Weakened by Meyer's miscues but apparently still viable, IHRA carried on into the 1990s under the leadership of Ted Jones, a former Carrier lieutenant. It still had its own championship series, its own periodical. It had television. For the pro contingent the rewards were not the same as at NHRA events, but then the competition was not as stiff either. The AHRA folded in 1984, not long after Jim Tice died from cancer, but the IHRA continued to offer an alternative, something to remind NHRA about the perils of becoming (in Densmore's words) "lackadaisical and inefficient; more pomp than circumstance." That was a key legacy from Carrier, but not nearly the most significant one. Carrier kicked off his association in partnership with a man named Carl Moore, who was in Tennessee state politics.

The twin cornerstones of their operation, in Bristol, Tennessee, and Rockingham, North Carolina, were cheek by jowl with the corporate citadels of the tobacco industry. In 1974 Carrier and Moore cut a deal with R. J. Reynolds Tobacco to underwrite the purse for a series of races and a season-long points competition; Reynolds pledged sixty thousand dollars initially. There was speculation that this "could open doors for other associations to do likewise," well-founded speculation, as it turned out.[21]

However one assesses the deal with RJR in hindsight, it certainly came at an opportune time insofar as many performers and promoters were concerned. Drag racing was in the doldrums. Attendance had taken nosedives before, most recently in 1966, when a good part of the cause was the nation's escalating involvement in Vietnam, but never like in 1974. 1974 will be remembered as the year of the first "energy crisis," the summer when there was no more cheap gasoline. Fearing poor attendance, a Las Vegas hotel canceled its sponsorship of an off-road race while announcing that "it is not in the best interest of our nation . . . to continue to promote a racing event where the energy resources are consumed for more than prime-life necessity."[22] Many drag strips were a considerable distance from any large metropolitan area; where they were close, air and noise pollution controls threatened. Noise complaints were what finally shut down Lions, a drag racing shrine, in 1972.

The pro racers towed back and forth across the country, 40,000, 50,000, or 60,000 miles a year. Not only was gasoline more expensive, so were all routine costs. There was talk, among the racers themselves, about prohibiting nitro to minimize attrition. Many people had a feeling that the 1974 PRO meet was a "crossroads," one that had not been successfully negotiated. Old-timers were talking about quitting. Major sponsorships like Mattel's had been dropped. Don Rackemann's Action Company signed Continental Baking to a funny car sponsorship, and Wonder Bread jumped in with both feet in 1973. It jumped out just as fast after only a year.

True, there were bright spots. Detroit was increasingly enthusiastic about the pro-stock class. Pro stock had its origins with racers such as Bill Jenkins—"Grumpy," he was called—who had never been happy with the metamorphosis of super-stockers into funny cars and with certain funny car racers, such as Don Nicholson, who were not comfortable with either the expense or the danger of fuel. As early as 1967, Nicholson had begun advocating limits on engine displacement and the proscription of nitro. He had met with opposition from strip management, notably from C. J. Hart, who (ironically, in view of his stance a decade earlier) proposed checking to make certain that there *was* nitromethane in the fuel tank of funny cars. But Detroit's abiding touchstone was product identity, and in 1969 a group whose leaders included Jenkins, Nicholson, and the North Carolina team of Ronnie Sox and Buddy Martin formulated a set of rules which stipulated gas-burning engines in stock chassis with stock bodies. The entire concept proved to be box office magic, and, as had happened several times before, Jim Tice took it to his bosom. NHRA inaugurated its pro-stock class not long afterwards.

At first Sox and Martin were dominant. If it was not their Plymouth in the winner's circle, it was Dick Landy's Dodge or somebody else on Chrysler's payroll. Not many spectators drove Chrysler products, however; most of them owned Fords and Chevys. Among the fans Grumpy's Chevys were the most popular of all, and he realized that he was in an ideal position to start taking advantage of the rules. In late 1971 he had a Pennsylvania chassis shop fabricate a tubular frame for a Chevy Vega body. The car was much lighter than the Dodges and Plymouths, but it still had doors that opened, and it still looked like the Vegas out in the parking lot. Despite the custom frame, NHRA approved it. Machines like Grumpy's Vega were soon capable of 9-second, 150-MPH clockings.

In the evolution of pro-stock rules "political" motives prevailed to a degree that made all previous episodes of the sort seem trivial. NHRA's executive vice president, Jack Hart, candidly admitted that it was essential to have close competition between Detroit's two most popular makes. NHRA permitted Chevys and Fords with wedge-head engines to run at a lighter weight than cars with Chrysler hemis, and that was all it took to send the likes of Sox and Martin into eclipse. Sox and Martin had reported $77,000 in earnings in 1970, but in 1973 Jenkins reported $200,000. Later the team of Wayne Gapp and Jack Roush would enjoy a period of similar fortune in a Ford, as would Nicholson, and Bob Glidden would eventually win more major NHRA events than any other drag racer in any kind of machine.

When Sox and Martin put their operation up for sale in 1974 Buddy Martin remarked that NHRA would do whatever was needed in order to cater to "Ford and Chevrolet fans sitting in the stands."[23] Politics was driving competition, he was saying. But, then, to say that was to say nothing new. What *was* new was the pervasive doom and gloom. Stars like Jenkins and Nicholson might be riding high, but a lot of others like Sox and Martin were finding it impossible to make ends meet; enthusiasm could carry only so far.

One result of the austere mood was NHRA's establishment of separate classes for both dragsters and funny cars restricted to alcohol. That could be an effective palliative for a cost crisis, and a number of future nitro racers were able to hone their skills in what was called "pro comp"—Dale Armstrong, Joe Amato, Jerry and Darrell Gwynn, Ken Veney. Veney would later become crew chief for Gwynn, when backing from a beer company permitted paying him a handsome salary, and another brewer's money enabled Kenny Bernstein to pay Armstrong handsomely. But that was all something for the Reagan years, the go-go 1980s. In 1974 the future of drag racing as a "business" seemed very clouded indeed.

Then Larry Carrier announced his "Winston-IHRA Challenge." And in April 1975 NHRA announced making a Winston deal of its own, announced it with a good deal more flair, with a gala at the Sheraton Universal Hotel in Universal City. RJR parted ways with IHRA in the mid-1980s but would keep putting more and more money into its NHRA Championship Series. Initially, the total was $100,000; by 1988 it had topped $1 million. "It's hard to imagine the sport colored in anything but Winston red," said Billy Meyer.[24]

Pessimism dissolved rapidly: "Six months into the action," wrote Dave Wallace, Jr., in mid-1976, "the box office has never looked better."[25] Much of this was due to a general recovery, but, as the years went by, people in drag racing tended to attribute the turnaround largely to the Winston presence, to the prestige of major corporate sponsorship. "I don't know where we'd be today without Winston," said Don Prudhomme in 1988.[26] By then a top-fuel or funny car racer who won the Winston championship and also did well at the NHRA World Finals could go home with something on the order of a quarter-million dollars. That was still not comparable to NASCAR—Winston bonus money made Bill Elliott "Million Dollar Bill" in 1985—but it certainly did a lot to ameliorate drag racing's minor league image.

In the 1970s, when people talked of the "enhanced reputation" that

Because NHRA rules permitted pro stockers with "small block" engines to race at a lighter weight than those with Chrysler hemis, Bill Jenkins had this 1972 Vega built with a chrome moly tube frame. It won six major NHRA events as well as the PRO meet in Tulsa. Note apparatus to prevent front wheels from coming up too high. (Photograph by Ed Sarkisian)

A defining moment in the history of drag racing, the event that would color it "Winston red." In 1975 NHRA's Wally Parks (left) signs a contract with R. J. Reynolds Tobacco's Richard Dilworth for Winston sponsorship of an annual championship based on accrued points, à la NASCAR. Initially called the NHRA-Winston World Championship Series, it would continue into the 1990s as the Winston Championship Series. (Courtesy National Hot Rod Assn.)

Winston had brought to drag racing, they may not have been aware of the contradictions inherent in such a perception. But years later Wallace (whose sense of irony rarely failed him) could still attribute drag racing's "newfound respectability" to its tie-up with R. J. Reynolds.[27] It was as if people had never heard about the surgeon general calling it "indefensible" that cigarettes were promoted "in a context of happiness, vigor, success, and well being." Not Surgeon General C. Everett Koop in the 1980s, Surgeon General William H. Stewart in 1969.[28]

NHRA had initiated a NASCAR-style season-long points chase in the 1960s but did not pay any money for the championship until 1974, after RJR had already expressed interest in getting involved. The program was pushed forward by Ralph Seagraves, who had inaugurated the NASCAR Winston

Cup Series. He installed such enduring fixtures as "free smokes." He bought newspaper ads, distributed press releases, arranged interviews. He made certain that Winston decals were "strategically positioned on race cars" and that Winston caps were "installed upon the heads of drivers and crewmembers for winner's-circle photographs." He made certain that one or another "Miss Winston" was always present in such shots. After a while it was hard for anyone to take photos of the action without getting a giant inflated pack of Winstons in the picture. Stars like Garlits and Muldowney might say privately that they did not like the idea of promoting smoking, but time and again they tacitly endorsed Winston. For its part the "Winston team" could always be found hard at work before, during, and after every event in the championship series: "Parties are thrown for media types and community leaders. Bright red banners are hung for miles around. Fences and walls and timing towers are repainted red, then lettered in white, ensuring maximum brand-name exposure to spectators—and, eventually, to millions of television viewers as well."[29] To millions of television viewers as well—it would never have been worth it to RJR without television.

Drag racing had been seen on TV in the 1960s, even the 1950s, and not solely rough-and-ready local productions; a CBS-TV program hosted by Eric Severeid covered the 1956 NHRA Nationals in Kansas City. But the networks had never shown any sustained interest. Then "subscription" TV had come along, and there sports always loomed large in the programming; subscription TV played a big part in the move of the Dodgers and Giants to California. There was also CATV, which initially stood for "community antenna television" but soon came to signify cable television. By 1976 segments of the NHRA Nationals and Winternationals could be seen on NBC, the Springnationals on ABC, and the other events were being taped for syndication by Diamond P Sports. Diamond P was the creation of Harvey M. Palash, whose initial involvement with NHRA was specifically in the realm of negotiating TV deals. Palash had worked in ABC's legal affairs office in the early 1960s, then gone into business packaging programs for radio syndication. Attending the Winternationals with Wally Parks in 1966, he noticed lots of people running around with cameras and asked Parks if NHRA had any provision for marketing television rights. The answer was no, and he seized the opportunity.[30]

Palash was soon ensconced on the NHRA board, a man emblematic of a new generation of high sheriffs. Parks had founded NHRA in concert with others like him. Even in the early 1970s almost all the board members were men who had been enthusiasts—Jack Hart, Ak Miller, Carroll Shelby, Nathan Ostich. Other former racers still moved into the NHRA hierarchy, notably Carl Olson, who had driven fuelers in the 1960s and early 1970s. But more typical of the new generation were Palash and Dallas Gardner. Palash was an attorney; Gardner had come to NHRA in 1973 as a financial consultant, then he was appointed treasurer, then executive vice president and general manager, then president. A graduate of Cal Poly in Pomona (in accounting and business administration), he had been comptroller at the Forum, working for Jack Kent Cooke. He had also been executive vice president of Ontario Motor Speedway. This gave him "motorsports experience" for the record, but it was certainly not the same kind of experience as Parks's. Wally had been enthusiastic about Muroc, and fifty years later he still loved drag racing's theater of machines; he remained a "motorhead," a buff. Gardner and Palash were both immensely personable, like Parks, but at bottom they were strictly businessmen.

Palash's executive director was John Mullin, who had been working for WBNS in Columbus and was assigned the task of making a feature out of the 1973 Springnationals, staged at nearby National Trail Raceway. Mullin had never been to a drag race, but he came up with a nice show. Three years later, when Diamond P began videotaping all the events on the Winston series, Mullin signed on with Palash. He worked with Opryland Productions in Nashville. In the early 1980s Opryland would grow into The Nashville Network, and TNN would become one of the prime outlets for Diamond P Sports.

Another outlet would be ESPN, founded in 1979 with funding from Getty Oil. Of necessity ESPN had to carry sports that were not seen regularly on the networks, but its management believed that "the appetite for sports in the country is insatiable."[31] ESPN aired its first videotaped drag race in 1983; six years later it was showing eleven NHRA events as well as several of the IHRA's. The most heavily involved of the networks by 1989

With Diamond P Sports video-cameraman Jeff Kelty on his left, Steve Evans tapes an interview in the pits at Indianapolis Raceway Park in 1986. Talking into Evans's mike is Graeme Cowin, a visiting funny car racer from Australia. (Smithsonian Institution collection)

was NBC, which aired three one-hour segments on "SportsWorld." Each hour cost NHRA $400,000, and it could only hope to sell enough advertising to hold deficits to a minimum. It was the same with ESPN: The cost was less, but so were advertising rates. "Drag racing is a drop in the bucket to ESPN," said Ted Jones. "ESPN pays a tremendous amount of money for NASCAR, whereas they don't for drag racing. . . . NASCAR doesn't beat us up in the ratings, but it's much easier to sell because it's live and there are more sponsors involved."[32] And it was the same story at TNN. In a good time slot drag racing got decent ratings (around .04, meaning it was being watched in about 3.5 million homes), but even so NHRA and IHRA still had a "vanity" relationship with TV. A glut of sports programming had inevitably driven down the price of commercial time. More specifically, the segment of corporate America which regarded drag racing as a good investment remained rather narrow.

Not that TV was not regarded as well worth the expense by the sanctioning bodies; without it there would be few major sponsorships at all, and certainly not Winston's. "We continue to view television as the key to both short- and long-range growth,"[33] said Brian Tracy, one of the high sheriffs. After 1986 every event on the Winston series was taped for television. By 1989 there were more than sixty NHRA and IHRA shows all told, including reruns.

IHRA's television productions looked amateurish compared to Diamond P's for NHRA, which were slick, if predictable, and distinctly devoid of anything negative aside from crash sequences.[34] Their formulaic quality was underscored when a New Yorker named Dean Papadeas produced a few shows for the home video market, which were a bit shy on polish but far more imaginative. If anything, Diamond P tried to make drag racing look staid. The team of commentators, Dave McClelland, Steve Evans, sometimes Brock Yates and Don Garlits, did its best to help NHRA put its best foot forward, but still there was, well, an image problem.

It was not so much the outlaw image that Wally Parks had been fighting since the 1940s, although drag racing still had its bad boys: Connie Kalitta had a penchant for getting disorderly, at one point losing his NHRA license for assaulting a track employee; several racers went to prison for trafficking in drugs (and there were persistent rumors that others kept getting away with it); several served time for fraud and tax evasion. But every sport had such problems, and drag racing could certainly present a number of articulate voices. Gary Beck impressed an uninitiated newspaper reporter simply by saying: "The mental preparation is very important, just as important as the mechanical preparation."[35] Gene Snow and Eddie Hill were impressive for their air of shrewd intelligence, Joe Amato for his boyish charm, others for their manic enthusiasm or imperturbable cool. Shirley Muldowney was impressive for her courage, Darrell Gwynn for his spunk and, later, for his courage, too.

The image problem had more to do with assumptions that drag racing was less "serious" than other forms of auto racing, that it was somehow more akin to such juvenile spectacles as "monster trucks" and tractors with seven engines.[36] But there was one man who did a lot to dissipate that misconception, Kenny Bernstein. Bernstein was a determined competitor in a race car, but what he projected most forcefully was an extraordinary business sense. People realized that Bernstein could make money however he chose. Among other things he fielded both a Championship Auto Racing team (CART) and a NASCAR team. If he was in drag racing, this must signify that drag racing also was serious business.

There were various ways to go about competing in the fuel wars of the 1980s. One could be independently wealthy and race all-out largely on one's own dollar. One could try and make do, even with limited personal finances, by keeping overhead down and running conservatively. Such people might have an eye out for sponsorship, but, particularly in top fuel, often not pursue it with anything like the same enthusiasm they had for the technology itself. "Top fuel consists almost entirely of drag racing's true independent operators," Carl Olson could say as late as 1985. "Many of them are self-funded and are more interested in doing their own thing than in seeking out sponsorship support."[37]

Unsponsored funny car racers were called privateers, and they had opportunities to ensure survival by match racing. The flopper contingent had been more strongly conditioned "to forget about drag racing as a sport and think of it in terms of show business."[38] Racers usually had an eye out for a deal. Most of them would settle for something modest, but a few were never inclined to sell themselves short. Prudhomme would not, but there was nobody less inclined to do so than Bernstein. If racers of the 1950s, 1960s, and 1970s were measured mainly in terms of performance (both in a technical and theatrical sense), Bernstein changed the game in the 1980s. Bernstein was always a staunch performer in the traditional terms; that was taken for granted. Where he also performed was in corporate boardrooms. As he himself put it, "I think when you can put a variety of sponsors in one room and put together a mutually beneficial program of merchandising and marketing, that's more of an accomplishment than actually winning a race."[39]

It seems safe to say that everyone who ever entered a car in a drag race, however much of a technological enthusiast, was looking to divest some of the direct costs. Though adamantly opposed to "professionalism," Dan Roulston of *Drag News* was writing about the importance of sponsorship as early as 1958. At first the reward for competitive success might be free parts. Beyond that one might get paid to run some particular brand name, though there were only a few who could command this favor. What quickly developed, however, was the custom of contingency awards, "decal money" posted not by track management or the sanctioning body, or even by

Winston, but, instead, by manufacturers of oil, tires, all sorts of hardware. *If* one was using a specific product, and *if* one won, the manufacturer would come through with some cash.

The system was sticky for many reasons, one being the difficulty of telling what was inside an engine. According to a 1972 survey, virtually every manufacturer claimed that "prominent racers used their products but displayed competitor's decals because of the higher awards or contractual obligations." "Contingencies turn some racers into whores," said one fuel racer.[40] Yet contingencies remained integral to the awards structure of drag racing.

Beyond that there were angels (or pigeons, in Keith Black's term), who would pay just to get in on the action. More common were people who put out money because of some supposed advertising benefit. Businesspeople sometimes budgeted quite a bit for this; in the early 1970s, for example, Vel Miletich and former Indy 500 winner Parnelli Jones, partners in a Torrance Ford agency, sponsored a funny car and paid driver Danny Ongais a handsome monthly retainer. A step beyond that was sponsorship by the maker of some automotive product marketed nationwide, or, better, by one of the automakers or an automotive subsidiary such as Ford's Motorcraft or Chrysler's Mopar Parts. Ever since the Mattel program for Wildlife Enterprises, however, the ideal had been to move horizontally, that is, into non-automotive realms.

How far one could move was largely determined by known demographics. For a long time the fundamental spectator composite was "the 18-to-34-year-old male."[41] Fast foods and beverages had seemed like a natural since the days of the A&W Root Beer Special and the Nesbitt's Orange Special. Don Prudhomme and Herm Peterson made deals with the Olympia Brewing Company in the 1970s, and Tom McEwen established his long-standing tie with Adolph Coors. But the big breakthrough came with Anheuser-Busch, specifically Budweiser, a product sold in astronomical quantities. The Bud deal was Kenny Bernstein's doing.

Bernstein was born in Clovis, New Mexico, in 1944. His father was head of a chain of variety stores, and Kenny began working in stockrooms when he was nine. As a teenager he lived in Lubbock, Texas. He got a motor scooter, then a quarter-midget. He fixed up a '56 Thunderbird. He did the usual street racing and started going to the drag strip at Amarillo. After moving to Dallas, he got involved with the Anderson brothers, top-fuel racers, and recalled that he flunked out of Arlington State because he skipped finals to go to Bakersfield. He took his "first ride in a real fast race car" in a fueler belonging to Vance Hunt and also drove for Prentiss Cunningham and the Carroll brothers, solid contenders in the tough Texas dragster league. On the West Coast he got to know Ray Alley, whose dragster experience dated back to places like Saugus in the early 1950s, and he drove a funny car for

Alley from time to time. Bernstein was an enthusiast; he did not dream of making drag racing his business. A business he knew was the clothing business, and he earned money selling high-fashion women's wear.

He also knew something about the car business, and when he turned thirty he was running a towing service in Dallas. By then it was normal for him to have more than one iron in the fire, and in 1976 he and Steve Evans bought a rocket machine, which Alley subsequently drove to clockings well over 300 miles per hour. But that was only an investment; he thought he had quit driving for good. After selling his towing business, Bernstein opened a restaurant in Lubbock, the Chelsea Pub, and eventually expanded into New Mexico, Louisiana, Tennessee, and Florida.

In 1978 a friend with a funny car persuaded him to go to the NHRA Nationals. They failed to qualify, but the experience set Bernstein to thinking about getting his own funny car. "I am by no means rolling in money," he told a reporter, "but I can now go racing with what it takes and we all know that is a lot of bucks."[42]

Bernstein liked to call his things "King." His "Chelsea King" funny car had the best of everything. It was hauled around in a forty-five-foot transporter containing a fully equipped shop behind a Peterbilt tractor. Bernstein's total investment was something like $260,000. In 1979 he won the IHRA championship. He finished in the NHRA's top ten, but a long way from the top, and in 1980 he did not even make the top ten. Because his other interests consumed so much of his time, he needed someone to devote full attention to the car; he needed a crew chief, pretty much a new concept in drag racing. Even more than that, Bernstein felt that he needed more money than he could provide from his own business, however well it might be doing, however nice a tax shelter the race car provided; he needed corporate sponsorship.[43]

He initially considered restaurant suppliers, companies like Land O' Lakes cheese, with which he did a lot of business, but the demographics were not right. What he needed was a company with a consumer product that was aimed particularly at the eighteen- to thirty-four-year-old male, a company that wanted television exposure in the context of an activity like drag racing. A couple of Austin beer wholesalers, friends of Bernstein's, got him through the right door at Anheuser-Busch headquarters in St. Louis. He was prepared with the proper answers. Without "numbers from television," he said, "there's a problem." Asked to come up with a marketing plan, he said he would do "anything they ask me to do."[44]

At first there were two funny cars running under Budweiser colors, Bernstein's and Roy Harris's. Harris had learned the ropes working for "Jungle Jim" Liberman, a showman par excellence, but in St. Louis they liked Bernstein's act better, and in 1981 he alone carried the Budweiser name. Bernstein now had the right equipment, and he had the bucks. He needed someone to make him a world beater. His friend Ray Alley might

Dale Armstrong (right) *confers with Lanny Miglizzi of L&T Clutches. The key to the L&T's "lock-up" feature was additional fingers on the pressure plate, the piece of hardware between them. On the next table is a blower and fuel injector. (Smithsonian Institution collection)*

have done that, but Alley did not want to travel. So Bernstein hired Dale Armstrong, a racer highly regarded for his skill and intelligence, and a topflight funny car contender himself. With Armstrong ensconced as "vice president in charge" Bernstein was left free to do what he liked best: "working with the media, doing public relations, doing publicity with the car."[45]

Armstrong was born in 1942 in Edmonton, ending up twenty-one years later in southern California, after having mastered all sorts of machinery, beginning with Soap Box Derby racers. He became a weekly competitor at Lions and first attracted attention racing a low-budget funny car called "Canuck." Armstrong was brilliant at figuring out where the rules were "soft." In the 1970s he raced mostly in pro comp, for which all sorts of different setups were allowed: He ran a funny car on alcohol, a funny car chassis with a Model-T roadster body, a blown dragster on alcohol, an unblown dragster on nitro. He tried nitro in conjunction with nitrous oxide. He juggled weight and displacement. He tried a four-speed transmission. He designed a funny car body that bore no resemblance to any known auto. "Every little thing counts," he said, "all the minute changes, and it's a combination of maybe ten minute changes that gets you five hundredths."[46]

Dave Wallace, Jr., suggested that the only limits on Armstrong's imagination were financial: "It was often said that if this guy ever hooked up with some smart millionaire, our sport might never be the same."[47] Armstrong's "smart millionaire" offered him something like $300,000 a year and gave him his head. Armstrong tried everything he had ever tried before and lots more. He built engines with a small bore and a long stroke, with dual fuel pumps and dual magnetos, then triple magnetos (and heads with twenty-four spark plugs). By means of an on-board data recorder he ascertained that the clutch was still slipping on the top end when it was thought to be fully engaged (something nobody suspected), and with Lanny and Tony Miglizzi he developed a two-stage unit that sent records toppling. He designed a two-speed blower drive that would have yielded even quicker and faster times, but NHRA disallowed the idea because its insurance carrier wanted to see the brakes applied, not more speed. Armstrong was accused of concealing a system for injecting nitrous oxide, which NHRA no longer permitted. "Dale's always got something going on," deadpanned Bernstein.[48]

Armstrong put in hundreds of hours on a dyno originally built to test aero engines. He made progress working the bugs out of the McGee Quad Cam from Australia, an engine whose apparent promise had remained unrealized for some fifteen years. Renting Lockheed-Georgia's wind tunnel for days at a time, he figured out ways to clean up a funny car's airflow and particularly to increase down-force. There were more of those outrageously configured bodies, particularly the Buick Regal Batcar, which Armstrong dreamed up with Gary Wheeler. NHRA authorized its use for one year; anybody serious about competing against Bernstein had to get something like it, perhaps from Bernstein himself, who, it happened, had carbon fiber Batcar bodies for sale.[49]

Budweiser was widely perceived as manipulating unlimited hydroplane racing for its own ends; it had "ruined" that activity, in Keith Black's word. "They have six [unlimited] hydroplane races per year put on by Budweiser," Black said. "So are they going to do anything to antagonize Budweiser?"[50] People wondered whether Anheuser-Busch had similar designs on drag racing. One thing was quite certain: Bernstein got more time on TV with his old partner Steve Evans than any other racer, more time by far. Whether this was right or wrong was strictly a matter of opinion. After all, Budweiser was more heavily involved with NHRA itself than any other sponsor except RJR. Other beer companies, Miller especially, had considerable presence at the races in the 1980s, but they rarely got as much attention from Diamond P. When Dick LaHaie won the Winston top-fuel championship in 1987, with Miller sponsorship, he was never featured much on TV, despite the intrinsic appeal of a former budget racer making good and with his daughter as crew chief. When Bruce Larson, a venerable privateer, beat out Bernstein for the funny car championship in 1989 he got only a fraction of the Diamond P time that Bernstein had received in former years.

One need not carry a conspiratorial interpretation of this too far. Bern-

Kenny Bernstein is seen here leaving the line at Firebird International Raceway near Phoenix, at the beginning of his first full season in a dragster, 1990; Dale Armstrong watches anxiously from beyond the fuel injector. Two years later Bernstein's sponsors would reap a publicity bonanza when he drove to a speed of 301.78 at Gainesville, Florida. (Courtesy Susan Arnold)

stein was, after all, invariably gracious and articulate, and his personal agent, Susan Arnold, did everything in her power to make certain he remained a "sellable commodity."[51] Bernstein and Budweiser were very powerful politically—but, ultimately, not invincible in competition. After winning the NHRA's funny car championship for four years running, as only Prudhomme had ever done before, Bernstein fell to third place in 1989. And when he fielded a dragster in 1990 he finished well behind racers who competed with only a fraction of his budget. So there *was* more to it than just money.

Bernstein had not been in a dragster since the days of slingshots. There were several reasons why he switched. Safety in case of fire was one of them. Second, Armstrong was convinced that the restrictions on body contours and spoilers pretty much assured that top-fuelers would always have a small but significant performance edge over funny cars. But the most important factor in Bernstein's mind was this: "Sponsors have the last say on everything. . . . In the old days, the dragster didn't give you a billboard. But it's different today. It's 300 inches long. The bodies are pretty big. . . . You can get a lot of identification on there. It's a lot different than it was years ago."[52]

A billboard. Besides Budweiser, Bernstein's new dragster was festooned with at least twenty other sponsors' names, including Planters Peanuts, an RJR subsidiary, and Racepak computers, a company in which Bernstein, Armstrong, and Ray Alley were partners. Space was priced like cuts of beef, according to its relative visibility, especially to television cameras. A firm in Ann Arbor documented "the amount of clear, in-focus exposure time" that sponsors received and provided them with a dollar value based on going rates for TV commercials; in 1990 it estimated that telecasts of NHRA events reached more than 83 million viewers and that "nearly 400 sponsors enjoyed national television exposure via the NHRA, sharing more than $34,000,000 of average exposure value."[53] The lion's share was for Budweiser, and for Winston; Bud was there because Winston was, and Winston was there because of television.

When tobacco industry lackeys said that they did not really expect anyone to take up smoking just because a certain brand name was associated with a particularly exciting activity, they made the people who paid out money for sports sponsorships seem pretty dim-witted. Art Buchwald had assessed the situation as far back as January 1971: "The Great Minds of the cigaret and advertising industries are hard at work trying to figure out ways of publicizing cigarets on television now that cigaret commercials have been banned. The tobacco companies are already going ahead with plans to sponsor automobile races." Twenty years later a physician at the Baylor College of Medicine stated that "motor racing, tobacco, and television are now so closely associated as to be inseparable."[54]

Besides a diversity of sponsors, Bernstein had diverse business interests. There was King Entertainment. King Racing Components. King Sports, Inc. King Racing, Inc.—the NASCAR team. The CART team. The "Bernstein File," part of a weekly ESPN program, "Speedweek." Not everything was a money-maker, but from all his ventures combined he reported a $20 million gross in 1989. What Budweiser budgeted for Bernstein annually was not public knowledge, but there was every likelihood that it was not far shy of what Augie Busch III made in salary and bonuses as CEO of Anheuser-Busch, $1.46 million.[55] When people interviewed Armstrong they tried to probe his "speed secrets." When they interviewed Bernstein the questions were about sponsorships.

By 1990 there was a growing apprehension that the well might not be much deeper, that there were not many more big companies that could be induced to sponsor drag racers. Bernstein said he did not believe it. NASCAR had attracted all sorts of sponsors whose products were not targeted at the eighteen- to thirty-four-year-old male. Since Bernstein actually owned a stock car team, he was in a good position to get NASCAR sponsors to look at drag racing. In late summer of 1990 he organized a benefit for Darrell Gwynn, with whom he had once shared his Bud largesse. NHRA versus NASCAR, the Darrell Gwynn Benefit Softball Challenge was a heart-warming event; it was featured on TV, of course. It raised quite a bit of money for Gwynn, who seemed to be looking at a bleak future. But possibly Bernstein was also thinking that there might be some practical payoff, just as the high sheriffs were thinking when they wined and dined executives from McDonald's at the NHRA Nationals.

One reason for NASCAR's stronger sponsor appeal was that it was usually seen live. Drag racing was almost always taped, yet drag racing seemed ideal for live TV. Dave Densmore admitted to having once "developed an adversarial relationship" with John Mullin:

> Through both word and action, he seemed convinced that the sport of drag racing was made for his television coverage. It was my contention that the reverse was true. I frankly resented the "anything for Diamond P" attitude that seemed to pervade the company [NHRA] at everyone else's expense. Race schedules were adjusted to accommodate TV, cameras became a starting line distraction and microphones a finish line inconvenience.

Densmore knew that the Diamond P package had been essential to attracting and sustaining corporate sponsorship. Yet NASCAR (and CART) did better than NHRA. The solution? "Give John Mullin the ball and let him run with it." Build on the episodic nature of the drags: There could be commercials after every race, every minute or so. Time between rounds would need to be decreased, fields might have to be cut from sixteen to eight cars, and there might have to be other technical concessions in order to make drag racing "totally subservient to television."[56] An exercise in sarcasm? Perhaps. Yet everyone knew that television was the key to further growth, and, if growth were seen as an overriding purpose, then why not make everything "totally subservient"?

Kenny Bernstein might very well have liked the scheme outlined by Densmore. One should bear in mind, however, that Bernstein was not typical, even among top-rank pros. Others, even in 1990, had more diverse motivations than his. And the makeup was diverse, too; there was scant reflection, anyway, of the profile that dominated the spectator demographics. Attending the season finale NHRA Winston Championship Banquet in Ontario, California, in 1990, one would see the racers with the most points in top fuel all standing together on stage receiving plaudits; there would be two women and an awful lot of gray hair. Only a couple of smokers, several teetotalers. People who kept racing through good times and lean. Indeed, if one started asking questions of any of these men and women, about why they did what they did, one would receive immediate confirmation of Eugene Ferguson's observation about "the strong romantic and emotional strain" that so often underlies technological pursuits.

12

MEN AND WOMEN

*I worked at a newspaper office for a
while, I waited on tables, and I worked
in a doctor's office. But what I really
wanted to do was drag race.*

SHIRLEY MULDOWNEY, 1987

*When I got married I told my wife, "I
have a terrible disease." She asked,
"Will the kids get it?" I said, "I hope
so."*

TONY MIGLIZZI, 1987

n the first chapter I noted George Basalla's concept of technology
as a manifestation of "the various ways men and women throughout
time have chosen to define and pursue existence." In that context it
is worth considering a composite profile of drag racing's 1990 competitors.
While two-thirds of the spectators were less than thirty-four, hardly any
top-fuel drivers were that young, and quite a few of them were past fifty. No
matter how old, nearly every one had been involved in drag racing for most
of a lifetime. Among drivers in the other pro categories the average age was
a few years younger, but at least half these people had been around for
twenty years or more. Overall, the average age was close to twice that of
1950. Consider, too, that at the beginning women were all but entirely
absent from the competitive ranks; indeed, they were rarely to be seen at the
drags, even offstage. In 1990 there were female drivers in nearly every kind
of machine, including the very fastest, and in the stands and pits alike there
was a strong indication of family involvement. Couples often competed as a
team, and a second generation had emerged which defined and pursued
existence just as its parents did, mostly sons but not an insignificant number of daughters.

In 1976 Ruth Cowan posed a question that in many ways remains the most interesting of all questions confronting historians of technology: "Was the female experience of technological change significantly different from the male experience?" Cowan went on to ask rhetorically about a "differential impact" with regard to "the introduction of the railroads, or the invention of the Bessemer process, or the diffusion of the reaper."[1] The differences are, of course, obvious. If one asks about the invention and diffusion of drag racing's theater of machines, the answer is the same, but the situation would change over time. In few aspects of American life was gendering the same in the 1980s as in the 1950s, and it was not the same among those who sought to define and pursue existence in terms of drag racing.

Those who gathered at Santa Ana, Paradise Mesa, Saugus, and other such places throughout the country were nearly all youthful males. There was little thought that the girls who came along, when they came at all, would have any active participation. In this context, at least, their existence was entirely circumscribed by the males with whom they were associated. For a decade or more this did not change much, except that women were cast more overtly in the role of window dressing. True, Doris Herbert was in a position of considerable influence as editor of *Drag News,* and, if it was perhaps tempting for NHRA officials to push her around because she was a woman, she could still publish material suggesting that some women were not happy with the inequities that men had built into the social structure of drag racing. Simultaneously, Herbert gave other women space to profess the virtues of a hobby that the whole family could share.

At first a woman accompanying her husband or boyfriend to the races was pretty certain to be confined to domestic chores, but not inevitably. C. J. Hart's wife, Peggy, was involved in the operation of Santa Ana right from the beginning, and Hart reported that she also shared "the same enthusiasm for racing."[2] Peggy took turns driving their Cad-powered roadster, clocking 113 when the fastest dragsters were not much faster. Eventually, she did all the driving and competed in a gas coupe, then, when she was past forty, in a dragster. But the potential danger troubled her husband, and it was C. J. who decided one day that "we weren't going to run anymore."

Reports of Peggy Hart's exploits were nothing if not patronizing. A *Hot Rod* writer/photographer posed her applying "drag race war paint" while waiting in the staging lanes; "She claims it gives her added 'go,'" said the caption. But Peggy Hart got publicity in any event; people were at least made aware "that drag racing is not an all male world,"[3] and by the late 1950s one could find women's names sprinkled throughout the lists of winners published weekly in *Drag News.* Mostly they were in stockers, where the so-called powder-puff derby became a staple. But there was also Lynn Sturmer, a regular at U.S. 30 in Gary in a Chrysler-powered gas dragster. And there was Fran Deggendorf of Dubuque, Iowa, whose spouse, Joe, was an outspoken opponent of paying appearance money to "parasite show offs," the touring fuel racers. He also gave Fran turns at the controls of his gas dragster at places like Minnesota Dragway and Cordova. Later the Deggendorfs raced a Chassis Research slingshot as partners, with Fran taking pride in managing to get "as greasy as the best of the hot-rodders."[4]

A more common, or more commonly expressed, attitude among women was that it was best to leave the racing to the boys—simply to accept the situation when "a dragster sits in the garage, the family car sits outside in the rain, things go unrepaired, [your] husband and friends track oil on the rug, and the subject of the day [is] 'Garlits did it again.'" "I feel the female gender can accomplish more in this manner than [by] donning a pair of coveralls or knowing how to jockey a wrench or a 'hot' car and losing our God-given femininity," wrote Marianne Beckstrom in 1960.[5] Patience and understanding rather than participation, she advised.

Not all women agreed, and by 1960 drag racing had acquired its first participant of "the female gender" who would become a strong threat in topflight competition, Shirley Shahan of Tulare, California. Shirley, who "came from a sort of racing family," had married H. L. Shahan when she was seventeen. "H" had a '58 Chevy, which Shirley drove "for kicks" a couple of times; she was adroit with a stick shift, and in 1959 she won super-stock at the inaugural U.S. Fuel and Gas Championship (another woman, Carol Cox, won the competition among super-stockers with automatic transmissions, a separate class). In 1963, when H bought an aluminum-bodied 427 Chevy, Shirley became the regular driver. Soon they switched to a Plymouth, with backing from Detroit; it was a Plymouth PR man, Sam Petok, who thought of dubbing Shirley the "Drag-on-Lady." Occasionally, H converted to nitro and fuel injectors, and Shirley could run 10.30s with the car set up that way. The money was so good in match racing that H quit his job to concentrate on taking care of the car. Shirley took care of their three children, drove at the drags on weekends, and still held down a regular job at the Southern California Gas Company.

Then in 1966 she won an impressive victory in super-stock at the NHRA Winternationals. The Drag-on-Lady became a hot property. She and H toured the nation's drag strips for several years, eventually getting into the new pro-stock class with an AMX. Fans could buy a Drag-on-Lady poster from Fram filters that made no mention of H; yet, like the other women who preceded her, Shirley's existence was ultimately defined by her husband. Years before, she had said that "'H' and I are working for the day when he can open his own racing shop, and then I'll retire from driving."[6] Sure enough, at the end of the 1972 season, H made a deal to build engines for a circle track team, and, Shirley said, "that pretty well wrapped up my days as a driver." Many years later, and long since divorced, she remarked that her first big victory at Pomona "was about five years too soon. . . . If it had happened later, in the women's lib era," things might have turned out otherwise; her racing career might have been different.[7]

On the cover of its June 1966 issue Popular Hot Rodding *celebrated Shirley Shahan's super-stock victory at the Winternationals. The event was covered by ABC's "Wide World of Sports."*

By 1972 there was another Shirley on the national circuit, Shirley Muldowney, who would become the most famous and successful woman ever involved in auto racing, indeed in *any* sporting endeavor in which men and women met as equals. In the interim a lot of power relationships had changed. When Shirley Shahan first began attracting attention in a super-stocker, Shirley Muldowney was already driving a slingshot. It was not terribly fast, not a fuel burner, but there was a brewing controversy. Although Shahan was not interested in driving a fuel dragster, another Californian, Sue Smiderle, said she was. Smiderle had worked her way up to a supercharged gas coupe that turned 10.40s and 140s. She had discussed the possibility of driving with the owners of a local fueler of some note, Frank and Walt Sandoval. While even her staunchest male partisans could be grossly condescending—"I personally have watched this girl tear down an engine and put it back together again, and I can tell you right now that this little gal really knows her way around an automobile"—she did have considerable support.[8]

The UDRA proclaimed its opposition to the "female driver thing," but leaders of that organization shot from the hip. Through long experience the NHRA was inclined to consider legalities. "You know, at first we were opposed to the whole idea," said Bernie Partridge, NHRA's West Coast regional director, but then NHRA decided differently: "Our lawyers told us we couldn't turn them down, it would be discrimination . . . and there is nothing in the rules saying a woman can't drive a dragster. So, consequently, if she passes the written test and the driving test, there is no legal way we can stop her."[9] It turned out that the Sandovals and Smiderle were not all that serious about fuel racing together anyway (nor was John Wenderski's widow, Dianne, who had likewise announced her intentions to drive a fueler), but such was not the case with another woman who was contemplating the same thing, Paula Murphy.

Murphy brought a background to drag racing whose diversity was unusual for anyone of either sex. She was born in 1929 in Cleveland, attended Bowling Green State University, then graduated from the University of Cincinnati with a degree in physical education. She had intended to become a teacher but took a job as a social worker for the Community Chest instead. In 1953 she moved to North Hollywood with her husband and three-month-old son. While working as a secretary, she met a woman who was interested in sports car competition. Paula had raced K-class catboats on Lake Erie while in college, and soon she was driving a 1500cc Alfa Romeo in "ladies' races." Later she moved into much faster machines, including a bird cage Maserati, and in the interim she and her husband parted ways.

In the early 1960s the Sports Car Club of America phased out ladies' races, establishing a policy of licensing women for competition under the same set of requirements as men. Murphy drove a Lotus, then a Ferrari, taking a win at Riverside, the premier road course in southern California. She also raced on various California ovals, and she became a mainstay of the "Economy Run" sponsored by General Petroleum, which tested drivers' ability to coax good gas mileage out of current Detroit autos (no mean feat in the 1960s). She was getting a name and offers to perform various publicity stunts; with two codrivers she set a "nonstop" record from Los Angeles to New York to Tijuana to Vancouver and back to Tijuana, some ten thousand miles; she ran a few 100-MPH laps at the Indianapolis Motor Speedway in an Andy Granatelli Novi to become the first woman ever to tour the Brickyard in a racing car. She set a "women's land speed record" at Bonneville in one of Studebaker's new Avantis and a much faster record at the controls of a Walt Arfons jet.

While at Bonneville, she got to know Bob Tatroe, who drove for Arfons at the drags. She had tried many different kinds of racing, but never drag racing. Murphy's first ride was a 1965 Olds 4-4-2 prepared by Dick Landy and sponsored by the Southern California Oldsmobile Dealers Association. By 1966 she was running mid-12s at about 110 miles per hour. These were good times for stock class, but she was anxious to go faster. The funny car craze was sweeping the country, and soon she had contracted with a journeyman fuel racer named Jack Bynum to prepare a flopper; the tube-framed, Mustang-bodied machine he assembled was on the crude side but quite typical of the period. By then Murphy had quit her job at a Van Nuys company involved in developing the Saturn rocket and gone to work in public relations for Granatelli's STP. (Her father, a retired engineer, looked after her son.)[10]

When Murphy first applied for a license to drive a blown-fuel funny car both UDRA and NHRA were negative, but she already knew that Jim Tice's AHRA, as was often the case, would be open to a novel idea. As she went through the standard test at Lions, a series of progressively faster runs, Terry Cook observed that "she handles the car better than some of the males who are loose on the drag strips with their 'Funnies.'" But there was plenty of opposition. Ostensibly, male concerns centered on the damage to drag racing's "image" should a woman get involved in an accident. Cook had an interesting comment:

> In the past few years the "Funnies" have taken over drag racing. . . . ALL rules were cast aside, and a free-wheeling "anything goes" policy was the guide. . . . Now where were all these people who are concerned with the "good" of drag racing when the circus was formulating?
>   . . . Their super-concern over the question of Paula and her Mustang seems small when you consider that they are the ones who paved the way for cars of this nature.[11]

This was, recall, just when drag racing was slipping into its epoch of "hoked-up hippodrome acts," in Brock Yates's phrase. Murphy was at least as "serious" as most people interested in funny cars, and there was no denying her rich competitive experience. After AHRA conferred a license UDRA did too, and even NHRA relented in the fall of 1966.

Then, the next summer, NHRA reneged. There were so many accidents and injuries involving blown, fuel-burning machines. Imagine the bad publicity "if it was a *girl* who got hurt," exclaimed Jack Hart, Wally Parks's lieutenant.[12]

Such concerns may have sounded valid in sanctioning councils, but to others they seemed disingenuous. In July 1966 a St. Louis woman named Ginger Watson *had* crashed in a fuel-burning dragster, when her parachute failed to open after a 169-MPH run at Alton (an AHRA strip), and she had been hurt. But she recovered and was soon back on the match race circuit, turning even faster times.[13] In Washington state a few years before, a woman had been killed in a roadster when the throttle stuck. There was bad publicity, to be sure, though with not nearly so much impact as later incidents involving the death of spectators or the death of a teenage boy taking his first ride in a blown fueler; Paula Murphy was three years older than Don Garlits, and she had been driving race cars for more than a decade.

Nevertheless, NHRA rescinded her license, along with those of several other women, including Muldowney and Barbara Hamilton, who had been licensed to drive a supercharged gas coupe since 1964. The ruling, which excluded women from more than a third of NHRA's ninety-two competitive classes and from all the faster machines, was rationalized so:

A great amount of evidence exists regarding the stamina and strength required to properly handle the more powerful cars in drag racing, and there is little doubt about the added stresses that exist with engines equipped with superchargers or burning special fuel mixtures. . . .

Protests have been voiced and lawsuit threats against NHRA have been made, but the organization's officials believe their decision in the sport's behalf has been a reasonable one.

Adding the trivializing note that seems to have been inevitable whenever any man opened his mouth on the subject, Wally Parks was quoted as saying, "It isn't that we don't like the ladies, we simply want to see them around for a long time to come." The measure was deemed "protective rather than discriminatory."[14]

On July 10, 1967, when Barbara Hamilton received a curt letter from Jack Hart returning her entry fee for the upcoming NHRA Nationals, she was certainly not ready to cave in. For three years, she wrote back, "my record as a driver and participant [has been] completely unblemished and I have driven with absolutely no incidents. . . . I am only interested in racing my own Willys, minding my own business, and being left alone."[15] Litigation loomed but never materialized. Hamilton argued with convincing pas-

Paula Murphy in her first funny car in 1966. Note the presence of an authentic Mustang door handle, in the interest of fostering an illusion that this was an actual automobile. (Smithsonian Institution collection)

sion, as did Murphy. And Murphy had a trump, a deal with a sponsor—one of the nicer deals in drag racing, as a matter of fact. She raced as "Miss STP." As she recalled, it was pressure from STP that got her license reinstated, and those of the others. In the interim, several months, she remembered being "out of business, at least with NHRA. . . . I lost a lot of money because we had [match race] dates."[16]

Murphy lost a lot of money. So, presumably, did certain promoters, who must have made this known to NHRA, too. Most of the concerns about image had turned out to be, well, imaginary, but, much more important, it was becoming clear that women could be a tremendous draw at the gate; women could bring out spectators who had never before attended a drag race. Murphy's agent, Ed Rachanski (a onetime funny car racer himself), could book three and four matches for her weekly. She also ran in open competition, particularly for IHRA, because Larry Carrier provided her with a guarantee. Other people were starting to catch on, such as Della and Bernie Woods of Lake Orion, Michigan, sister and brother, both in their late twenties. They bought a 426 hemi from Garlits and fielded a fuel-burning funny car, too, as the "Bernella Racing Team."[17]

Promoters now knew women were valuable, but they did not know how to exploit them most effectively. Julio Marra, the impresario at Capitol Raceway, midway between Washington and Baltimore, staged an event in 1970 called the "Miss America of Drag Racing." Tom McEwen drove the "Hurst Golden Shifter" girls up and down the strip "as they bowed and waved to the spectators from atop their perch on the Hurst parade car." There was a "miniskirt contest," and powder-puff races in stockers. There was also a match race between two up-and-coming super-stock competitors, Judy Lilly from Colorado and Shay Nichols from Texas, and another involving a local woman, Carol Burkett. There was a match race between Paula Murphy and Della Woods, with best clockings of 202 and 7.56. And Muldowney made two passes in her twin-engine gas dragster, nearly equaling the times of the fuel funnies. A *Drag News* reporter suggested that there weren't many men "who could stand the embarrassment" of being beaten by somebody so "slightly built and very attractive."[18]

Shirley knew this better than anybody, but people like Julio Marra were beginning to catch on, too. When a Pennsylvania promoter presented his idea of "Women's Liberation Night" a few months later it was Paula and Della against machines like Lew Arrington's "Brutus."[19] The ladies lost this time, but that was irrelevant; by 1971 the luxuriant foliage of the money tree was becoming quite apparent.

With Bynum as her mechanic Murphy raced floppers for seven years. In 1973, at the age of forty-four, she took a ride at Bonneville in a rocket-powered machine, the "Pollution Packer," owned by a Minneapolis millionaire named Tony Fox. She made a run in this same rocket at the Pomona Winternationals, a prime showcase for the NHRA, which had once fretted

After falling out of favor for several years, twin-engine gas dragsters enjoyed a resurgence in the late 1960s. Here Shirley Muldowney poses with hers; the chassis was from the California shop of Don Long, whom Tom Hanna called "the best technical craftsman I ever met"; the Chevrolet engines were built by her husband, Jack. (Photograph by Ed Sarkisian)

incessantly about women "in the more powerful cars." Later Ky Michaelson, who built Fox's rockets, prepared one to carry the STP banner for Murphy. She debuted it at another NHRA track, Sears Point, near San Francisco. When she deployed the parachute after her first full pass it tore away, and she went off the end of the strip at high speed. Her neck was broken. A man would die in a Pollution Packer a few months later, but Murphy survived. After a long recovery she did a little more drag racing, in stockers, but mostly it was back to stunts for Paula. NASCAR permitted her to turn a few fast laps at Talladega in Richard Petty's STP Pontiac. In 1976, with backing from Pontiac and National Car Rental, she drove "around the world" with Johnny Parsons, an old-time Indy 500 standout. Then she hung up her helmet, taking a job as a buyer for a Palmdale aerospace firm. She lived within earshot of a drag strip, Los Angeles County Raceway, but rarely attended. For former performers "just spectating" was often emotionally difficult.[20]

When Bernie Woods gave up drag racing his sister did, too. Della opened a beauty salon. But later she got back into funny cars, in partnership with her husband. In addition to match races she competed on the NHRA's Winston circuit, and in 1990 she was still at it part-time, even after a bad crash and fire, at the age of fifty. Other female drivers came and went in the

1970s and 1980s. Judy Lilly won several major events. Shay Nichols graduated into pro stockers. Carol Burkett became a solid campaigner in an alcohol funny car, and her times still ranked her in the top 15 percent of that class as of 1990, as did those of another woman from the Washington, D.C., area, Carol Henson. Among the better competitors in alcohol dragsters were Lena Williams, Mendy Fry, Heather Sanders, and Paula Gage from the West Coast, and especially Amy Faulk from Memphis. In a super-stocker Faulk had won the NHRA's Winston championship in 1979, and she ranked with the top ten alcohol dragsters a decade later.[21]

Women were in and out of various rides in top-fuelers. Lucille Lee finished fifth in the NHRA standings in 1982, and eight years later Lori Johns came in fourth after winning three major events. An Australian named Sue Ransom sometimes drove Chris and Phil McGee's quad-cam machine from Down Under. As of 1990, the two quickest elapsed times ever turned by Europeans who drove top fuelers, only a handful of people, had both been clocked by women, Liv Berstad and Monica Oberg. The sanctioning bodies had long since banished rockets, but jet cars were still around, their design and safety much improved in thirty years. NHRA had licensed at least five women to drive jets, including Marsha Smith, a black woman from Indianapolis, but the most stalwart performer was Agnes "Aggi" Hendriks, from British Columbia, who had clocked 5.40s and 280s in a machine called "Odyssey."[22]

From Peggy Hart to Aggi Hendriks did mark something of an odyssey. But there were always mountains of ambiguity about gender. Articles in popular magazines would often juxtapose material on women who were actually involved in competition with cheesecake shots of the likes of "Miss Power Pipes." On the one hand there was George Hurst's "Miss Hurst Golden Shifter," Linda Vaughn, who mostly smiled, waved, and hugged and kissed men.[23] On the same hand as Vaughn there was Pam Hardy, who was part of the act put on by funny car racer Jim Liberman, "Jungle Jim." Liberman had started out in California driving Lew Arrington's Brutus. Having built his own machine, he resettled in West Chester, Pennsylvania, more central to the match race action. There it was that he discovered Pam Hardy one day in 1972 and persuaded her to go on the road with him. He was twenty-seven; she was eighteen. As "Jungle Pam," her most visible function was to pour liquid under his tires for his burnouts and guide him as he backed up. Hardy was very visible indeed. She "made an Andy Warhol art form out of starting line preparation."[24] She wore short shorts, high boots, and net blouses but never a bra and glasses only if she wanted to see clearly. In the years Liberman and Hardy traveled the circuit together (they split up in early 1977, and Liberman died in a highway accident a few months later) they were just about the most popular performers in drag racing, among "males eighteen to thirty-four," anyway.

Pam Hardy was defined by Jim Liberman, Linda Vaughn by George Hurst. On the other hand there was Shirley Muldowney, who likewise put on quite a performance but, as a competitor, was every bit as fiercely driven as Don Garlits or Don Prudhomme and, as a person, ultimately chose to define her own existence. Belgium Benedict "Tex" Roque, Shirley's father, had been a cab driver, a musician, a boxer. If a boy bothered her, he told his daughter, she should pick up a board or brick and "part his hair."[25] While still in high school in Schenectady, Shirley Roque met Jack Muldowney, a mechanic whose passion was drag racing in his '51 Merc. She found she loved to race, too. She dropped out when she was sixteen, married Jack, and their son, Johnny, was born the next year. She worked as a carhop but lived for the competition—at the local drag strip on Wednesdays, out on Depot Road the rest of the time. "I was always in trouble," she remarked thirty years later. Drag racing at least gave her a focus, however, and could she drive, whatever it was that Jack came up with. There was a Cadillac-powered Ford roadster (the same as Peggy Hart's first ride), several Corvettes, then a gas-burning slingshot in which Shirley turned 9.50s at 155, then a supercharged engine, then the twin-engine machine that ran 7.50s at 200. Shirley and Jack Muldowney were regular winners at Lebanon Valley and all the nearby tracks, and in 1970 they just missed qualifying for the NHRA Nationals.[26]

Shirley would say that part of her ability lay in her "light touch." "You've got all this incredible power, but you can't manhandle it. That doesn't get the job done. Women, I think, have better reflexes and better concentration." Shirley's observation about "touch" is true. Frank Hawley, an excellent driver himself as well as the headmaster of a Florida establishment called the Drag Racing School, enumerated something of what it takes to handle a dragster: "Because a dragster is so powerful, so fast and in some sense a very violent vehicle, a rookie driver believes that a great deal of strength and quickness is required to hold these cars under control and in a straight line. . . . Quickness is required, but violent, aggressive moves aren't necessary and are detrimental to a smooth, straight run."[27]

Driving a dragster, despite what people like Jack Hart might have once claimed, was never a matter of "stamina and strength." Hawley pointed out that anyone who conceived of driving in terms of "muscle" was going to be out of control, or close to it, much of the time. But Shirley had a lot more than just the right technique for winning; she also had the right stuff. "Concentration, judgment, and awareness," said Hawley; "the winning drag racer has to have all three working together while accelerating at a breathtaking rate." And Shirley had something even beyond that. "I think the difference between me and the other guys," she once observed, "is that a lot of them really don't have, truly don't have, that kick-ass attitude."[28]

So here was Shirley Muldowney, who said, just like Don Prudhomme had said a few years earlier, that she wanted to be a professional driver. By

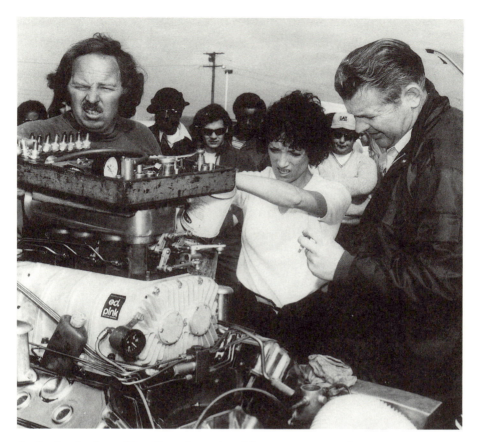

At the beginning of her first championship season, 1977, Shirley is seen in the Pomona pits flanked by Connie Kalitta and Keith Black. (Courtesy Shirley Muldowney Enterprises)

1970 she was commanding pretty good appearance money in Jack's twin, but there were rumors that NHRA intended to phase out blown gas dragsters (and, indeed, this happened a year later). One day they met Connie Kalitta, who was racing a funny car at the time because he had a Ford engine and the Detroit brass had told him the top-fueler "wasn't selling any Falcons or Mustangs for them."[29] Kalitta sold Jack and Shirley an old Mustang-bodied flopper, and soon they had more dates than they had ever dreamed of. A new funny car followed, then another, but by the time Shirley debuted the third one she and Jack had split up, and she had moved to Michigan, where Kalitta lived. For more than five years Kalitta became her "teammate, crew chief, and confidante."[30] Until he had his license suspended the two of them even toured a pair of cars together, the "Bounty Hunter" and "Bounty Huntress."

Shirley was, of course, a star performer on the match race circuit. But she also did fairly well in open competition, winning the IHRA's Southern

Nationals at Rockingham in July 1971 and making a tough field at the NHRA Nationals in September. The next year she had a serious fire at West Salem, Ohio. Shirley had always felt more comfortable in an open cockpit, and certainly she preferred to have the engine behind her; she wanted to switch to top fuel. While she was quite accustomed to wrangling about licenses, there had never been a female driver in a top-fueler, and getting NHRA to accept the idea was no easy matter. When the high sheriffs finally relented in June 1973, one of the drivers who acted as an official observer during her licensing runs was Garlits. Briefly, Shirley alternated between her funny car and a dragster owned by a friend of Kalitta's, but she decided to give up floppers forever after another fire at the 1973 NHRA Nationals. She and Kalitta ordered a new dragster from the Logghe shop.

Shirley's first full season in top fuel was mostly uneventful. Although she ran match races galore (with Tommy Ivo doing her booking), she lost in the first round at both of the big NHRA meets she entered. In 1975, however, she went to the finals at the Springnationals in June and the Nationals in September. In 1976 she won the Springnationals and the World Finals at Ontario Motor Speedway. In 1977 she was top qualifier at Pomona, then she won the Springs again, Le Grandnational in Quebec, and the Summernationals in New Jersey—three of NHRA's eight major races—and she won the NHRA's Winston championship.

Much of her success was owing to Kalitta's savvy, but some of it apparently had to do with her "kick-ass" attitude. At first very few top-fuel racers had taken her seriously; they simply were not prepared for somebody who could be so tough, who could drive "just like a man," as Jungle Jim put it.[31] Her mother was not surprised: "Whatever she went for, she got," said Mae Roque. "She didn't let anything stand in her way." But people like Gary Beck and Garlits had to get used to her tongue lashings. She once threatened to shoot Garlits's crew chief, Herb Parks. Her relationship with Kalitta was explosive. She would tell him: "If my Dad was alive now, he'd kick your ass from here to China. Make mincemeat out of you, crush you like a grape." And she provoked some wonderful responses. Said Richard Tharp, whom she succeeded as NHRA's Winston top-fuel champion: "I used to like Shirley six days a week and couldn't stand her on Sunday. Now I can't even stand her during the week. I'll tell you, I'm not above punching her out."[32]

Her rivals complained that "*all* Shirley does is drive." True, she would sit off to the side with her puppy dog, Skippy, signing autographs, while others did the mechanical work. But Shirley said that this arrangement was ideal, because when she came to the line she was not distracted by second thoughts about decisions made back in the pits. Nobody complained about other people, about men, who "only" drove, such as Tom McEwen. And furthermore, Shirley said, she knew exactly what her crew was doing: "because I'm the wallet, you know? I know what the parts are, I know what

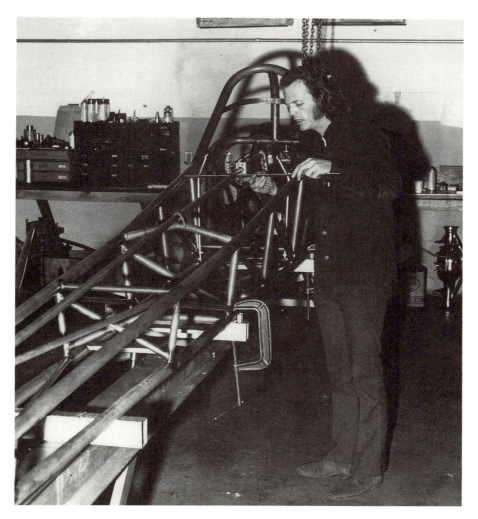

Fabricator Ron Attebury is seen here measuring to install the steering on a new chassis for Muldowney. Welding up a chassis on a jig like this assured that all components would be in proper alignment. (Smithsonian Institution collection)

they do, and why." So men then complained that "you never know what to expect" from Shirley. She could be "as sweet and charming as any woman alive or as brutal as a Marine Corps obstacle course."[33] Garlits, Kalitta, most of drag racing's competitive standouts, had exactly the same temperament, and with them that was not considered remarkable at all.

Some of the confrontational rhetoric, on all sides, was pure hype. From a commercial standpoint Shirley was a godsend. Garlits may have had other reasons for supporting her at the beginning, but it certainly did not escape his notice that "the battle of the sexes" was a surefire hit at the box office. "The ladies finally had someone they could root for," said Shirley. Garlits, with whom she ran dozens upon dozens of match races for $2,500 and $3,000 guarantees, added that previously "the men had to drag the women to the drag strip."[34]

Years before, at a small-time track somewhere, someone had taken white shoe polish and scrawled "Cha Cha" on the side of her car while she and Jack were distracted. Shirley never particularly liked the nickname—she thought it made people expect "some kind of exotic character, a dancing gypsy or something"—but it was certainly catchy, and publicist Ben Brown told her she'd be crazy not to capitalize on it. Shirley went along. At some point in 1976, however, Cha Cha disappeared from her dragster, and she began insisting that she wanted to be "just Shirley."[35] For a time men had fun calling her Shirley "Don't Call Me Cha Cha" Muldowney, but eventually the Cha Cha was forgotten. It was an important step in a process of self-definition.

After Shirley won the 1977 NHRA championship she and Kalitta split up. Their personal relationship had become impossible, yet Shirley knew how important Kalitta's skills were to her racing success; Paula Murphy had been hindered, Shirley said, by not "finding help that really knew how to run her race car": "I think she had the *ability,* but it takes more talent to *maintain* that car than it does to drive it."[36] The next year there was every evidence that Shirley's domination in open competition had been substantially due to Kalitta. She could still book match races for top dollar. But, with Rahn Tobler as her crew chief and her son, Johnny, as his helper, she finished out of the NHRA top ten. Tobler was twenty-three, John three years younger, though neither was inexperienced. Rahn had grown up in Inglewood, hung around Lions, then moved to Houston, where he began helping Steve Stephens and Dick Venables, local top-fuel racers, in 1971. Later he worked for Marvin Graham from Oklahoma City, who had finished as high as third in the NHRA standings. Tobler and John Muldowney had both been with Shirley during her championship season.

Tobler admitted that it was tough to learn "our own way of doing things."[37] But he started to get a combination sorted out in 1979. Shirley qualified first at two NHRA races, ran top speed four times, and finished fourth overall. The momentum carried into 1980. Shirley and Rahn showed up at Pomona with a brand new car, just completed in the northern California shop of Ron Attebury. She qualified handily and had an easy time of it in the first round. In the second she defeated Dave Uyehara with low elapsed time of the meet, 5.83, and in the third she defeated John Kimble. By then the local press already had a pretty good story: a woman versus a Japanese American who had been born in a Wyoming concentration camp, then a woman versus a black man from Compton.[38] It got even better. In the final round she defeated Kalitta. Shirley and Connie were nothing if not ambivalent about one another: Later Kalitta would call her "a gutsy little bitch," while she maintained that her favorite description of Kalitta was "Charles Laughton gone to seed."[39] But after the race they pulled off their helmets and head socks and gave one another a proper embrace, and they gave Diamond P Sports one of the best moments ever captured by its now omnipresent videocams.

Shirley won four of the eight events she entered on NHRA's Winston Championship Series (there were ten in all at the time), and she took her second NHRA title. Nobody else had ever done that, not even Garlits. She also won a title on the AHRA's "social security circuit," something rarely accomplished by anyone other than Garlits. She made a deal with Pioneer Stereo which paid her $150,000 in 1982, when she won the NHRA title for the third time. She was featured in TV commercials. She received awards never before won by a woman.[40] And Hollywood made a movie about Shirley Muldowney, *Heart like a Wheel*. The movie was produced by Chuck Roven, her manager since 1978, and directed by Jonathan Kaplan, who knew a good story line when he saw it: "Most Hollywood films with a feminist slant are about women making choices about which New York townhouse to live in. Shirley, on the other hand, is a working-class woman who made some real choices. She could have been married to a gas-station owner today, taking care of a lot of kids."[41]

Hollywood did not quite know how to market *Heart like a Wheel*, but it did get plenty of good notices and was, in fact, a minor cinematic gem. Kaplan and screenwriter Ken Friedman built the central tension expertly on what a woman had to do to define and pursue existence in an endeavor dominated by males whose consciousness had scarcely been touched by the women's movement. Summarizing a study of champion race drivers, Frank Hawley later enumerated some of the essential qualities as "cool, intelligent, dominant, sober, shy, tough-minded, slightly suspicious, shrewd, self-assured, self-sufficient, and controlled."[42] Shirley Muldowney possessed most of these qualities, and Bonnie Bedelia captured them with skilled clarity. Still, Roven and Kaplan must have been terribly sorry that they did not wait for a couple of years to make their film about Shirley, for the most dramatic episode, by far, had not even begun.

Late in the afternoon of Friday, June 29, 1984, Muldowney stormed through the traps on a qualifying run at Le Grandnational Molson at Sanair in Quebec. She clocked 247 miles per hour. Just as she pulled the chute, she saw the tube come out of the left front tire. The same thing had happened recently to both Beck and Amato, with no serious consequences. But Shirley's wrapped around the spindle and locked the wheel, flinging her across a ditch beside the track (there was no guardrail past the finish line) and directly into an embankment. The car disintegrated, the roll cage breaking away from the engine, as it was designed to do, and also breaking off at her thigh, as it was not supposed to do. Later, nobody could recall ever having seen such a violent crash. Nobody could even find the cage at first. Finally, a spectator several hundred feet away yelled, "Here she is!" Tobler, who had

Dave Uyehara, seen here checking a fuel mixture with a hydrometer, drove drag-sters and funny cars for many years while also operating a chassis fabrication shop. This photo shows the typical layout of a fifth-wheel trailer, with nitro in plastic containers arrayed low along the wall and spare wheels and tires hung up high. (Photograph by Leslie Lovett, courtesy National Hot Rod Assn.)

been romantically involved with Shirley for some time, was in sheer panic. Shirley was covered with mud, oil, and blood, but she was conscious. Rahn bent over her, and she whispered, "I'm in a lot of trouble."

Both hands and three fingers were broken, one thumb was almost torn off, she had a double compound fracture of her right thigh, her pelvis was broken, her right ankle was broken, and her lower left leg and ankle were smashed beyond belief. Barely attached, her foot was in her lap. It took six hours just to get her wounds clean enough for surgery.

Shirley spent six weeks hospitalized in Montreal, then three weeks more in Detroit, before Rahn took her home to Mt. Clemens in a wheelchair. Mae

John Kimble, owner of a Compton trucking firm, is seen here in early 1978. A year later he suffered a devastating crash at Fremont; a year after that he was back, making it to the semifinal round at the Winternationals in a brand new car. (Smithsonian Institution collection)

Tommy Ivo had a last fling in the movie business and in a dragster at the same time, when he worked as a stunt driver in Heart like a Wheel. (Photograph by Leslie Lovett, courtesy National Hot Rod Assn.)

Roque showed him how to cook, and besides that he also became Shirley's "housekeeper, nurse, chauffeur, and psychiatrist." Her right leg was in a cast up to the hip and her left leg could not be permitted to touch anything. Rahn rigged a harness to lift her in and out of her wheelchair. The pain was unremitting, even six months after the accident.

Initially, Tobler had wanted to get out of racing, to have them start anew, and Shirley agreed that this might be the right thing. But she had plenty of time to think, and she decided she wasn't ready to quit. After that she said: "I was a different person. Now I had something to shoot for." Tobler called Roger Penske and found out who had taken care of Rick Mears and Danny Ongais after they suffered leg injuries in accidents. Soon Shirley was being treated by a specialist in Indianapolis, Terry Trammell. This time she spent eighteen weeks in the hospital. There were reconstructive operations, bone grafts, skin grafts, therapy every single day. She progressed from a wheelchair to a walker to crutches. Shirley was now certain she could drive again. She and Rahn began talking to potential sponsors, and eventually they made a deal with Keith Harvie, president of an operation called Parts Automotive Wholesale, one of whose subsidiaries was Milodon Engineering. Though barely mobile and still in pain, Shirley was going back on the NHRA circuit in 1986.

She checked out her new car at Firebird International Raceway in Arizona, nineteen months after the crash that had crippled her. In the early 1980s most top-fuel drivers had switched from foot-actuated clutches to centrifugal units with no pedal, such as the L&T made by Tony Miglizzi and his son Lanny. Shirley liked the pedal, for it permitted her to take advantage of her superior reaction times. But now there was no choice about that; her left foot was close to useless. Following her accident Al Swindahl and the other chassis builders had come up with a better design for the driver's "tub," so a fueler was no doubt safer than before. But reporters naturally asked Shirley why she would get back into one of those things after what had happened the last time. She said, "It's what I do."[43]

The men who ran NHRA had once feared that a woman suffering a bad accident would ruin drag racing's image; the irony was that Shirley's crash and comeback resulted in more favorable publicity than any other episode in drag racing's thirty-five-year history.

Shirley and Rahn raced in Harvie's P.A.W. colors for four years. They got married. They raced for another year with backing from Larry Minor and Otter Pops, Paul Pope's "fast freeze" product. By then Shirley was fifty, and (as one man wrote) "tough in a way that no man probably will ever be able to understand."[44] Although she never regained her old dominance, she never finished out of the NHRA top ten either. Her competitive instincts were sometimes as good as anybody's, her car was sometimes as quick as any, and she even held the national ET record for a brief time. But she and Rahn were both inconsistent.

Seeking a way out, they decided to hire Don Garlits as a "special advisor"—Garlits, whom she had called an old fool, a creep, and who had himself aired his resentment at seeing her "sit in the cab of that truck filing her nails while those turncoat men flog her car for her." After Garlits and Muldowney got together in 1989 a *Sports Illustrated* writer visited them on a sweltering day at the NHRA's Cajun Nationals in Baton Rouge and wrote an inspired story, which began:

> So it has come to this. Shirley Muldowney, her long hair streaked with gray, hobbles over to the man wearing thick glasses and hands him a plate with a turkey sandwich and potato chips on it. "Let me get you some apple juice here. Do you want to come in where it's nice and cool?"
> . . . The man shakes his gray head. "No thanks," he says with a pleasant smile. . . . Don (Big Daddy) Garlits is content to sit in a pink plastic chair under an awning attached to a trailer truck, drink from a pink plastic cup and watch four men tear apart the engine of Muldowney's pink dragster.[45]

Once men like Garlits had made a big deal out of women like Shirley "not getting their hands dirty," not getting intimate with the workings of their machinery. But such women were hardly alone anymore. In the pits, when they were not signing autographs, Eddie Hill and Gene Snow busied themselves poring over computer readouts. So did high-paid crew chiefs like Dale Armstrong and Tim Richards. The men who employed Armstrong and Richards, Kenny Bernstein and Joe Amato, rarely if ever involved themselves with mechanical matters. Ironically, one crew chief who *did* get dirty hands was Kim LaHaie, a woman three years younger than John Muldowney, who was teamed with her father, Dick; the LaHaies finished third, first, fifth, third, and third in Winston points between 1986 and 1990, the second-best average of anybody in drag racing.

Don Garlits and Shirley Muldowney, it turned out, had a lot more in common than they ever dreamed. Garlits, who had begun a career as a bookkeeper at eighteen, stopped and redefined himself in terms that were absolutely novel: professional drag racer. It did not matter that this would make ordinary people look at him "kind of queer." He motivated himself by devising an image of the "rich guys" as the enemy—be they over in St. Pete or out in California—but he excelled because he was cool, intelligent, dominant, sober, shy, and all those other things that a psychologist determined, many years later, made for a champion race driver. Muldowney, who had begun a career as a housewife at sixteen, redefined herself in terms likewise novel: a drag racer who was both a woman and a champion in the ultimate type of machine. (Before Amato made the list in 1990 Muldowney and Garlits were the only top-fuel racers ever to capture the NHRA's Winston championship more than twice.) Both of them had "something to

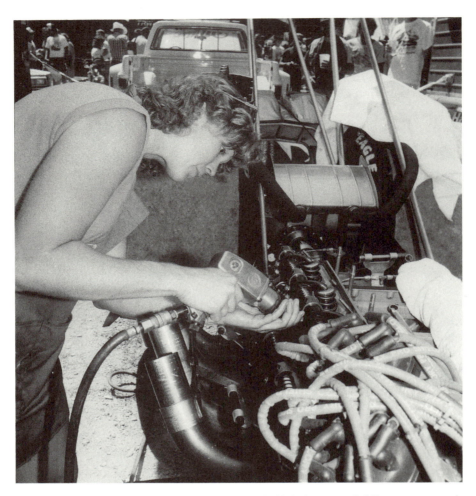

In the pits at a track near Denver in 1987 Kim LaHaie is seen wielding a pneumatic wrench. Kim and her dad, Dick, were on their way to the Winston World Championship. (Courtesy National Hot Rod Assn.)

prove"—yes, of course. But it was not so much to prove themselves winners over and over as it was to prove that they could "pursue existence" as they chose.

One thing was much different for Muldowney than it was for Garlits. She did not need to devise an enemy. The enemy was real, and she could have been blocked altogether from attaining what she wanted. As Virginia Scharff writes, "The auto was born in a masculine manger, and when women sought to claim its power, they invaded a male domain."[46] Grudgingly, men with the power to bestow or deny sanction yielded; political relations shifted. Attitudes changed, if only slowly and certainly not much. If there had not been key members of the establishment (and, in this context, that included Garlits) who saw that they themselves stood to gain from permitting Shirley her own choice, it is not at all certain that she would have been able to define herself as she sought to do.

It might be said that something else was different for Muldowney. Though he regarded patents as a waste of time, Garlits invented things, hundreds of things over the years. Shirley was not an inventor, not in a mechanical sense. What she did do was invent herself in a technological context that explicitly denied any "'natural' desire for calm, peaceful, un-challenged dailiness."[47] After a time she found it almost impossible to conceive of herself in any other context. Drag racer: "It's what I do." In this she was nothing like unique. Many of her contemporaries had trouble imagining themselves doing anything else either. This was partly why Shirley, at fifty, was not exceptionally old for a top-fuel driver. "It takes a special kind of love to keep at it," said Chris Karamesines in 1990, by then past sixty.[48] Apparently, it took even more for some people to give it up.

Even though he finished just out of the Winston top ten (Muldowney got by him at the last event of the season), in 1990 Karamesines had his best year ever on the NHRA's championship trail. The other front-runners included Eddie Hill, fifty-four; Gene Snow, fifty-three; Gary Ormsby, forty-nine; Dick LaHaie and Frank Bradley, forty-eight; and Joe Amato and Kenny Bernstein, forty-six. Don Prudhomme, who figured to be a factor but crashed three times, was forty-nine. The only anomalies were Darrell Gwynn, twenty-eight when he took his last ride at Santa Pod, and Lori Johns, a twenty-four-year-old from Corpus Christi.

Scanning a list of other leading top-fuel racers in the latter 1980s, one finds that Dan Pastorini was thirty-nine when he finished seventh in the Winston points in 1988, and Hawley was thirty-four when he finished ninth in 1990 in Gwynn's car. Then it was back to middle age: Pat Dakin was born in 1949; Jim Head in 1948; John Carey in 1946; Earl Whiting in 1943; Gary Beck in 1941; Larry Minor and Dennis Forcelle in 1940; Jimmy Nix in 1939; Kalitta, Hank Endres, and Jack Ostrander in 1938; Bill Mullins and Richard Holcomb in 1934; and Garlits in 1932. Drivers of the fastest of all race cars averaged a full generation older than in the 1950s.

This was a tenacious bunch of characters; by and large their skills had not slipped a bit. That fact may have been of no more utility than to make other fiftyish people feel good. But the dominance of this gray brigade suggests something more. Sociologically, drag racing was largely a ritual in male bonding at the outset. In 1990 a lot of sustenance appeared to come from family ties. Joe and Jere Amato seemed inseparable. So did Eddie and Ercie Hill. Dick LaHaie's wife, Claudia, and most other spouses were around the races most of the time. LaHaie's daughter, and usually his son, were integral parts of his operation—so was Bradley's son, and Ormsby's, and Muldowney's, and, of course, her husband. Darrell and Jerry Gwynn raced as a team, as did Connie Kalitta and his son Scott, who was the same age as Darrell.

In the 1970s teenagers broke into funny car and top-fuel racing all the time. Jeb Allen and Frank Hawley at seventeen, Billy Meyer and Fred Mooneyham at eighteen, Mike Dunn at nineteen. The second-generation phenomenon was already emerging: Mooneyham's father, Gene, had campaigned the 554 coupe in the 1950s; Dunn's father, Jim, had been around then, too. By 1990 it was unusual for anyone new to penetrate the top ranks, but of those who did many had family ties; Chet Herbert's son Doug, for example, and Cruz Pedregon, whose late father, Frank, had been a California standout in the 1960s, teamed up at one time with Dale Armstrong.

Not that family involvement did not have a long tradition. Pat Garlits was on the road with her husband in the 1950s, and their two daughters were a familiar sight at the drags all the while they were growing up. Tom Hoover began campaigning a top-fueler in 1964, when he was twenty-three, and soon his mother and father, Ruth and George, caught his enthusiasm; George Hoover was in his late fifties then, and he was still working as his son's crew chief in 1990.[49] Tom Hoover had been among the first top-fuel racers to switch to a funny car; another was Jim Dunn, who also was one of the first to race as a team with his wife, the first of many. Over the years Jim and Diane Dunn would occasionally come up against their son in competition, but during 1990 all three of them campaigned a funny car together.

In pro stock, for which engine and chassis setups were treated as confidential information, as in NASCAR, there were family operations built around the aim of protecting secrets. Warren, Arlene, and Kurt Johnson's was one, but in a class by itself was the Glidden family, Bob and Etta, sons Robert ("Rusty") and Billy. Bob Glidden was the most successful racer in NHRA history by far, having won close to eighty major events when no other competitor had won more than forty. He made a fetish out of secrecy: "No one sees any of our engine parts but our family. If your last name isn't Glidden, you have no business looking at them." Glidden sometimes sounded as if he did not quite trust the boys, but he made it clear that Etta was integral and essential to all aspects of the operation.[50]

The relationship of people like Bob and Etta Glidden was indicative that gender roles had changed, and they had. But they had also stayed the same. Many women around drag racing still seemed almost exclusively defined "by their husband's achievements." In 1984, in the wake of Muldowney's accident, a group banded together to form the Drag Racing Association of Women and raise money to defray her medical expenses. DRAW became quite successful in a more generalized charitable role—and as a way of giving a lot of women a better sense of purpose. "For more than 20 years I spent my weekends at the races sitting in the station wagon, reading paperbacks," said Linda McCulloch. "There's more to life than that. There's not a day goes by that I'm not glad I belong to DRAW." While Muldowney was unfailingly grateful for what DRAW had done on her behalf, she also believed that DRAW was a good thing because "a lot of ladies are interested in what their husbands or boyfriends are doing, but otherwise don't have an important function at the races."[51]

Yet the importance of family participation was treated with utmost reverence, and couples often traveled together no matter what "function" each person had. Most drag racers who were not simply adventurers—and there were only a few of those left—were so consumed by technological enthusiasm that family life had to be redefined: "My family is racing." "Drag racing is our family." One can multiply quotes like these endlessly.

When NHRA, at the behest of its underwriters, sought to declare the staging lanes off-limits to the young children of competitors, the outcry was instantaneous: "What is going through the minds of those within our drag racing organization who are standing for such unfeeling, uncaring, and cruel stabs at the family structure?" There was a periodical called *Racing for Kids* which gave heavy play to the family aspects of NHRA activities. Everyone loved the Ceraolos—father Tony, a shop teacher, and his preteen daughters, Kathy and Tonia—who took to the national circuit in 1981, just the three of them, and managed to stay in the hunt for a while. They had to push too hard, though, and were forced to drop out when they ran low on funds. But rarely had there been a more popular team, one reason being that Tony had instructed Kathy and Tonia in "wrenching," in mechanical matters.[52]

Another girl who was learning to wrench around the same time was Kim LaHaie. Like several other top-fuel standouts, the LaHaies were from Michigan, and Dick had been competing since the 1960s on a circuit promoted by the one surviving chapter of the UDRA. But not until 1980 did he join the NHRA's Winston tour as a regular performer, at age thirty-eight. Kim joined

At the 1981 Gatornationals Tony Ceraolo qualified with a 5.76, two-tenths of a second quicker than he had ever run. Then, as he was staging for this first-round race, his transmission broke. Here his daughter Kathy tries to find words to say to her father. It was Ceraolo with whom Virginia Bonito, quoted in the preface to this book, had gained her experience with top-fuelers in the 1970s. (Photograph by Leslie Lovett, courtesy National Hot Rod Assn.)

Jim and Alison Lee, husband and wife from Warrenton, Virginia, were fixtures on the top-fuel scene from the mid-1960s to the mid-1980s. Here, at the NHRA Springnationals near Columbus, Alison is checking spark plugs, while Jim removes a rocker arm cover from their Ed Pink hemi. (Smithsonian Institution collection)

him a couple of years later, when she was twenty. She was just going to sell T-shirts, things like that, but soon Dick discovered that "the fellow I had with me didn't know half the stuff Kim did." Within a year she could perform all the basics, and within two she could do anything her father could do. What was different about Kim LaHaie? She had always been curious about mechanical things, just like her father and her brother, Jeff. "What I do seems unusual to a lot of people," she said, "but to me, it's just something that comes naturally." For Kim "what I do" meant hands-on mechanics.[53]

Actually, there had always been such women in drag racing. Before LaHaie there was Alison Lee, who teamed with her husband, Jim, for some twenty years. Alison's background was quite different from Kim's. She and Jim raised horses at a place called "The Plains" near Warrenton, Virginia. They hired drivers for their fueler, which they called "Great Expectations." But Jim and Alison always worked side by side in the pits, and he said the only difference was that she was a bit faster. "There is nothing inevitable about the masculinity of technology," Scharff observes, "even where the automobile, often considered a kind of metallic phallus, is concerned."[54]

The Lees were wealthy, and Great Expectations was always a machine to be reckoned with. But they never won big, and after decades of hard work they finally grew weary. Said Jim Lee: "We were only trying to make the car support itself, to keep its inventory in good shape. In the end, the money just wasn't there for what we put in it. We just got tired of feeding and feeding and feeding. We got tired of the long thrashes on Saturday nights. We figured there had to be something else we could enjoy."[55]

When the Lees sold their dragster and all their spare parts in the mid-1980s, when they said good-bye, it seemed like losing part of the family. But at least there were others like Alison to make people understand what choices were possible for women. In addition to her mechanical skills, Kim LaHaie could drive as well as her dad. Shelly Anderson, daughter of Brad Anderson, a standout in the alcohol ranks, could both drive and wrench, too. And there were still other women who had the same all-around range of skills which had always been the measure of drag racing's best male competitors.

Having been socialized as they were, it is not surprising that second-generation racers like Kim LaHaie, Mike Dunn, and Shelly Anderson were often quite successful. "Being around the sport from the time they're old enough to walk does make a big difference," observed Jim Dunn; "Shelly could take a race car apart and put it back together when she was a teenager," added Brad Anderson.[56] Darrell Gwynn, who finished near the top of the Winston standings each of the five years he drove a top-fueler (he was second twice), comes to mind, of course. An even better instance is Pat Austin, who was dominant in alcohol funny cars during the same period (and one of the few full-time pros in that realm), with his father, Walt, in charge of the mechanics. In the "industry"—among the people who manufactured components and were as intimate with the racing scene as the competitors themselves—the second-generation standouts were legion. The hope that Tony Miglizzi confessed to his new bride came true. Their son, Lanny, became the force driving L&T Clutches; he did, indeed, get the "disease."[57]

Why, one wonders, would somebody use such a term? Or similar terms like *addiction,* like *being hooked*? This sort of language suggests the possibility of a technological activity harboring some sort of power of its own. That in turn suggests technological determinism, which is considered a dubious concept. Still, the choice of terminology is significant. So is the difficulty that some competitors had giving up an extremely strenuous, not to mention dangerous, activity, even when they grew old, even after they had suffered tremendous losses. So is the enthusiasm of sons and daughters socialized in the activity, who would provide the core of a youth movement bound to occur in the 1990s. In the next chapter, as I begin to wind up this narrative, I propose to consider this matter of technological enthusiasm more closely.

13

ENTHUSIASM

*My chief goal in life is to give the spec-
tators a good show.*
    CHRIS KARAMESINES, 1967

*Very few racers . . . have ever really
made any money.*
    GEORGE HOOVER, 1981

opular perceptions of technology are often distorted by a pair of
assumptions: that it advances according to some evolutionary
imperative and that its most compelling motivation is economic self-
interest. These are logically inconsistent, and they are also incapable of
standing very well alone. I have already marshaled a certain amount of
evidence contrary to the first and will suggest more in concluding this
narrative. As for the second, by considering people with a profit motive, I
may have fostered a misapprehension. Eugene Ferguson's observation
about motivation bears repeating yet again. Self-interest may guide us all,
but the guises that self-interest takes can far transcend the quest for economic gain.

Worth noting, though not central to the argument, is the minority of
competitors in top ranks for whom money was evidently beside the point.
Most of them considered racing explicitly as a recreational activity, but,
even if not, it seemed that they could spend whatever it took. In the 1950s
and 1960s there was Frank Cannon, Tim Woods, John Mazmanian, Lou
Baney (sometimes called "Lou Money"). There was Ted Gotelli, another
Italian-American who, like Baney, had a variety of business interests and

could always afford the best of everything. "You can't run a big fueler on beans," said "Terrible Ted."[1]

In the 1970s there was Jim Annin, who seemed much like Roland Leong. "I knew a lot of guys who had funny cars, and I decided I'd like to do it myself," said Annin. He contracted with Keith Black to build his engines and with Pat Foster for a chassis. He hired the versatile Mike Snively, who had worked for both Black and Leong. When Annin switched to a dragster Snively put it right into the 5s. Annin apparently had all the money necessary to bankroll his racing operation as a diversion, as "more of a hobby than a business venture." On the same order was Barry Setzer, for whom Foster drove at one time. Setzer's mechanical guru was Ed Pink, Black's chief rival; his chassis man was John Buttera. Said publicist Ben Brown, Setzer's machinery was the ultimate because of his "combination of parts, people, post-race maintenance, and schedule." It went without saying that money was the key to the equation.[2]

In the 1980s an affluent life-style was, of course, a major facet of Kenny Bernstein's image. But Bernstein was not the wealthiest drag racer, far from it. Consider Larry Minor, whose family business, headquartered in California's Imperial Valley, was called "Agri-Empire." Agri-Empire employed some two thousand people and was reported to encompass 140,000 acres of farm and range land all over the western states. "To forget the pressures of business," Minor got into off-road racing, then into drag racing.[3] He hired top talent, men like Gary Beck, Ed McCulloch, Bernie Fedderly. Beck and McCulloch were standout performers who had conceded that they could no longer make ends meet on their own, even with help from sponsors; Fedderly, like Dale Armstrong a Canadian, was almost as savvy a crew chief. A sight to see were Minor's pits in the mid-1980s at a major NHRA event, with a dozen mechanics ministering to a pair of dragsters and a funny car simultaneously.

By the time he added the funny car to his racing empire Minor had garnered sponsorships from Miller beer and Oldsmobile, but he still budgeted $500,000 for his newest acquisition alone. He was Keith Black's best customer. His first dragster had been ruined in a crash when still brand-new, and so was his first funny car, but that was no problem, not really. Beck won NHRA's Winston championship in 1983 flying Minor's banner, as did Dick LaHaie four years later, and McCulloch finished near the top of the funny car rankings year after year. Minor helped back Shirley Muldowney for a season in the twilight of her racing career. He offered to share with Garlits the expense of building a machine with the sole aim of going 300 miles per hour. He himself took a ride occasionally, and for a while he had the second best elapsed time in the record book, second only to Beck's in Minor's number-one machine.

Joe Amato was not a landed baron like Minor, but he certainly had all the money it took to compete against him. He said: "Racing is my real lifelong ambition. I've always figured if I got a big enough business, that I could afford to race."[4] He meant race top fuel, for he competed in the alcohol ranks long before stepping up in 1982. Amato had dropped out of high school twenty years before that to take over his father's A&A Auto Parts in Moosic, Pennsylvania. With his brother and two other partners he opened more retail outlets, more than a dozen eventually, and also got into wholesaling. By the late 1980s Amato's Keystone Automotive Warehouse in Exeter covered 350,000 square feet and employed 550 people. One year he reported his gross as $100 million.

Amato's cars always carried signage indicating major sponsorship—TRW, later Valvoline—though most of his competitors thought it likely that his sponsors only began to cover his expenses. He was one of a handful of racers who regularly "rented" strips for test sessions. Asked about R&D, his crew chief, Tim Richards, said: "I would say our efforts are similar to Bernstein's team. . . . For all the things we try, the ratio might be one in twenty that works, so the money is an important tool."[5] Like Minor, Amato backed other racers; these were mostly in the alcohol ranks, but he was also involved with Bruce Larson during Larson's championship funny car season in 1989. Amato himself won three Winston top-fuel championships in a span of seven years.

Amato, Minor, and Bernstein were all newcomers compared to Paul Candies. For a quarter-century Candies was "The Wallet" to Leonard Hughes's "The Wrench." Like Roland Leong, Candies and Hughes retained drivers as full-time employees. They were the first to have an aluminum KB hemi, in 1974. They switched, as the spirit moved them, from funny car to dragster and back, and they were the only team to win NHRA's Winston championship with both types of machine. Candies was CEO for Otto Candies, Inc., a marine transport firm based in Des Allemands, Louisiana, which dominated harbor services at New Orleans. Like the others, he usually had backing from commercial sponsors, such as Ford Motorcraft and Stroh's. But he sometimes incurred large deficits, too, and he covered these out of pocket. "My job," he said simply, "has been to provide the financing."[6]

Not like Daddy Warbucks, but not strapped by any means, were several other fuel racers who owned substantial businesses and were perennial championship contenders in the latter 1980s, such as Gary Ormsby with his Toyota and Suburu agencies in Auburn, California, and Eddie Hill with his Honda-Kawasaki dealership in Wichita Falls, Texas. Their budgets were smaller, and the deals they made for outside financing were very important to them. Although they hired first-class help, people like Ormsby and Hill were operating closer to the edge, especially when parts attrition started running high. And it did. When Hill had a blowover in a spanking new Dave Uyehara car at the 1990 Winternationals he borrowed Darrell Gwynn's spare chassis. Like Amato, Ormsby ordered his chassis from Al Swindahl two at a time (at about $40,000 a pop); in 1989, because of crashes, he needed three just to get through a season.[7]

Top-fuel racers who sustained themselves in championship competition largely on their own dollar (as Keith Black would have said) included Kalitta, with his airfreight service in Michigan; Gene Snow, with his oil ventures in Texas; Jim Head, with his general contracting firm in Ohio; John Carey, with his diesel engine emporium in Delaware; Richard Holcomb, who manufactured corrugated iron pipe in Stockbridge, Georgia; Rudy Toepke, who owned a trucking firm in Dardanelle, Arkansas; and Earl Whiting of Montesano, Washington, whose enterprises were called Olympic Fiber, Inc., and Northwest Log, Inc. "Self-funded" competitors like this could disappear after a couple of seasons of drenching red ink, though some of them hung on year after year, or, in Kalitta's case, for most of a lifetime. None of them won an NHRA championship, or even came close, but they were all enthusiasts; they would all tell you that they loved the technological challenge, however costly it might be. Carey designed a capacious rolling workshop, and for a time he fielded three cars at once, just like Larry Minor. The expense, he said, was "unbelievable." "If this wasn't fun, I wouldn't be doing it . . . this sport leaves a pretty good dent in your wallet."[8]

Some of drag racing's upper crust had inherited wealth, others were self-made men, but rarest of all were those who put themselves "ahead of the game" (Don Prudhomme's phrase) through racing. Notable besides the Snake and Garlits—and one should not forget Bob Glidden in the pro-stock ranks—was John Force, who was perhaps drag racing's ultimate hustler. Force cut dozens of sponsorship deals, large and small, and his funny cars were an ever-changing panorama of commercial signage. When he felt he could afford it he hired Austin Coil as his crew chief; Coil did not come cheap, but many insiders regarded him as Dale Armstrong's closest rival. Force had grown up "by the railroad tracks in a mobile home" and was making his living as a truck driver when he signed his first contract with a sponsor.[9] He loved to remind people of his humble beginnings, to make sure they understood that "I ain't a potato grower" (an allusion to Minor, whose father had founded the San Jacinto Packing Company when Larry was a child). As a rule, however, people who already had plenty of money seemed to find it easier to attract corporate funding, perhaps because they had some previous experience with boardroom types, perhaps because potential sponsors could feel confident that they would not prove an embarrassment if they needed to go deep into their own pocket to keep up.

In any event, competitors with cushy sponsorships were, at most, only a handful. In the 1980s an enthusiast named Bill Pratt began publishing a periodical called *The Drag Racing List* for "hard-core" fans, ranking racers worldwide according to their best elapsed times. In 1990 he enumerated about 90 top-fuel dragsters—75 percent of them in North America, 10 percent in Australia, the rest in England and the Scandinavian countries. There were about 100 fuel funny cars. Pratt also listed something like 450 pro stockers, 250 alcohol dragsters, 300 alcohol funny cars, 80 jets, wheel-standers, and the like, and 320 pro modifieds (a hybrid sort of funny car/pro stocker). All told, he listed some fifteen hundred racers in twenty distinct classes, but in its full array of more than two hundred classes NHRA had more than twenty thousand competitors, and there were thousands more who raced primarily at IHRA or independent tracks. Among fuel dragsters and funny cars the number bearing evidence of substantial corporate sponsorship was about 20 percent; it was a minuscule proportion among all the rest.

To be sure, there had been dozens of corporate brands linked to drag racing: R. J. Reynolds and Budweiser, of course; also Skoal, Kodiac, Levi Garrett, Export-A, Winfield, and e.z. Wider; Coors, Miller, Stroh's, Old Milwaukee, Schlitz, Simpatico, Olympia, and Hamms; Valvoline, Pennzoil, Castrol GTX, Quaker State, Mobil 1, STP, Amalie, Kendall, Crown, Amsol, Huzoil, Exxon Superflo, and Sunoco Ultra; AC-Delco and Exide; Black & Decker, Sentry, and Raybestos; Lava, Jet-X, and TR-3 Resin Glaze; Buick, Oldsmobile, Pontiac, Dodge, Plymouth, Mopar Parts, and Ford Motorcraft; Super Shops, Chief Auto Parts, Nationwise Auto Parts, Summit Racing Equipment, Competition Specialties, and Parts Automotive Wholesale; King's Hawaiian Bread and Wonder Bread, Wendy's and In 'n' Out Burger; Coca-Cola, Pepsi, Bubble Up, Mountain Dew, Hawaiian Punch, Snickers, Jolly Rancher, and Otter Pops; K-Mart and 7-Eleven; Pioneer Stereo and JVC Car Audio; Mattel and Revell. Even English Leather and Faberge Super Brut. And even the U.S. Army, Navy, Marines, and Air Force.

Many of these names appeared on a succession of race cars, some on several of them simultaneously. Yet over the years the vast majority of drag racers had never had any significant amount of outside funding. Moreover, a corporate brand name emblazoned on a fueler or funny car was not necessarily indicative of a financial commitment for better or worse. Because sponsors were primarily interested in television exposure, contracts with racers sometimes stipulated that there would be no payment in the event they failed to qualify or advance into late rounds. Looking around the staging lanes at Pomona in 1990, Shirley Muldowney suggested that only Bernstein's Budweiser deal, Prudhomme's with Skoal, and a few others really amounted to much. Plenty of racers with Fortune 500 names painted all over their cars, she said, were actually getting less than what Pioneer Stereo had given her in 1982, $150,000 a year, and some were getting next to nothing.

Most significant were those cars on which the most prominent billing was the owner's *own* name. At the level of the Winston championship points chase this was almost inevitably a losing proposition. Or, if it were conceivable that someone in such a position could stay out of the red, it is still safe to say that motivations could not be "precisely calibrated in economic terms." To haul a fueler around the country and flog it (the expression was both ubiquitous and apt) week in and week out was extraordinarily hard work. By the latter 1980s only a handful of competitors had more than a prayer of winning a Winston series event: In the eighty-odd events staged between 1985 and 1989 there were only fourteen different winners in top fuel and eighteen in fuel funny cars. In each category the same half-dozen competitors won about 80 percent of the time. All of these had corporate sponsorship, but, more to the point, there were unsponsored racers who made it to every major event—from New Jersey to the Pacific Northwest, Minnesota to Louisiana—and rarely even got past the first round.

Through contingencies and various quid pro quo arrangements all racers sought to minimize their out-of-pocket costs. They learned ways to take advantage of the American tax system: Expenses were usually deductible if someone could show "an intention to make a profit" (even if this intent did not appear "reasonable"), or they could sometimes be charged off to a separate business as advertising.[10] Even so, very few competitors could keep the books balanced over the long haul. McCulloch, who switched from dragsters to funny cars in 1969, recalled running match races in the early 1970s for $1,500 guarantees, one hundred dates a year, at a time when "there was really very little engine maintenance": "Just make three laps, check the bearings, and go to the next race." But flush times faded in the rampant inflation of the Carter era, and when Minor hired McCulloch he had been out of racing for several years.[11]

Whatever their financial situation the majority of competitors had been socialized into the activity at a young age, often as children. McCulloch first became involved in 1957, when he was fifteen, but Jeb Allen began going to Lions every Saturday when he was seven. Each of his two older brothers raced a dragster, and his dad had been a chum of Keith Black's since school days. Jeb got his top-fuel license in 1971 while still attending Bellflower High. ("They're really easy on ditching," he observed.) He went on tour in the spring of 1972 and won the NHRA Summernationals a month after his eighteenth birthday. Resourceful and hard working, he would win the Summers twice more. Match races, easy money, were impossible to come by for most people with dragsters, but not for someone who had made his name as "the world's fastest teenager." In return for publicity English Leather gave him enough to pay some of his bills. Although he pushed his equipment hard in open competition, he tried to keep overhead to a minimum. Racers like Billy Meyer and Bernstein were beginning to show up with semitrailers crammed full of spare parts, but, to NHRA's Wally Parks, Jeb personified the competitor who had succeeded without "high-roller rigs and the extras they afford."[12]

Allen won the AHRA Grand American Series in 1976, the IHRA's Winston championship in 1980, and the NHRA's Winston championship in 1981, after finishing third to Beck and Muldowney the year before. He remained in the top ten in 1982, but then everything began coming unraveled. At the start Allen had told an interviewer that he was going to join the pipe fitters' union, like his dad, "to have something I can fall back on." He had been around drag racers most of his life, and he knew that there were a lot of people who had "been on top and now they're broke." By 1983 Allen was forty thousand dollars in the hole. Bookings grew scarce because he had lost his reputation for reliability. Then his creditors pounced, and, like others he had known, he too was broke, at the age of twenty-nine.[13]

"That's racing," people remarked. Everyone knew that a competitor's fortunes could turn quickly. Yet there was nothing that Jeb Allen loved more than putting on a good performance, and he never stopped dreaming of a return to the limelight.

Occasionally, a prominent racer would retire and simply exit the scene, once and for all, apparently with no regrets. Such was the case with John Wiebe, who had come off a Kansas farm in the mid-1960s and abruptly returned to farming a decade later. But the tendency was to treat retirement as something transitory. Texan Jody Smart, who finished third in both the NHRA and IHRA standings in 1983 but sold his equipment to John Carey soon afterwards, said he would return in a minute if circumstances looked favorable. So would nearly every other racer who had ever been on the circuit.

In 1986 Prudhomme's major backer, Pepsi-Cola, dropped his option. The Snake had not been without a well-heeled sponsor since the formation of Wildlife Enterprises. His rig was ready to go, but he announced, "I don't plan to run until I can get a sponsor deal where I can afford to run properly."[14] With NHRA and even RJR people mediating, U.S. Tobacco came to Prudhomme's rescue in 1988. For a whole year, however, he had followed his old boss Keith Black's admonition about the inadvisability of running on "one's own dollar." So it was, too, with Bruce Larson, the Winston funny car champ in 1989. The money Larson had received from Sentry Tachs and Gauges was the largest line item in that firm's budget. When a new management reconsidered its priorities and bowed out of drag racing a year later, so did Larson. Before he made his Sentry deal Larson had managed to survive as a privateer for twenty years, and he stood as good a chance as anyone of competing on his own dollar. Well into his fifties by 1990, he elected not to try.

Yet, for every such instance of that sort of caution, there were instances of racers who were ready to challenge the odds. The risk could be rationalized. Bernstein might estimate his average cost per run on the order of five thousand dollars, but he was factoring in an R&D budget as well as paying several employees and amortizing his investment in an enormous spare parts inventory. Expenses could be substantially less for someone with minimal overhead: A New Jersey racer, Nick Bonifante, figured his cost per run at just one thousand dollars. That was with only routine expenses, however. An engine explosion could wipe out $20,000 worth of equipment, a crash could cost much more than that, and "Nitro Nick" had both types of mishap. His outside funding was modest, and he had to be fighting a losing battle. But drag racing had plenty of Nitro Nicks, even in 1990, all of them driven by some variety of technological enthusiasm.

Some people loved the idea of "performing" with their machines, and they treated this as being worth almost any price. It was not worth any price to Prudhomme, but it did mean a lot to him. His wife remarked, "The fans might have missed him, but he missed them a whole lot more."[15] When Tom McEwen found himself without a sponsor for the first time in twenty years he reacted differently; he built an exhibition machine, a funny car that looked like a '57 Chevy hardtop, and kept on trouping while awaiting the appearance of someone "with enough money who wants us to be competitive again."[16] Deep down nearly every competitor may have resonated with Chris Karamesines's comment that racers were "just like actors and singers."[17] Or athletes. After his days as a weekend hero in the NFL came to an end Dan Pastorini bought a dragster from Gene Snow and dubbed it "Quarterback Sneak." Only moderately successful at wangling sponsorship, or in the Winston chase, he sought to make ends meet by match racing and selling T-shirts to fans (a fair source of income for "name" racers). "I love competition," Dante said, "I thrive on it."[18] Presumably, he thrived, too, on the roar of the crowd.

Another competitor already familiar with the roar of the crowd was Jeff Bernstein, a studio musician and sound system specialist, who got into drag racing with his wife, Susan, because, as Susan put it, "we were bored to death and I . . . made up my mind to do something exciting." Neither of the Bernsteins yearned for center stage, the driver's seat, but there were others for whom suiting up and getting buckled in was the biggest thrill of all. After Paula Martin, a skydiver and water ski racer as well as drag racer, was seriously burned in her funny car, she promised she would return: "Some people have a biochemical requirement for kicks," she said. "I will race my car again . . . because of the sheer and unequivocal feeling of joy I experience when I go fast."[19]

While nearly everyone connected with drag racing had tried driving at one time or another, many lacked the basic skills, and many more had nothing like the right stuff. Yet the vast majority of those who owned fuelers and funny cars were also enthusiastic about driving them. Even Kalitta, after several horrendous crashes and undoubtedly out far more cash than he had ever pocketed, remained certain that he would keep driving until he got bored, a situation he suggested was not imminent. Jim Bucher, a top-fuel standout in the early 1970s, kept driving when he knew his days were numbered because of a rare disease. Richard Holcomb died of heart failure at fifty-six, not unexpectedly. Ernie Hall was stricken at fifty-five, shortly after completing a qualifying run at a Winston series event in Seattle; many years before he had told his wide-eyed six-year-old, Carolyn, "I'll race 'til the day I die."[20] Gary Ormsby, who won the Winston top-fuel championship in 1989, then barely lost out to Amato in 1990, had apparently known the whole time that he was dying of cancer.[21]

Some racers liked putting on the dog, but others took pride in being thrifty. It worked, up to a point. From the inception of the Winston Championship Series in 1975 Frank Bradley managed to finish in the top ten two-thirds of the time; he was second in 1976, fourth in 1987, sixth in 1988, seventh in 1990. Although he was a regular on the AHRA's social security circuit, rarely did he have any substantial amount of outside funding. At one point he established a base at Mike Lewis's Maple Grove Raceway, camping out all summer between race dates. Dick and Kim LaHaie won the Winston championship in 1987 with backing from Miller beer, but even before that they had been able to finish as high as third with minimal sponsorship and a modest budget; when Miller bowed out of drag racing in 1990 Dick kept going, although he insisted that he "wouldn't let his ego get in front of his wallet."

Racers were always talking about "ego." Ego was the "one thing" that kept them going, a *Drag News* commentator suggested in 1970.[22] That was not true for everybody, but it was certainly a big factor. When not racing themselves, Prudhomme, Pastorini, and others stayed in the limelight by doing TV. When Garlits finally quit the NHRA circuit he signed on as a commentator for Diamond P Sports. And, no doubt, some racers had colossal egos, including some of the best—Jerry Ruth, for example. Ruth was from Seattle and called himself "King of the Northwest." While he never made a run at the NHRA championship, he was as comfortable financially as any racer on the national circuit. He prospered by running a conservative tuneup, rarely traveling long distances, cultivating amicable relationships with people who counted, and aiming "to hang on to as much of his earnings as he can while still being a winner."[23] Though never heavily sponsored, Ruth derived his income solely from drag racing for nearly twenty years, and he might have continued if he had not gotten into a mess involving cocaine and drawn a long term in federal prison.

There were others who became so used to the hustling it took to survive that they hustled themselves right into jail. Tom Madigan, who had been around drag racing for thirty years, first as a driver and later as a journalist, insisted that most of the veterans had "criminal minds." He could cite a lot of evidence, yet by the 1980s there were probably no more bad apples in drag racing than in baseball, while there were untold numbers who lived ordinary lives, for the most part, but harbored an abiding passion for high performance machinery.

Ernie Hall ran a business outfitting fire engines in Cornelius, Oregon, and he also loved drag racing. A teenaged Ed McCulloch was an early protégé of Hall's, but mostly he went it alone. For a time he joined the Grand American circuit, but after the AHRA expired he kept his equipment up to date and still made a few race dates every year. He had his purposes, but showing a profit was not among them.[24] Like Hall was Ray Stutz, a fire fighter in LaVerne, California, a stone's throw from the Pomona fairgrounds. In 1984 Stutz competed at a handful of races, taking in $16,000 while claiming expenses of $28,000 (and surviving an IRS audit). That same season Dave Braskett and Dennis Taylor made seventeen runs in their "Lone Eagle" at a cost of $15,000, likewise about twice what they took in. "The sport is addicting," Taylor observed, a point others made time and time again.[25]

In 1990 forty-seven top-fuel racers competed in the Winston series, although more than half of them entered only one or two events, and only a dozen made the entire circuit. Of that dozen several were hobby racers in some sense; the other thirty-five were that in every sense. They never thought in terms of amortizing their investment but only about trying to get off easy on their expenses. A few of them claimed they did. Ron Cochrum and Howard Haight, friends of Ray Stutz's from Pomona, entered four events in 1984, running well enough to qualify at the NHRA Nationals. They spent ten thousand dollars and reported exactly breaking even. Georgian Bob Strauss had once been an IHRA regular, but by the latter 1980s he was racing only a few times a year. In October 1988 he ran his all-time best at the Texas Motorplex, 5.27, 273 miles per hour, and claimed his debit came to only $1,100. "A guy who loves billfishing could spend that much on a weekend trip," he remarked.[26]

Strauss insisted that he used only fresh equipment, having decided that this was cheaper in the long run than paying for the damage when some tired internal component failed. "Both my crankshafts are brand new," he pointed out, "$2,500 apiece." More common were racers who bought used parts that others felt were near their fail-safe margin. Some well-financed teams were very conservative about this—when Bruce Larson was chasing the championship in 1989, his crew chief, Maynard Yingst, would put only a dozen runs on a crankshaft—so there were always plenty of secondhand parts available.

While a broken crankshaft was catastrophic, there were components like fuel injection systems which never failed or even wore out. But setups were always changing, and anybody trying to maintain an outdated combination was fighting an uphill battle. A poignant case in point involved Roger Coburn of Bakersfield, who had fielded winning top-fuelers for most of the 1960s and 1970s. In partnership with James Warren and Marvin Miller, Coburn won the March Meet at Famoso three times in a row and won week after week at southern California strips like Lions and Irwindale.

Opponents dubbed the team "The Ridge Route Terrors," the Ridge Route being the road over the mountains between the San Joaquin Valley and the Los Angeles Basin. They tied for fourth place in the NHRA Winston series in 1976, after having won the Winternationals, then, on a rare trek east, the Gatornationals too. Chris Martin of *National Dragster* wrote that there was probably no track in the West at which they had not held the record at one time or another: "They were that good."[27]

Warren once remarked: "If I couldn't drag race, I don't know what I'd do."[28] But, finally, he grew weary after two decades of thrashing alongside Coburn. Miller, whose "Rain for Rent" (portable sprinklers for San Joaquin Valley farmers) had helped sustain the operation, backed off. Coburn was on his own, with obsolete equipment. It seemed he could only get mediocre drivers. He increasingly resembled, as Martin put it, "Muhammad Ali in the twilight of his career." A construction worker, Coburn had once said that, if he and Warren could ever win the NHRA championship, "maybe by then we'll have enough money to open a little business. Maybe a machine shop." It was a line straight out of the movies. In the middle 1980s Coburn was still trying, but he could not match times he had run ten years before. "I've lost the handle on it [the motor combination]," he said, "but I don't know, I think we're gettin' it figured out. That's the whole thing that's motivating me . . . to get the car to run well again."[29] Warren and Coburn had been highly respected by everybody in drag racing, and no one was anxious to tell Roger that, given what he was working with, it just was not possible. Nor did anyone discount the possibility of a successful comeback by Coburn someday, somehow, perhaps even with James Warren driving.

After John Mulligan was killed, his partner Tim Beebe returned to racing with a funny car, but he too seemed to have lost the combination. Old timers like Jack Chrisman and Mickey Thompson had similar experiences, but people like Coburn, Beebe, Chrisman, and Thompson had at least known what it was like to be winners. There were, on the other hand, plenty of racers who did not. Take Rodney Flournoy, who began racing in 1977 in a well-worn Don Long funny car that Jerry Ruth sold Rodney's father, Eddie, for $5,500. Ruth had met Eddie Flournoy in the mid-1950s, long before Rodney was born, when Ruth was running a Studebaker gasser at Santa Ana and Flournoy was crewing for Jazzy Jim Nelson, among others. Eddie had been around, and he could rely on sound advice from friends like John Kimble. In 1981 his son went to the final round in an event at Orange County. He never again did that well, yet the Flournoys kept at it, with an array of equipment that was probably even sadder than Roger Coburn's.

"I don't want to see my kid and his car just sitting around," said Eddie at the 1987 Winternationals. "Once in awhile we make a few bucks with it." But events like Pomona were so tough, he added, and he knew it was "a 90 percent chance we won't qualify." Really, they were there just "to have fun."[30] Once in a while some independent fuel racer, perhaps just there to have fun, actually won a big event, though never anybody as underfinanced as the Flournoys.

Fielding a funny car was more expensive than a dragster, because of the cost of the bodies and because racers generally felt constrained to use fresher engine components, in the interest of protecting both the driver and all that brittle and combustible fiberglass. To offset this there might be a match race once in a while. For the vast majority of people with dragsters, however, open competition—no guarantees—was the sole option. Racers would show up at Winston series events with the odds stacked against them even higher than 90 percent. There was, for example, Arnold Birky of Santa Rosa, California, who had raced dragsters since the 1960s but was perhaps best known for a 1988 engine explosion at Pomona, reputed as the most awesome ever. Neither driver Bob Neal nor anyone else was hurt, but one piston left the premises altogether, and other parts disappeared into thin air. As they went out the gate, it seemed unlikely that Birky and Neal would ever return, yet they were soon regrouping, buying cast-offs from Ormsby and eventually even replacing their old hand-me-down chassis from Jeb Allen.

By the latter 1980s NHRA was paying first-round losers five thousand dollars, and that was tempting to someone with a relatively modest investment. But it generally took 5.10s to qualify; Neal seldom made a complete run all the way through the lights without something going awry, and never once had he run quicker than a 5.38. Qualifying runs would consume a $1,500 drum of nitro, just for starts. What could be on Arnold Birky's mind? "You've heard it from others, I'm sure," he said, "but I love racing. . . . I love the sound of the motors, the power of the cars. I love the people involved. How they are always willing to help. They are the greatest in the world. . . . If I wanted to quit, which I don't, it would be really tough."[31] The sound of the motors, the power of the cars, the people. The element of camaraderie might be particularly important to drag racers insofar as their activity defied all efforts to endow it with the chic of the NASCAR or CART circuit. It might be something special to people like Birky, who were derisively called "leakers" and yet admired for their tenacity. In that tenacity, however, one could often get an impression of something obsessive.

One heard the expression "hooked" time and again, but there were many other ways to express the same idea. In 1965 Lou Baney's mechanic, John Garrison, said, "You never can get ahead of it, it won't give you an inch." And listen to Pete Robinson, also in 1965: "I love to race, and as long as I can do it safely . . . I intend to race. I guess it is just in my blood."[32] After two serious crashes Pete came right back, and it happened again, the third crash fatal. Another competitor who suffered a series of mishaps, but survived them all, was Jeg Coughlin, a parts wholesaler from Columbus. While recovering from his injuries, he would always order a new car. Early in the 1982 season an ad appeared in *National Dragster* for Jeg's state-of-the-art Al Swindahl fueler:

FOR SALE—*QUICK!*—BEFORE JEG DECIDES TO GO RACING AGAIN

Jeg's home track was National Trail Raceway, where in 1984 Doug Kerhulas, from Bakersfield—a protégé of Warren and Coburn—had his wing struts break and foul his parachute just as he cleared the traps. With one of the shortest shutoff areas of any major track National Trail had a spring-loaded net to catch cars unable to stop otherwise. Kerhulas hit the net at high speed. A cable snapped back, inflicting head injuries that left him critically injured and with some long-term impairment. Yet he was back in competition inside of a year, with Don Prudhomme's cousin Mark driving. And that was nothing compared to the saga of Dave Edstrom, a Minnesota fuel racer who lost his eyesight as a result of diabetes in 1974, then returned with a machine called "Blind Faith," performing the mechanical work himself, his son driving.[33]

Such people could hardly bear the thought of being left out. It was an emotion shared by losers and winners alike. When Garlits was temporarily inactive he yearned "to see the action, smell the familiar nitro odors, and experience those electric, throbbing moments." When Prudhomme returned after making his Skoal deal he remarked: "I feel like I've come home. I've missed the sound of those engines and all the technology."[34] Arnold Birky told a classic yarn: "I know this racer who used to have a dragster, and he wanted to quit and make it stick. He moved to a small out-of-the-way town, and, to this day, almost totally refuses to speak on the phone to anybody connected with racing. He won't go to the drags, read *National Dragster,* be around any kind of hot car . . . that's how bad it gets."[35]

The phrase "technological progress" used to roll off everybody's tongue with consummate ease, but no more. As negative perceptions mount, historians of technology have become increasingly inclined to dwell on the dark side, to point out that costs often outweigh benefits. But scholars run a risk in universalizing such perceptions. They risk losing sight of the fact that technological practitioners can find their pursuits every bit as exciting as they themselves find their own work. This excitement is abundantly evident in books like Tracy Kidder's *The Soul of a New Machine* (1981) and Tom Wolfe's *The Right Stuff* (1979), and, as I have noted, historians like Brooke Hindle, Thomas Hughes, and Eugene Ferguson have described it in such terms as exhilaration and enthusiasm. Ferguson writes:

Enthusiastic technologists not only have built the world we live in but . . . by and large, they themselves have hustled the support required to be able to do so. To build nearly anything, and particularly something that is difficult or hazardous, is an intensely interesting process for those who have to solve the many problems that are inevitably encountered.[36]

To concur in this observation is not necessarily to abandon a critical stance towards the "progress talk" that infects popular discourse about technology. The point is that people whose lives are concerned primarily with words on paper need to realize how compelling an experience the pursuit of technological progress can be. For an academic to write contemptuously about "the widespread feelings of individual impotence that somehow found compensation for many Americans in their love affair with cars" does little to enhance understanding.[37] There is significant meaning in what racers say about being "hooked." The allusion seems, yes, to be to some sort of "elemental force" (in Hindle's phrase) intrinsic to technology. Positing the existence of such a force is not the same as claiming that technology advances autonomously, but it certainly does say a lot about how it gains momentum—how *any* technology gains momentum, not just the esoteric and essentially benign pursuit being addressed here—and it is clearly important for us to know how technology gains momentum.

Many of the people mentioned above sought simply to keep up by learning the techniques of the pacesetters. They were not innovators, except insofar as they were cast in that role by the nature of their efforts to wring high performance out of limited finances. While emulators always outnumbered innovators, nobody doubted that it took an innovative flair to become a pacesetter. "What may be new today is old stuff tomorrow," said Tony Miglizzi, who spent a lot of time on R&D work with Dale Armstrong.[38] In part, men like Armstrong, Tim Richards, and Austin Coil were paid expressly to devise novelties, and they devised plenty of them, often to the point of alarming their less affluent competitors along with the high sheriffs, who had long since indicated that they would permit the exercise of only so much "ingenuity."

Coil refined his skills as a co-owner of the "Chi-Town Hustler," a popular match racer in which Frank Hawley also won NHRA's Winston funny car championship two years running. Coil talked of an endless succession of novelties such as "a short stroke V10," which, he thought, "might produce considerably more usable power than a V8 of comparable displacement."[39] After he went to work for Force he contrived electronic timers that eventually aroused suspicions that when Force went down the track he was not doing all the "driving" himself. NHRA fretted about such devices for several reasons and, as we shall see, eventually forbade their use. As had happened before, political forces stood directly athwart the path of technological progress, yet a frontline crew chief was always facing the necessity of gaining an edge.

By 1980, when the NHRA's Winston series had clearly become drag racing's premier competitive stage, Coil was convinced that this chase would be a losing proposition for any racer except the winner. Of the twenty championships between 1981 and 1990 (ten in top fuel, ten in funny car) Coil won three (two with Hawley and the Hustler, another with

Force), Richards and Amato won three (four, counting the role Richards played in Larson's success), and Armstrong and Bernstein won four—half the total.[40] Garlits won twice, as did dragsters owned or backed by Minor. Candies and Hughes had one win, as did Gary Ormsby and his crew chief, Lee Beard.

Ormsby remained devoted to the pursuit of novelty even after losing a small fortune in his ill-fated experiments with aerodynamics: "We still believe in developing new technology and taking some chances," he said. "We're not gun-shy." Beard added: "We never try to copy anybody. We've never looked at something on another car and said, 'That's the piece that beat us.'"[41]

The winners, no doubt, were also innovators. But the reverse of this proposition was not necessarily so. Among those racers who were most enthusiastic about novelty some seemed to hold to it as an end in itself, to pursue novelty at the *expense* of having a shot at winning. Walking among the cars pitted at Pomona, Englishtown, or Indianapolis Raceway Park, one was bound to see various paeans to nonconformity, often in the form of neatly lettered inscriptions:

> Don't Go Where There's a Trail . . .
> Rather, Go Where There's No Trail
> And Leave a Path.

Or,

> If a man does not keep pace with his companions,
> perhaps it is because he hears a different drummer.
> Let him step to the music which he hears,
> however measured or far away.

If sufficiently dauntless, even the unsuccessful innovator was held in utmost esteem. Jocko Johnson was a sterling example, but there were countless others who had heard the sound of a different drummer. Tony Nancy remarked that innovation was one of his primary objectives. The "basic urge" behind his mid-engine Wedge of the mid-1960s, he said, was "to try something different and new—maybe even oddball."[42]

*E*ventually, mid-engines became absolutely conventional, of course, and by the 1980s a top-fueler hewed to all sorts of other conventions. At any given time there were three key determinants of "normal" design. There were mandatory materials specifications and other safety standards derived from a dialogue among competitors, manufacturers, and sanctioning bodies, the forum provided by the SEMA Foundation, Inc. (SFI), a spin-off of the Specialty Equipment Market Association.[43] There were also mandatory but essentially arbitrary rules imposed by the sanctioning bodies. The NHRA stipulated the maximum engine displacement, tire width, and airfoil surface, the minimum weight, and many other matters; SFI set forth specs pertinent to matters such as the size of chassis tubing and the strength of containment devices. NHRA's constraints were in large measure political, SFI's derived more explicitly from "technical" considerations. Beyond these there were only tacit conventions, a prevailing consensus among competitors about empirical matters. And, even though many options had been closed for either political or technical reasons, there were plenty of choices still open.

Consider, for example, the mix of factors pertinent to the running gear and chassis of a top-fueler. For many years the NHRA rule book never specified the number of wheels. Four had been conventional, of course, but there had been dragsters with three wheels: Kenny Ellis had built tricycle slingshots in the 1960s, and Jeff Jahns raced a machine that not only had three wheels but was also a mid-engine sidewinder that started from jacks. And there had been dragsters with six: Beyond those with two pairs in the rear Art Chrisman designed a twin-engine slingshot with two independent rear axles, one with the wheels and tires set wide apart and overlapping the other pair. There were several machines that had both the front axle and the rear powered. Tommy Ivo's Showboat was the most famous, but Manuel Coehlo had actually captured the Lions ET record with such a dragster in the 1950s.

Except when both axles were powered, all dragsters were designed to have very little weight on the front end, and the convention became a pair of spoked wheels with tires about the size of a bicycle's. After Shirley's accident in Canada NHRA mandated disc wheels. When Garlits debuted Swamp Rat XXX he ran on flat Kevlar belts instead of tires, but soon substituted small cast-aluminum wheels with inflated tires about thirteen inches in diameter. The idea was to minimize frontal area. Most other racers emulated Garlits, but a few preferred the old way, and in 1990 the two setups remained one of the obvious points of design divergence.

After 1987 the rules specified two front wheels with a minimum of twenty-six inches between them, but the distance between the rear wheels remained optional, like engine placement and attitude; there was every reason to suspect that cross-mounting had not been tried for the last time. Mickey Thompson had once observed that, ideally, there would be only one rear wheel. What he had in mind was directional stability, and that is presumably what the Sandoval brothers were seeking in the 1960s when they designed a fueler with the tires only a few inches from one another. When the idea was revived in the 1980s, however, it was with different considerations. Richard Holcomb was making a run at DeSoto Memorial Speedway, a Florida strip owned by Art Malone, when an engine explosion damaged his chassis. Holcomb's crew chief, Clayton Harris, could see that it would need to be "back-halved" (i.e., reconstructed aft the cockpit), and Malone remarked that any car ought to run quicker and faster "if there were some way to pull the tires in out of the air."[44] With Connie Swingle, Harris worked things over, leaving only seven inches between the rear tires. Among other positive results Holcomb reported an absence of "shake," a problem of dynamic imbalance in the rear axle and tires which occasionally gave everybody fits. Holcomb became the third man to run in the 4s, after Eddie Hill and Gene Snow, at the Motorplex in October 1988. There was no immediate rush to emulate his setup but every likelihood that it would happen eventually. When Garlits speculated about a fueler unconstrained by any "foolish rules" he envisioned a three-wheeler with a rear end like Holcomb's.

Since the rules said nothing about chassis width, it remained permissible to design a frame much narrower than conventional from the cockpit all the way forward, and Jim Head actually did so. But Head was a man who frankly admitted to being "more interested in inventing and engineering and coming up with a better mousetrap than . . . winning races."[45] As of 1990, there were no other cars like his. Nor were there many that were not just about 300 inches long, the maximum NHRA permitted. There was a minimum, too: 180 inches. This left a large area of choice, however—ten feet, to be exact—which was settled only in the realm of tacit convention.

Initially, there had been no consensus about length. Scotty Fenn regarded a chassis as "nothing more than a clutch between the engine and the ground." "If the chassis doesn't utilize the power delivered to it," he said, "then it is not accomplishing its mission."[46] Fenn's theory about optimum wheelbase—that it should be equivalent to the rear tire circumference—meant that his cars were very short. While he never made it clear why he advocated precisely that length, the general idea was obvious. A short dragster would readily pull the front wheels off the ground when leaving the line, transferring all the weight to the slicks.

There was an opposing school that minimized the import of dynamics and maintained that it was preferable to have a longer wheelbase, which put a greater proportion of the *static* weight on the rear tires while *discouraging* wheelstands. Static weight distribution clearly affected elapsed times, and, as Roger Huntington wrote, "with the engine in a given position in relation to the rear end, a dragster can get a greater percentage of its weight on the rear end by merely lengthening the wheelbase." Although gas coupes (and later funny cars and pro stockers) provided clear-cut evidence for the viability of Fenn's concept, as improved tire compounds permitted utilizing more and more horsepower, the trend with dragsters was towards longer and longer wheelbases.[47]

In the interest of optimizing static and dynamic constraints, there was continual tinkering with the arrangement of components—how high or low the engine was mounted, how close to the rear end, how far the cockpit was behind the axle or, later, in front of the engine—and there were strong and conflicting opinions about what was "right." There were also strong

Clayton Harris, after a long roller-coaster career with top-fuelers (and bearing the scars of one of his crashes), is clearly a happy man in this 1989 photo, as he goes about his work in his role as crew chief for Richard Holcomb. (Smithsonian Institution collection)

Among the novelties displayed by Jim Head's 1990 machine were the narrow chassis and front axle, large front wing, vertical stabilizers on the wing stand, and aerodynamic pods in front of the slicks, the latter integral with a ground-effects tunnel aft the engine. (Courtesy Jim Head)

opinions about whether a chassis should be flexible or rigid. Woody Gilmore emerged in the mid-1960s as an outspoken exponent of flexibility, but after the advent of the mid-engine configuration it became apparent that a flexible chassis was more susceptible to shake, and a new convention emerged.[48] Along with extra diagonals to ensure rigidity, torsion-bar front suspensions gave way to wheels solidly mounted with A-arms. But a completely rigid structure tended to break welds, and, eventually, people began to perceive that "the chassis was telling them it did not want all those extra crossmembers."

Talking in terms of what some piece of hardware was "saying" was not unusual. Garlits, who happened to be running one of the few chassis he ever bought from another fabricator, sawed out "all the rick rack." Swindahl did the same thing to a chassis of his. What both men figured was that a chassis "wanted" to be rigid behind the cockpit but flexible in the long expanse of tubing up front. Swindahl began using thinner tubing of larger diameter, as did Dave Uyehara, because it was stronger and more flexible at the same time.[49]

By the mid-1980s the chassis of a state-of-the-art top-fueler was typically about twice as long as a funny car's, had no diagonal members ahead of the cockpit, and was more flexible than anything Woody Gilmore had ever dreamed of. But there were a few people who kept wondering about alternatives. Although design conventions had been tested before, mostly with discouraging results, they were about to be tested again by a fabricator in Kenosha. Dennis Rollain had started out in partnership with John Buttera but remained behind when Buttera went off to California. In 1984 Rollain struck a deal with an old friend from Itasca, Illinois, Dave Miller; he would provide him with a chassis if Miller would outfit and race it. Although Miller claimed he was expecting a long "flexie flyer" pretty much "like everybody else's," Rollain had something else in mind. He delivered a machine with diagonals between every upright member and a 200-inch wheelbase. Simply put, Rollain did not believe that a dragster had "to be long and flexible to go fast."[50] His creation was replete with one of those professions for which nonconformist racers had a penchant:

> There is no anguish
> known to the human race . . .
> like the pain of a new idea.

Watching a conventional top-fueler in the lights, Rollain noted, one could see that "the front wing is holding the front down and the back is holding the back down, while the middle of the car is being flexed." With Swindahl and Uyehara that was the whole idea, to get the cars to arch up between the wings, but Rollain thought this flexing must consume a lot of energy. Hence, he designed Miller's "shortie" much like a funny car, with the entire body acting as one big spoiler. When Rollain and Miller first

When Californian John Rodeck debuted this machine in 1974 spectators re-marked on its resemblance to a go-cart. Rodeck, who also manufactured alumi-num engine blocks patterned on a Chevrolet design, campaigned his "shortie" only briefly. (Smithsonian Institution collection)

In 1974 Don Kohlor had Michigan fabricator Wayne Farr build a chassis for a machine even shorter than Rodeck's, which had a body and hood that mimicked a Model-T Ford. Kohlor campaigned the "Defiant One," only slightly lengthened, as a top-fueler until the latter 1980s. (Photograph by Tom Schlitz)

Dennis Rollain poses with the machine he built for Dave Miller in 1984; the wheelbase was 200 inches, when virtually every other top-fueler was 260. (Smithsonian Institution collection)

showed up at an NHRA event a nervous official suggested that they first "rent a track and go up and down . . . a few times." The car worked all right, but, when Miller started entering Winston series races, he got mixed reactions. Remembering his own misadventures with a shortie, Garlits told him, "I personally wouldn't sit in that position." On the other hand, Gene Snow was supportive, and Kalitta, who helped Miller sort out his fuel system, said, "You know you really have your hands full and I love it."[51]

In 1985 Miller qualified for some tough fields and won a few rounds. Even though his car was a little on the heavy side, it was competitive. The big problem was that he was neither especially savvy technically nor did he have especially deep pockets, and, eventually, the costs became too much to bear. Why would anybody on a limited budget try to compete in a machine that departed so dramatically from the consensual norm? Partly it was because Miller relished the image of nonconformity; a little sign displayed prominently whenever he was working in the pits inquired WHY BE NORMAL? Rollain had confidence in the concept, yet he also liked to affect a mad scientist demeanor. He enjoyed being known as somebody who was not afraid to try a new idea.

As for the idea itself, it was not proved at Dave Miller's hand, nor was it disproved. The question remained: If a funny car with a 125-inch wheelbase could go almost as quick and fast as a dragster, why did a dragster need to be more than twice as long and springy as a whip? Who better to put this question to rest than an old-time funny car racer now in the dragster ranks.

Gene Snow campaigned this shortie in 1987 in several variations. At one point it had no wing at all; at another it had after-parts resembling the 1971 Leland Kolb machine pictured on page 226. (Photograph by Leslie Lovett, courtesy National Hot Rod Assn.)

Gene Snow was as shrewd and skilled as anyone in racing and as resourceful as well, in all senses of the term. If he lacked a generous sponsor—and in his dragster years he usually did—there seemed to be sufficient money in the coffers of the Snow Oil Company. Snow had always had a penchant for novelty; the very fact that he would go against the tide and switch to a dragster in 1980 was suggestive of this.

Indeed, while Snow had been a winner in several different kinds of machine, his repute was primarily as a tireless innovator. He liked that. Sometimes he was successful, as in changing to direct drive when all funny cars had torque converters, or in doing away with suspension when they were invariably sprung. Sometimes success was questionable, as when he eschewed fiberglass and molded a Dodge Charger body out of Gelco with the color blended in. And sometimes things did not work out, as when he tried substituting turbocharging for the standard Roots (positive displacement) blower or substituting tanks of compressed air. "I'm never happy with the norm," he said. "I always want a new deal."[52] Intrigued with

The heliarc process was developed in the aircraft industry for welding chrome moly tubing in a cocoon of inert gas to prevent occlusions. Note the precision bead that Jim Davis has drawn where the spring mount and radius rods are attached to the axle. Every other chassis fabricator had turned to torsion bar front ends by the mid-1960s; Davis remained enthusiastic about using a quarter-elliptic spring. (Photograph by Jerry Mason)

Miller's shortie, he ordered one of his own from Gene Gaddy, a Texan who built all his chassis. Unlike Miller, Snow had the wherewithal to stick with something novel for a lengthy shakedown period, and he did well enough with this car.

At exactly the same time, however, exciting avenues of innovation were opening up with regard to the use of electronic devices to "manage" mechanical functions. Chassis fabrication—the structural engineering facet of drag racing technology—had been the province of specialists for many years; heliarc welding was a skill unto itself, and very few racers made their own frames. Yet, when it came to what made the cars go, that was different. The racers still considered this their own domain. A rule of thumb, even with the most enthusiastic innovator, was not to try too many new things all at once. Because Snow wanted to focus his attention on the promising new interface of electronics and mechanics, computerization, he put any further chassis experimentation on hold.

Computers were thought to be the forte of kids with little or no mechanical aptitude. Contrary to the stereotype, computers proved seductive to a good proportion of the middle-aged mechanical brigade who dominated fuel racing. As these people began stretching, however, they stirred fears that what they were doing somehow transgressed permissible bounds and that someone would have to "draw the line." In the latter 1980s enthusiasts who sought to push technology unfettered by constraints they regarded as extraneous were going to have to confront the reality that "purpose" had always been a matter of opinion and always would be.

14

CHOICE

*You have to try new things all the
time. . . . You try lots of things.*
  GARY BECK, 1984

*It's that idea of taking what one engi-
neers in one's mind and seeing if it
works.*
  AUSTIN COIL, 1989

*Do you know how many different com-
binations there are?*
  JIM HEAD, 1991

verything about drag racing was a human invention of one
kind or another, technological or cultural. While there were always those who regarded the sole purpose as getting quickly
from point A to point B, that aim was deflected the first time anybody
"legislated" anything not in keeping with that purpose, such as proscribing
nitro. Was this about safety? About expense? A preemptive strike at incipient professionalism? All three, probably; even when one knows that technological choices are driven by nontechnological agendas, it can be difficult to determine who is really being served. The point is this: By the 1960s
people who regarded the purpose as solely technological could not pursue
their aims with anything even approaching autonomy; by the 1980s someone like Larry Carrier could treat the notion that "there is nothing to [drag
racing] but going from point A to point B in the quickest amount of time"
as foolish on the face of it. Carrier was making his point in conjunction
with a decision to drop top-fuelers from the IHRA fare. Top-fuel enthusiasts were appalled, but, to the last one, they would have agreed that
the sanctioning bodies were right in barring the door to rockets, which
got from A to B with a dispatch that no conventional dragster could hope

to match. Everybody wanted political bounds set somewhere.[1]

As drag racing evolved, it provided a microcosm of virtually all technological pursuits, with technology just one ingredient in a heady brew of
contextual forces. But one impulse remained a constant: a passion for novelty. There were always enthusiasts who believed that "you have to try new
things all the time."[2] In forty years one choice after another was
legislated—that is, precluded by the rules—and still the horizon never
narrowed significantly. A step ahead, there were those who could "assemble and manipulate in their minds devices that as yet [did] not exist."[3]
When Don Prudhomme said that "drag racing today is a science, but there's
really no substitute for an intuitive driver," he was half-right.[4] The best
drivers did have a sixth sense, which could make the difference between
winning and losing. And yet drag racing was no science; at the design stage
it was always a matter of individual abilities "to weigh the imponderable
and sound the unfathomable." People who made rules could only react,
while "the opportunities for a designer to impress his particular way of
nonverbal thinking upon a machine or a structure" were, as always, "literally innumerable."[5]

Mechanical ability, wrenching, was likewise fundamental; even in 1990
building an engine was mostly an art, and there was certainly an art to all
the tricks of tuning it up, such as "reading" spark plugs or assessing the
condition of bearings by their color, not to say standing at a Bridgeport mill
fashioning a new part directly from a mental image. Chemistry, on the other
hand, required a different facility for mental manipulation. Chemistry was
filled with abstractions. Although hot rodders understood that thermo-
dynamics was basic to any internal-combustion engine—alcohol was different from gasoline, and nitro was much different—at first there were
"secrets" to which only a handful of racers were privy. Later people actually did try to keep knowledge secret, but in the 1950s a secret was simply a
technique. Knowledge was shared in a craft milieu, as Emery Cook did
when he showed Don Garlits that a 25 percent blend of nitro was tame.[6]

As the concept of "liquid horsepower" exerted a growing appeal, the
first scientific instrument with which drag racers had to become familiar
was a hydrometer. Enderle initially sold these under the name "Nitro An-
alyzer," but racers already realized that there were other chemicals that
might be even more propulsive than nitro—hydrazine (N_2H_4), for example. Hydrazine had been developed in Germany as an oxidizing agent for
blending with nitro and was later used as a propellant for rockets such as
the Titan II. In an internal-combustion engine the idea was to scavenge
oxygen and thereby "excite" the nitro. Holly Hedrich and Jack Chrisman
both experimented with hydrazine in the 1950s, as did Jack Hart and Dean
Hill, who subsequently became NHRA's executive vice president and resident fuel expert, respectively. Hedrich recalled that a flathead Ford on
hydrazine and nitro would put out 25 percent more power than on straight

At Famoso in 1963 the "Freight Train" versus the "Shoehorn." These two radically different configurations suggest something about the range of technological choices open to dragster designers. (Photograph by Bill Turney)

nitro but was pretty sure to self-destruct after only a couple of minutes on a dyno.[7] That is, the "unwanted results" were extremely problematic.

Even so, there were those who could not leave hydrazine alone, and in the 1960s it was reportedly responsible for startling performances by Karamesines and Don Nicholson, among others. Because of hydrazine's reputation, both men denied using it or else equivocated. Another racer confessed but said that "we just haven't found the right combination for mixing this fuel," admitting that, as chemists warned, it was "extremely hazardous."[8] In 1966 an engine explosion at a strip in the Northwest injured several people. The finger of blame pointed to hydrazine. Experts reiterated that "they haven't got it completely predictable for rocket use yet, so how reliable will it be when sloshed together by a racer?"[9] Racers were notoriously skeptical of experts, but the negative evidence was overwhelming, and, eventually, everyone agreed that the risks were unacceptable. In the rules one choice was closed, but only one.

Soon there was a wave of enthusiasm for another oxidizer. This one could not be "sloshed together," for it was gaseous, nitrous oxide (N_2O). In 1972 Clayton Harris hooked up with a wealthy angel, builder Jack MacKay of Columbus, Mississippi, and parlayed a fueler with a nitrous system into a string of elapsed times in the 6.10s. *Drag News* named Harris "Driver of the Year." NHRA affirmed the "legality" of nitrous, and supercharger mogul Marc Danekas began marketing injection systems with the query "What is 300 h.p. worth?" Nitrous was appealing because it was inexpensive and widely available (dentists had used it as an anesthetic, after all). And there was considerable experience with nitrous in internal-combustion engines; it had been used as an emergency power booster in World War II Spitfires as well as in various kinds of air and auto racing subsequently. Then, in 1982, Prudhomme's funny car turned a 5.63 at Indianapolis Raceway Park; the quickest anyone else had ever run in a flopper was in the 5.80s, and the fuel dragster mark had stood at 5.63 for nearly seven years. Prudhomme was apparently using nitrous, and it looked like anyone who wanted to keep up, in any kind of machine, was going to have to emulate his combination. What happened instead was reminiscent of the saga of Pete Robinson's jumping jacks; Prudhomme's performance set in motion a train of events that would leave nitrous excluded from fuel racing altogether.

Nitro engines were run so rich that a lot of raw fuel ordinarily blew out the headers. Nitrous enabled burning all of it; there was no hint of header flames, a sign that unburned fuel was exhausting. That might well be worth 300 horsepower, but nitrous had destructive tendencies, too.[10] Competitors as well funded as Prudhomme could run the risk, but most others were not eager for it. NHRA had often been accused of jamming things down the racers' throats (as Scotty Fenn would have put it). When NHRA reversed its position on nitrous, however, it was not an arbitrary measure taken against the wishes of the majority of fuel racers, as had been the case a quarter-century before, with nitromethane. Rather, it was at their behest. To repeat, everyone wanted political bounds set somewhere. With nitrous oxide, as with hydrazine, there were convergent opinions regarding purpose. Not every racer agreed, but most of them felt that the price to be paid for high performance was too high. Another area of Don Jensen's "clean slate" had been ruled out of bounds, and yet the options that remained open were still literally innumerable.

On the mythic "day that drag racing began" back in 1949 one of the featured cars had a supercharged engine burning pump gasoline, and the other was naturally aspirated on a 3:1 blend of methanol and nitromethane. Both were Ford flatheads. The first NHRA Nationals, in 1955, was just about the last major event ever won with such an engine. The winner in 1956 had a Chrysler, and during NHRA's gasoline-only years there were a pair of Olds-mobiles, a Pontiac, and a Lincoln. One Olds was naturally aspirated; the other was supercharged. Preferences varied not so much on the basis of what engineers would call "best practice" as according to individual per-ceptions of optimum combinations of power, weight, and durability and according to subjective (one might even say irrational) quirks such as loy-alty to specific makes. Bob Gorman ran one of the fastest competition coupes of the mid-1950s with a Studebaker engine. Marvin Schwartz and Earl Canavan ran Lincoln engines on fuel. Certain racers had a persistent affection for four cylinders, and Ed Donovan built a dragster that turned

180s with an Offenhauser, an engine designed for circle track racing.

During the epoch of Howard Johansen's "Bear," twin-engine gas dragsters typically had Chevrolets, but, when twins enjoyed a final surge of popularity around 1970, many had Chryslers. There were a few valiant attempts to harness the power of two blown Chryslers on fuel, but more successful were twin-engine fuelers that were naturally aspirated; Nye Frank's "Pulsator" of 1965 was a contender, as was Gene Adams's "Double Eagle" of 1970. All twins were heavy. A wealthy young road racer named Jim Busby briefly fielded a slingshot with a pair of dual-overhead-cam (DOHC) engines that Ford had designed for the Indy 500. Each engine weighed only 350 pounds, and the entire car came in at 1,250 pounds while putting out 1,450 HP. Even so, single-engine blown fuelers had a better power-to-weight ratio.

Such ventures as Busby's were, however, suggestive of a different approach. A dragster with just one unblown engine, a lightweight Chevy V-8, could be awfully quick off the starting line: Bob Tapia earned the nickname "Giant Killer" one night by defeating an Arfons Allison with such a machine. Another one was campaigned by Dennis Rollain and John Buttera, but perhaps the ultimate in lightweight design was the creation of a former motorcycle racer named Bud Morehouse. Using an aluminum Olds F-85 engine in a 40-pound frame, Morehouse came up with a dragster that weighed just over 700 pounds ready to race. Many lightweights had small-displacement engines, around 300 cubic inches, and ran in "junior fuel," a class popularized by C. J. Hart at Lions. But Roy Steffey and the Logghe brothers from Michigan, Bob Noice from California, and Ade Knyff from Massachusetts all won top-fuel purses with large-displacement engines in such machines.

In the latter 1960s, as competition between M&H and Goodyear led to improved tire compounds and better traction for fuelers with blowers, naturally aspirated top-fuelers faded from contention. Then, in the 1970s, various shops tooled up for production of aluminum engine blocks that would normally be fitted with blowers—the Donovan 417, Don Alderson's Milodon, Keith Black's KB, the Rodeck, the Arias. While all these found a niche, for nearly a decade the KB dominated. Black's hold on the market was eventually loosened in the 1980s by Joe Pisano's JP-1, but a JP-1 was not really much different from a KB, a V-8 with the cylinder bores on the same centers as a 426 Chrysler.

A myriad of choices had eventuated in almost all fuel dragsters and funny cars having essentially the same power plant. They had 490 to 500 cubic inches, a positive-displacement supercharger providing 30 to 35 pounds of boost, and an injection system set up to deliver fuel at the rate of about one gallon per second, 75 percent of it going directly into the combustion chambers—yielding air-fuel ratios way beyond the conventions of stoichiometry. There was a dual-pattern camshaft with roller tappets and

Bones and Curt Carroll were among the few fuel racers ever to try running twin supercharged Chryslers. Plagued by excessive wheelspin, the Carroll brothers sought to compensate by setting up a wing at this extraordinary angle of attack. (Photograph by Jere Alhadeff)

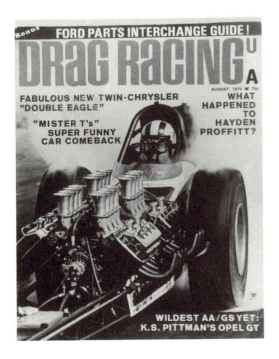

In early 1970 driver Don Enriques put the naturally aspirated "Double Eagle" in the 6.70s, only two-tenths off the elapsed times of the best blown fuelers. Drag Racing USA featured this machine on its August 1970 cover.

two valves per cylinder. Cylinder heads, pistons, rods, and rocker arms were machined from aluminum billets rather than castings, the crankshaft was machined from an alloy steel billet rather than a forging. Nitro blends were usually around 90 percent; because of the way nitro burned, slowly, dual magnetos were set with about 55 degrees spark lead.[11]

Such an engine, weighing about 650 pounds and developing 7 or 8 horsepower per cubic inch (some said 10),[12] was a most wonderful machine. Yet, compared to engines used in other racing constellations, it also seemed wonderfully unsophisticated. In CART, in Formula 1, the engines had four valves per cylinder and dual overhead cams, a design that went back to the Offenhauser and even further. Theoretically, it was superior. But in drag racing the theoretical advantages had not been actualized, despite efforts to develop a DOHC head to fit KBs and JP-1s and despite efforts to develop entire four-valve DOHC V-8s. In the 1980s Joe Schubeck created the "Eagle," which he envisioned as "a 'foundation' motor for the next generation of fuel cars."[13] That remained only a dream, however, and he had not dreamed nearly as long about the "next generation" as three enthusiasts from Sydney, Australia.

Around the time that Donovan and Black were tooling up to produce aluminum engine blocks on Chrysler patterns, Hedley McGee and his sons, Chris and Phil, were working on the McGee Quad Cam. Hedley was a leading cam grinder in Australia, where motorsports were as popular as in the United States. American drag racers had been coming down since the 1960s, but American hardware was not easy for Aussies to come by. As Chris McGee explained: "We had to, right from day one, build a lot of parts. Stuff from America . . . always had a lot of import duty on it." Designing and tooling up to manufacture a complete engine was a terribly involved undertaking, and it took years to begin to sort out problems with the prototype. In the 1980s Gary Beck helped with development, as did Dale Armstrong, and Amato and Bernstein acquired McGee engines with the thought of conquering all the Chrysler clones. But Hedley would not live to see the day, and, even though his sons were both still passionate about the idea, in 1990 the McGee was not yet regarded as a viable alternative for fuel racing. There was reason to suppose that the McGees were out far more money than they would probably ever recoup, just like a lot of racers.[14]

Among the designers and manufacturers of drag racing hardware there were countless success stories, about people who had improved upon parts once salvaged from junkyards or who had invented and developed components that were altogether new; a typical issue of *National Dragster* had close to four hundred advertisers. But there were many other stories more like that of the McGee engine, about inventive efforts that were reasonably well conceived but faltered at the developmental stage. For the McGees the dream was to banish rocker arms, push rods, and the whole "train" of components necessitated by locating the camshaft inside the block; for another enthusiast it might be to devise a better method of supercharging. The superchargers used on most racing engines were driven by exhaust pressure, but these turbochargers did not give the crucial throttle "snap" of a Roots (i.e., GMC-type) blower, which a drag racing machine required. While the strictly technical problem was perhaps soluble, there was something else, a matter of theater: Turbochargers also functioned as mufflers, and most people felt that a big part of the appeal of a blown fueler was its "apocalyptic roar."[15] One of the innovators who investigated turbocharging, Gene Snow, said that he dropped his efforts partly for that very reason.

Drag racers had tried all sorts of setups with Jimmie blowers, such as arranging them in two stages or having two of them feed into the same manifold. With the supply of GMC units becoming exhausted, specialty manufacturers had begun casting them anew out of aluminum and magnesium, and they increased the size time and again until NHRA finally imposed a limit, rotors 19 inches long (GMC units that hot rodders first adapted to their own purposes had rotors about a third or quarter as long). There was some fine-tuning of design, such as grooving rotors for Teflon strips, but the basic design remained the same, and what was curious about this was that supercharging (i.e., boosting the pressure of the fuel charge) was not what had initially been intended at all. On a two-stroke diesel engine a blower did just that; it blew air, at zero boost, into the combustion chamber as an aid in exhausting.

Norm Drazy was a mechanical engineer employed by the Garrett Corporation in Phoenix. He also did some drag racing as a hobby and recalled years of "making big changes to the fuel system, camshaft, and cylinder heads that made only small changes in performance."[16] He suspected a "bottleneck" (had he been current in the historiography of technology, he might have referred to a "reverse salient"), and gradually he began to focus his attention on the method of supercharging. Whereas a turbocharger accelerated air to a high velocity, then diffused it to a low velocity, a Roots unit trapped a specific volume of air and discharged it at pressure. A Roots would "spool up" to peak RPM instantaneously; a turbo could not. The trade-off lay in the power a Roots consumed itself; high-pressure backwash in the manifold always sought the low-pressure rotor spaces. It sought to turn the rotors, and the whole engine, backwards—truly a reverse salient.

There had to be a better way. For fifty years, maybe more, hot rodders had been tooling up to make engine components from scratch, but very few had ever addressed a matter of such fundamental engineering design, and nobody had ever designed such a complex device strictly for drag racing. Drazy began poring over the printed literature. There was something called a screw compressor, invented by a Swedish engineer, A. J. Lysholm, in the late 1930s and patented in the early 1950s. It was pretty much standard with pneumatic tools and air-conditioning equipment. Unlike a Roots, such a device compressed air within the lobe spaces rather than at the

Nick Arias, Jr., seen here in 1983 with his grandson Nick III, designed a hemispherical-head V-6 as well as two different V-8s. A lifelong enthusiast, Arias also appears on page 5, in a 1950 Santa Ana shot. (Courtesy Nick Arias, Jr.)

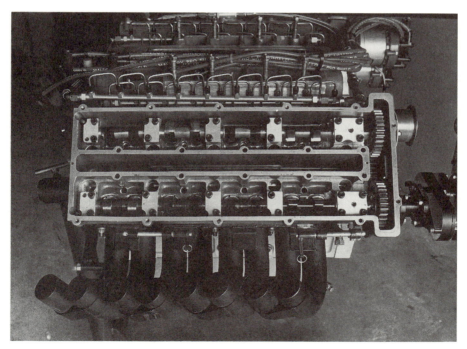

With the cams exposed a McGee had an air of complexity, but actually the absence of rocker arms and push rods made it simpler than the Chrysler design. (Courtesy Petersen Publishing Co.)

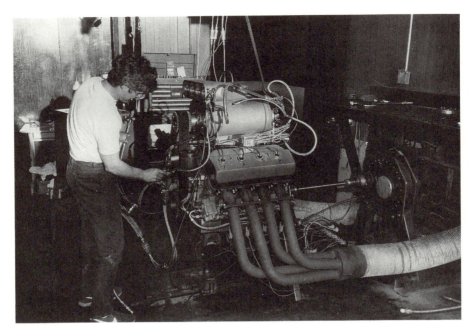

At a Los Angeles dynamometer shop in early 1988 Norm Drazy tests a PSI atop a Chrysler engine. (Courtesy Pat Alexander)

instant the lobes parted in the manifold space. "There is less lost work and a better inlet charge," Drazy observed. "The result is cooler air and less total horsepower used." A screw compressor, he estimated, should be nearly twice as efficient.[17]

Drazy could not afford to pay someone for prototyping while he was trying out ideas, so in 1979 he began outfitting his own machine shop. With Pat Alexander, another Garrett engineer whom he later married, he also established Performance Systems Incorporated (PSI). In 1985 Drazy resigned from Garrett, after seventeen years. Following "months of work at the drawing board and the computer," he finished designing the rotors: One had four lobes; the other had six and turned 50 percent faster (a Roots blower had two identical rotors with three lobes each). By 1987 he had completed design of the housing, gears, and shafts as well as a new fuel injector.

Next Drazy built a test stand designed to measure and compare the performance of a conventional Roots blower with a PSI in terms of inlet and outlet pressures and temperatures, airflow, and volumetric and adiabatic efficiency. At nearby Firebird International Raceway he and Alexander tested a Roots, then turned to the PSI prototype. They got "one perfect run," but the second time a part broke, and the test stand self-destructed. (Fortunately, everyone had retreated to a remote location.) After attending to the problem, Drazy installed his PSI on an alcohol-burning

Chrysler engine and spent days at a southern California dyno facility.

Drazy presented his test data to the NHRA, which had previously put blower development under "moratorium." Then, while the high sheriffs deliberated, he put the PSI on a friend's alcohol dragster, and they took it to Firebird. Everything worked fine. For starts NHRA agreed to legalize the PSI for alcohol engines. Drazy and Alexander moved into production, advertising the entire setup for a little less than five thousand dollars. That was a lot of money to most racers in the alcohol ranks, who competed for purses that were about 20 percent of fuel payouts, but a trickle of orders began coming in, and the trickle grew larger after a machine with a PSI won the alcohol dragster class at the NHRA Nationals. While there were no PSIs on nitro motors—NHRA was still thinking that over—Tim Richards, for one, was ecstatic: "I think it's the future," he said. "I'm sold on him [Drazy]. . . . the guy knows what he's doing." Tests were scheduled for the Amato fueler.[18]

The first PSIs used billet aluminum rotors, but Drazy planned to switch to cast aluminum to reduce both cost and weight (makers of Roots units such as Gene Mooneyham typically used cast magnesium). Lighter rotors would also enhance the appeal to fuel racers; drivers of alcohol machines brought the RPM up to 5,000 after staging, but fuelers staged at an idle, and the engine had to develop peak torque the instant the throttle was opened. Drazy's first cast rotors failed under stress. He went back to the drawing board, switched foundries, and tested the new prototypes to 25,000 RPM. In the fall of 1989, at the Motorplex, PSIs propelled alcohol dragsters and funny cars to their quickest and fastest times ever, 5.80s, with speeds above 235. "Performance Systems Inc. was the hottest name in the industry," wrote Dave Wallace, Jr. PSI had delivered more than one hundred units. The company's biggest problem was filling orders. "What could possibly go wrong?"

Everything, it turned out. After the Texas race the Winston series moved to Firebird. During qualifying a PSI disintegrated, sending shrapnel flying into the stands. Wallace reported that "the extent of damage to the blower prompted NHRA to suspend alcohol qualifying on the spot—and ban screw superchargers from further competition."[19] Drazy blamed the failure on the low elongation of cast aluminum. "What we were selling was a very, very strong China vase," he admitted. "And if it ever did crack, it would shatter." He would have to switch to magnesium, which would increase costs by fifteen hundred dollars. Even so, he insisted that a PSI would still be cheaper than a Roots, which needed to have the Teflon strips replaced every few runs. But again the PSI was under moratorium. For all that it could deliver in the way of efficiency Drazy's device simply did not perform adequately, not if one factored in divergent opinions about purpose; every racer wanted to set records, but everyone agreed that people in the grandstands should not be put at risk.

After investing so much, Drazy was not about to give up. Although he had probably missed out on fuelers forever, in all likelihood there would again be PSIs on alcohol machines someday. But Drazy was bound to lose a lot of time. Others would be able to develop devices similar to the PSI, knowing things he had not known. Performance Systems, Inc., would never succeed as it might have succeeded if Drazy had not stumbled in the developmental stage. Any ultimate success might be quite modest. If so, Drazy could not be altogether surprised. Just as Jeb Allen had known from the very beginning that there were lots of racers who had once been successful but had later gone broke, Drazy had known that "there are lots of experimental products which haven't worked well in drag racing."[20]

The PSI exemplified an innovation worked out with a much higher component of "science" than most others in drag racing, or at least a lot more formal engineering input and recourse to computer-assisted design. The flaws seemed to stem from a misapprehension of various arts, compounded by an inattention to strictures concerning how much was too much. Some PSIs were being overdriven twice as fast as the engine, and the margin of safety was insufficient. Strictly speaking, that was not Norm Drazy's fault, but he had to bear the blame when the purposes his customers had in mind diverged from what he had in mind.

The high sheriffs had often rationalized their mandates as safety measures when, in fact, they were serving other agendas, but not in this instance. They also learned ultimately to avoid making rulings without some semblance of a bilateral consensus. As with nitrous oxide, the impetus for limiting choices could come primarily from racers. The same thing happened with regard to Dale Armstrong's "blower transmission," which enabled a driver to change the speed of the supercharger (hence, boost) manually. Even though Bernstein indicated that he would sell these units to his rivals, Paul Candies wanted no part of the deal; all he and a lot of others could see were escalating costs. Bernstein had a lot of leverage on NHRA, but Candies was one of Wally Parks's closest friends, and, with a petition in hand, he had even more. Beginning in 1988, the rule book prohibited "multi-speed supercharger devices."

Armstrong was not happy. Innovation was central to his notion of purpose. "Where would we be," he asked, "if new ideas weren't developed or used?"[21] The fact was, however, that innovative minds were never stymied. The more one understood about a technological system, the more choices appeared feasible. Thus, even as Armstrong's two-speed blower was quashed, the field of computerized electronics was still wide open.

*O*n-board "data loggers" first started showing up in the early 1980s. The aim was to record such information as cylinder temperatures, fuel pressure, blower boost, and G-forces. Though often called "computers," they were only recording devices; stored information could be printed out or

A typical installation of a Racepak data logger. (Smithsonian Institution collection)

At the NHRA's World Finals at Pomona in 1989 crewman Henry Walther makes a printout from a data logger on board Chris and Phil McGee's topfueler. (Smithsonian Institution collection)

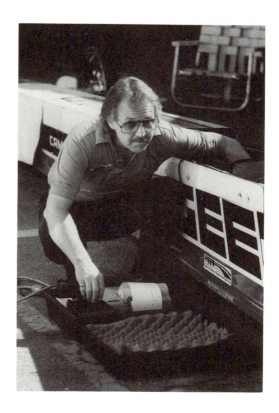

downloaded to a video display terminal. Several people worked on developing data loggers, but these were fraught with problems, almost as many as Dave Zeuschel had encountered in the 1960s when he tried training a movie camera on a battery of instruments arrayed aboard a slingshot, the idea being to film readings during runs.[22] The company that eventually came to the forefront was Racepak, whose device actually did perform a computational function, correlating the engine speed and the speed of the drive shaft. Racepak's origins were in a firm called Competition Systems, which had been founded by two hydroplane racers. Much of the developmental work was done by Armstrong, and in 1985 Bernstein, Armstrong, and Ray Alley established a partnership whose "main assets were knowledge gained over 18 months of working with Racepak . . . and an exclusive agreement with Competition Systems to distribute the product to racers."[23] Its market advantages included the computational capability, which turned out to be of immediate value.

As soon as Armstrong ascertained how to make a Racepak record reliably in an extremely hostile environment, he learned something fascinating. Since the late 1960s the clutches everyone used had been designed to slip for a certain distance, then lock up, putting all the power of the engine to the ground. Nobody doubted that they actually worked that way. But the Racepak readout showed that the engine was turning at least 10 percent faster than the drive shaft at all times during every run; the clutch never stopped slipping. With that information in hand Armstrong got together with Lanny and Tony Miglizzi and developed a two-stage clutch that was truly locked up in the final stage. After that Bernstein's times were out of reach. While working on the car in the pits, Armstrong kept the Rackpak covered with a towel; it was a secret, literally. But the "terrible towel" and the car's performances, in conjunction, attracted plenty of attention, and it was obvious that people like Richards and Coil were going to figure out what was going on. The three Racepak partners went public and began selling their device to competitors.

Technological momentum began to build at once. It began to build in a mechanical realm with the development of ever more sophisticated clutches. L&T soon had a rival, Applied Friction Technologies. AFT's Bob Brooks did R&D work on a three-stage clutch in concert with Richards. Coil devised a four-stage unit, and eventually Jim Head was using a clutch that he claimed had thirteen stages. At the same time Head and others had been experimenting with direct drive, and soon the two-speed transmissions that had been conventional since the latter days of the slingshot were superfluous; why shift gears when a multiple-stage clutch served the same function?[24]

Well, there *was* a difference. Drivers had shifted for themselves, normally punching a button that actuated a compressed-air mechanism attached to the transmission. But a multiple-stage clutch was actuated by a box full of preset electronic timers that phased in automatically. Noting that everything on a drag racing car was "a function of time," Head started talking about a "fully programmable controller": "At time zero it started its sequence, at .5 seconds it would do something, and at .7, .8, and .9 seconds it would do something else."[25]

From microprocessing data to preprogramming the action of a clutch—and fuel delivery system and spark curve as well—it was only a short step to cybernetic devices, with output channels linked to input channels through feedback loops. Road racers had units that could "read" track surfaces and automatically adjust the suspension, and one of the first applications that occurred to people like Coil was "spin control." With phasing of the clutch being an operation that was performed in the pits, misjudging the track surface meant that a car would go up in smoke (i.e., spin the tires) if any one stage came in too soon. What could be more natural than a sensor to determine if the tires were starting to break loose, and, if so, to trigger a mechanism to close the throttle just a bit?

Word had it that certain racers were on "full computer control" as early as 1986, and three years later Paul Candies's driver, Mark Oswald, said, "About the only thing a driver had to do now is just hold onto the steering wheel and keep the car straight."[26] Because drivers had formerly been required to manipulate a variety of buttons and levers, did this automation not represent improved efficiency? That was only a matter of opinion. What if such setups resulted in a bigger gap between the few racers with unlimited budgets and all the rest? People in the computer business said that their equipment did not, after all, cost very much. Not yet. As Garlits remarked, however, "We're going to see the day in the very near future when there'll be a young engineer just in charge of the electronic, computer end of the race car."[27] Indeed, some racers already had such experts on staff.

Then there was a fear of something called "fan backlash." Referring to the incursion of computers and cybernetic devices into other forms of auto racing, Brock Yates said, "We're taking a very human activity and making a technological game of it." What was at stake was a matter of "integrity."[28] It was not clear that this concern had much basis in reality; in the 1980s there was hardly anything that Americans loved more than a technological game. Yet one thing was certain about fuel racing: It was what made people come out to the tracks and what provided television time. As one commentator said, it was NHRA's "meal ticket."[29] Anything disruptive was a matter of deep concern. Steve Gibbs, NHRA's vice president for competition, worried aloud about how fine the line was "between using the computer as an analytical tool, and as a . . . device that directly controls race car functions."[30]

But it was not just NHRA. Racers were expressing similar concerns. Jim Head had been one of the first to experiment with data loggers, in 1981, and eventually his dragster became a "works" car for Tracy Holmes, proprietor of the firm that produced the first unit with any commercial viability, the "Mem-Re-Temp." Like everyone else who explored this new territory, Head was very enthusiastic. By 1986, though, he was suggesting that "the overall picture" needed a serious look:

> I'm finding myself on the other side of an issue that I've been close to since its origin in drag racing. I've been fooling around with this stuff for the past five years now and spent a lot of money on it, and now here I am saying let's take it off the cars. I hate rules and I hate to make more rules, but I think we should at least consider it.

The concerns of people like Head were framed not so much in terms of cost or backlash as in terms of reverence for traditional skills. "On-board computers should be used strictly for monitoring," said Sid Waterman. Paul Candies, who emerged as a major advocate of upholding traditional skills, spoke yearningly of "the days when you tuned the engine and the clutch and you were your own computer."[31]

The NHRA was on the horns of an exquisite dilemma. There was no chance of rationalizing "legislation" in the name of safety, as the organization had been able to do so often previously. Although the backlash issue seemed contrived, there was definite danger in putting too heavy a squeeze on racers who were not funded like Bernstein or Amato but who were essential to filling out the traditional sixteen-car fields. The perpetuation of traditional skills was not a trivial matter either, yet Ingenuity in Action was a venerable NHRA motto, and there was no disputing the ingenuity of the new electronic systems. Maynard Yingst excitedly described waking up in the middle of the night with an inspiration—then getting up, opening a black box in Bruce Larson's car, and rewiring some of the circuitry. Not surprisingly, the foremost proponents of electronics for "systems control" as well as recording were the front-runners, but others shared the same enthusiasm who were never a threat in the Winston points chase. John Carey had one of the most lavish arrays of equipment. So did Head, a man who admitted to pinching pennies in his business "so I can throw it all away on my race car." "Why should we back up?" asked Earl Whiting.[32]

But the decision was indeed to back up. When it revamped the rules for the 1990 season NHRA hit on a rather odd compromise. While data loggers remained legal, there could be no electrical circuitry otherwise, except for ignition systems and gauges. Oh, there could still be timers that served functions such as "managing" clutches, but they had to work on different principles, hydraulic or pneumatic. The top teams got right to work on substitute devices, and these seemed every bit as complex (and were as baffling to tech crews) as what they superseded. Perhaps the possibilities for control through some manner of "artificial intelligence" were reduced.

Perhaps, although some crew chiefs said that the new devices had the same potentials as before.[33]

Whenever the rules were revised racers grumbled about legislation as if it were something new, or as if there were no choices left. There were always choices, but it could be costly to adapt to new strictures. Although IHRA did not follow NHRA in mandating against electronics, it was far more casual about nixing entire competitive classes, first the top-fuelers, then, after reinstating top-fuelers, the funny cars. NHRA exerted a conservative influence whenever there was talk about the advisability of combining the two fuel classes, having just one type of "ultimate" machine. That would be extremely costly for half the fuel racers (whichever class was legislated out of the picture) and would drastically affect drag racing's "bread and butter."[34] Moreover, it may not have seemed like such a bad idea to have both classes to play off against each other; who could forget the United Drag Racers Association? Escalating costs rendered it unlikely that there would be an appreciable increase in the number of championship-caliber machines of either type, but there was no reason for the total number on the Winston trail to fall off drastically. Sheer enthusiasm ought to be sufficient to keep those like Jim Head involved.

By the 1990s Head had an image as a man to whom novelty was everything. Having grown bored with timers and data recorders, he was back to thinking about unfulfilled potentials in chassis design:

> Our sport has not come very far. The cars go faster, but look at them: They haven't changed in nearly 20 years! Sure, the wings are higher, but nobody—including Joe Amato—is so sure that's a good idea; they work, but they're not the plan for the future, right? And the clutches are basically the same; we've put some more levers in there and we've got some sexy control devices, but they're still pretty much the same. The blowers are a little better, but they're legislated. So what are we going to work on? The chassis! . . . If I don't run out of time or money or desire, it's going to pay off.[35]

Time or money or desire. Some of drag racing's most intrepid innovators ran out of time—Pete Robinson, for one—and a lot of them ran out of money. Very few ever ran out of desire.

In 1951 Wally Parks predicted that, with proper leadership, hot rodding would "take its rightful position in public esteem along with such ever-growing sports as baseball, basketball, football and track." In 1964 an enthusiastic journalist wrote that fuel dragster racing would "eventually supplant all others as *the* American favorite."[36] Obviously, these were unfulfilled dreams. Among all the auto racing events in America, only the Indy 500 and NASCAR's Winston Cup chase attained any widespread "esteem." If drag racing avoided being stigmatized as a nonsport, as Brock

The black box containing the clutch "management" system for a 1990 fuel funny car. (Smithsonian Institution collection)

Jim Head wears ear protectors as he warms up his engine in the Pomona pits in February 1990. Head's cars were never quite like anyone else's, and often they were radically different. Though he seldom won a major event, his innovative spirit endeared him to old-timers who treasured iconoclasm. (Smithsonian Institution collection)

Yates had once warned might happen, to major segments of the public it was still something more akin to "mud boggers" and monster trucks than to the sport of Daytona and the Brickyard. A drag racer, as Garlits had once put it, must have been somebody who was "promised a pony one year for Christmas" and it never came.

Nevertheless, drag racing had become something much more than anyone would have ever dared predict at the beginning. After all, it did make headlines on sports pages in Los Angeles if not in Washington, D.C., in *USA Today* if not the *New York Times*. NHRA's own *National Dragster* was a weekly paper of remarkably high quality. There was that wealth of TV exposure, and Diamond P advertised dozens of home videos in its lavish catalogs. Even more lavish were the catalogs of "NHRA Winston Drag Racing Official Products." NHRA's membership had been on a ten-year upswing and was pushing eighty thousand, and Yates was right when he called it "one of the finest racing organizations in the world."[37] Drag racing was becoming less parochial. In addition to Australia, New Zealand, England, and the Scandinavian countries—all of which had been involved for many years—there was burgeoning enthusiasm in Japan and in the rest of Europe. With Carl Olsen ensconced as vice president for international relations NHRA was actively involved in promoting events abroad.[38]

Stunning technological performances from a dazzling theater of machines were the standard fare of the Winston series. Because the pits were open for the price of a pass, fans who turned out for these events had a unique opportunity to rub elbows with drivers and crew chiefs, an admirable cast of men and women. Although Billy Meyer had flopped in his attempt to take drag racing to "the next plateau,"[39] his Motorplex was positively palatial, and other NHRA venues were comparable. If one had a taste for high drama and high-powered machinery, there was nothing to top big-time drag racing.

Yet drag racing was fraught with ethical ambiguities, foremost being its relationship with the tobacco industry; above all else drag racing relentlessly glorified a product that (as it says on any pack of cigarettes) causes cancer, heart disease, and emphysema and may complicate pregnancy and which also adversely affects the health of people who do not even use it. Whether that was a high crime, a misdemeanor, or no crime at all was alleged to be a matter of opinion. It was not a matter of opinion at all, though in fairness I have to add that there was no disputing the contention that one could not fault drag racing without likewise faulting most of organized sports.

Aside from their partnership with the purveyors of Winstons (and Skoal and its ilk) NHRA's leaders were by and large honorable people, and one of them was extraordinary. When Wally Parks was speaking from the heart, you could be absolutely certain of it. Parks had created it all, yet when he emphasized that there was still as much excitement at places like Inyokern as at places like Indianapolis Raceway Park—that there was much more to drag racing than the Winston series—one knew he meant it. When he said such things, indeed, he put his finger on something fundamental. If one's purposes did not include performing for cheering throngs and TV cameras, an enthusiast still had many choices.

There were all the homey venues that Parks said were really the soul of drag racing, all those places like Inyokern and like Samoa, Sumerduck, and Rock Falls Raceway. There was all the rich array of classes for enthusiasts of virtually any kind of machine. There was the "nostalgia" scene. Dozens of classics had been restored to running condition, beginning with Art Chrisman's No. 25, the Hustler, Mooneyham's 554 coupe, Cook and Bedwell's TE-440, Jack Chrisman's Comet, and Swamp Rat I. Bill Pratt listed dozens of stylish slingshots whose performance consisted of smoking the tires right through the eyes, just like in the 1960s. Parks called Garlits's performance in Swamp Rat I at the silver anniversary NHRA Nationals "the outstanding highlight of the event."[40] In the growing collection of the Museum of Drag Racing, besides more than half of all the Swamp Rats, there was Tommy Ivo's twin Buick, Bob Sullivan's first Pandemonium, the Bustle Bomb, Speed Sport I, the Magwinder, and a faithful replica of Dick Kraft's Bug, fashioned by Kraft himself.[41]

Though nearly all of the historic California strips had closed, one by one, Famoso was still going, and Pomona hosted two events annually, including the Winternationals. There were annual reunions among the denizens of Santa Ana, Irwindale, and San Fernando, where old-timers would go to bench race for hours on end.

At the outset I mentioned that many people who were around when drag racing began are still around. But, sadly, many familiar faces are gone: Ed Winfield, Vic Edelbrock, Paul Schiefer, Phil Weiand, Dean Moon, Howard Johansen, Ed Donovan, Dave Zeuschel, Keith Black, Joe Pisano. Romeo Palamides, Frank Huszar, and Bob Summers. Peggy Hart, Jack Hart, Jim Tice, George Hurst, Roger Huntington, Ruth Hoover. Kenny Arnold, Calvin Rice, Emery Cook, Jack Ewell, Bob Sullivan, Jack Chrisman—each of whom survived his racing days to die like most of the rest of us will. Red Henslee, Lloyd Scott, Gary Gabelich, Rod Stuckey, and Jim Davis survived every close call while racing but later died in other sorts of accidents. Mickey Thompson and Marc Danekas were murdered; Mike Snively committed suicide, with only one thing in his pocket, a handwritten list of his major victories. Lou Baney died just as this book went to press.

But Art Chrisman was building engines, Don Nicholson was on the nostalgia trail, and Marvin Rifchin still ran M&H. Bruce Crower could be seen frequently at El Mirage and Bonneville, wearing a driver's suit, as a matter of fact. Chet Herbert also put in appearances at Bonneville, as did Joaquin Arnett, Dick Landy, Art Arfons, even Garlits. Gary Cagle was

Surrounded by little more than salt, SCTA starter Elice Tucker waits for the signal to send a Bonneville roadster on its way. (Smithsonian Institution collection)

SCTA's chief timer. If some of these men no longer went out to the drags, they were still enthusiastic about hot rods.

Even though people who were there at the start would say they never dreamed that drag racing would become what it did, there was still a strong tendency to assume that its evolution was preordained. All one had to do to disprove this was to go out to Speed Week at Bonneville, an annual event that began at almost the same time as organized drag racing. There one would see some of Parks's early dreams fulfilled to perfection; there was an unending display of ingenuity in action, and there were no prize money inducements. Participants might well outnumber spectators, just as in the first days of drag racing. Performance was judged strictly by technological criteria; nobody thought in terms of theatrics. As *Hot Rod*'s Gray Baskerville put it, Bonneville (and El Mirage, still the site of six SCTA meets each year) was a place where people still spent "ungodly amounts of time, money and energy for the pure pleasure of going faster . . . for fun and for free."[42]

The quintessential Bonneville competitors in 1990 were Nolan White of San Diego and Elwin Teague of Santa Fe Springs. Both had tried drag racing; both were much more enthusiastic about the salt. And both had topped 400 miles per hour in homemade streamliners, the first to go that fast with internal combustion since the Summers brothers in 1965. White and Teague paid all the bills themselves. "No 18-wheel support vehicles," Baskerville wrote about Teague. "What suffices is the leftovers from a mill-wright's paycheck, the space afforded by a completely cluttered 1½ car garage, a parts bin full of IOU's, and the support of a battered blue Chevy pickup truck towing a rusty open trailer."[43] To attain 400 miles per hour Teague estimated that—thousands of person-hours aside—he had spent ten thousand dollars a year for fifteen years. For that kind of money he could have taken a fueler on the Winston circuit for a while, but he preferred the splendid isolation of Bonneville. Al Teague had other dreams.

Between the 1930s and the 1980s hot rodding progressed from Muroc to the Motorplex. Yes, in one sense it did—but not inevitably, not for everyone, and not irrevocably. What would happen if, sometime in the 1990s, RJR just walked away from drag racing, followed by Diamond P's television crews and then the other big sponsors? Would there be no more drag racing? I have spent a lot of time describing how people contrived to make drag racing into a paying proposition. But I trust readers will understand that many racers would still have been there even if there were no chance of remuneration. There was always plenty of talk about money, but people kept going even in the face of incredible personal sacrifice. In addition to the financial cost there was always the physical peril. Technological enthusiasm is a powerful emotion.

Is there something "in" technology, as Hindle suggests? Maybe there is. There is certainly something more than a hope or promise of economic gain: "If you have ever heard, in the night, the distant prairie call of a steam locomotive's whistle," Ferguson writes, "you must surely know that the products of men's minds and muscles do not exist because somebody saw in them a way to make money."[44] So it would be for many people with the sound of open headers, a whiff of nitro, the challenge of finding just a little extra horsepower, a little more speed.

It is not an enthusiasm everybody can appreciate, and in the grand scheme of things it would not make much difference if nobody had ever dreamed the dreams of hot rodders. But we need to remember that their dreams are, after all, not much different from the dreams of enthusiasts for superconducting supercolliders and manned space stations. The big difference is that those people claimed a much grander purpose; they managed, for example, to make President George Bush believe that Space Station Freedom, all $30 billion worth, was "essential to our destiny as a pioneering nation." Talk about finding a pigeon . . .

No drag racer ever tried to claim that his or her peculiar brand of technological enthusiasm was anything more than a matter of "satisfaction, meaning, and self," and for that kind of candor, even if for nothing else, all drag racers deserve to be called out to take a bow.

APPENDIXES

Selected Major Events, 1955–1992

Winners, Low Elapsed Times, Top Speeds Except for certain of the events in the last three series listed, elapsed times and speeds are the best for the entire meet; often these times were recorded by someone other than the winner. If it happened that best marks *were* by the winner, they are denoted by an asterisk (*). Where two names and sets of times are listed, the first is for top fuelers, the second for funny cars.

	NHRA Nationals	Bakersfield	NHRA Winter's	AHRA Winter's	NHRA Finals
1955	Calvin Rice 10.30*/151.00				
1956	Melvin Heath 9.99/159.01				
1957	Buddy Sampson 10.42*/152.54[a]				
1958	Ted Cyr 9.56[b]/161.67				
1959	Rodney Singer 9.12/172.08	Art Chrisman 8.70*/180.36			
1960	Leonard Harris 9.25*/171.10	Ted Cyr 8.60/185.00	Lewis Carden 8.84/165.00		
1961	Pete Robinson 8.48*/175.78	Jack Ewell 8.28/185.58	Jack Chrisman 8.99*/176.00	L. Mudersbach 8.83*/175.43	
1962	Jack Chrisman 8.60*/182.18	Don Prudhomme 8.21*/185.58*	Jim Nelson 8.50/176.47	Rod Stuckey 8.51*/188.66*	
1963	Bob Vodnik 8.50/182.18	Art Malone 7.99/194.88	Don Garlits 8.11*/188.66	Danny Ongais[c] 8.21/196.06	
1964	Don Garlits 7.67*/202.24	Con. Kalitta 7.84/195.64	Jack Williams 7.86/195.22	Ron Goodsell 8.03/190.26	
1965	Don Prudhomme 7.50*/210.76	Don Garlits 7.60/206.42	Don Prudhomme 7.56/206.88	Tom Hoover 7.47/203.16	Maynard Rupp 7.57/204.54
1966	Mike Snively 7.31/218.46	Mike Sorokin 7.34*/213.28	Mike Snively 7.51/210.76	Bob Hightower 7.58/205.94	Pete Robinson 7.19*/212.26
1967	Don Garlits 6.76/223.88 Doug Thorley 7.60/192.30*	Mike Snively 7.10/216.00	Con. Kalitta 7.17*/219.50*	Con. Kalitta 7.44*/205.46 Don Nicholson 8.23*/172.08	Bennie Osborn 7.03*/223.88*
1968	Don Garlits 6.67/233.76 Paul Stage 7.84/193.12	Ron Rivero 6.91/221.12*	James Warren 7.37*/236.00	Don Prudhomme 7.02/226.12 Ed Schartman 8.43/182.18	Bennie Osborn 6.93/225.00
1969	Don Prudhomme 6.43/231.76 Danny Ongais 7.22/207.37	Jim Dunn 6.75/230.76 Danny Ongais 7.56*/196.92	John Mulligan 6.81/225.00* Clare Sanders 7.79/198.00	L. Goldstein 6.89*/221.67 Dick Harrell 7.64/199.10	Steve Carbone 6.68/223.88

	NHRA Nationals	Bakersfield	NHRA Winter's	AHRA Winter's	NHRA Finals
1970	Don Prudhomme 6.43*/233.16 Don Schumacher 6.80/214.79	Tony Nancy 6.75/231.36 Hank Clark 7.69/199.10	Larry Dixon 6.74/223.32 Larry Reyes 7.30/203.61	Don Garlits 6.74/227.84 Tom Grove 7.23/204.08	Ron Martin 6.53/226.70 Gene Snow 6.86/214.79
1971	Steve Carbone 6.21/232.44 Ed McCulloch 6.54/226.70	Don Garlits 6.64/226.13 Jim Dunn 6.95/212.26*	Don Garlits 6.70*/223.32 Butch Maas 6.93*/214.28	Gary Cochran 6.37/222.76 Dick Harrell 6.84/214.28	Gerry Glenn 6.55/227.27* P. Castronovo 6.90/210.28*
1972	Gary Beck 6.06/234.98 Ed McCulloch 6.43*/229.50	Tom McEwen 6.35*/232.55 Ed McCulloch 6.96*/217.38	Carl Olson 6.49/231.95 Ed McCulloch 6.68*/220.04	Steve Carbone 6.40/228.28 Dale Pulde 6.57/219.50	Jim Walther 6.21/230.17 L. Fullerton 6.51/222.32*
1973	Gary Beck 5.96*/243.90* Don Prudhomme 6.27*/232.55	D. Salsbury 6.40/238.09 Tom Hoover 6.86/218.96	Don Garlits 6.51*/235.60* D. Schumacher 7.18*/220.58*	Don Garlits 6.06/239.36* D. Schumacher 6.40/221.13	Jerry Ruth 6.11*/232.55* Frank Hall 6.38/232.99
1974	Marvin Graham 6.01/242.58 Don Prudhomme 6.19/231.36	Carl Olson 5.94*/236.22* Ed McCulloch 6.41/226.70*	Gary Beck 5.84*/243.24* Dale Emery 6.30/232.55	Mike Wagoner 5.83/245.90 Ray Beadle 6.34/230.76	Don Garlits 5.88/248.61 Dave Condit 6.16/233.76
1975	Don Garlits 5.86/247.93* Raymond Beadle 6.14*/232.55	James Warren 5.87/243.24* Dale Pulde 6.08/235.60	Don Garlits 5.93/244.54 Don Prudhomme 6.21/233.16	Marvin Graham 6.10/237.46 Shirl Greer 6.32/227.27	Don Garlits 5.63*/250.69* Don Prudhomme 5.98*/241.53*
1976	Richard Tharp 5.79/247.93* Gary Burgin 5.97/238.09*	James Warren 5.95*/242.58* Jim Liberman 6.22/226.70	Frank Bradley 5.78/243.90 Don Prudhomme 6.02*/238.09*	Frank Bradley 6.03/229.00 Mike Van Sant 6.44/221.67	S. Muldowney 5.77*/249.50* Don Prudhomme 6.02*/240.00*
1977	Dennis Baca 5.80/250.69 Don Prudhomme 6.02/242.58	James Warren 5.75*/240.64 Ed Pauling 6.03*/231.36	Jerry Ruth 5.85/248.61 Don Prudhomme 6.03*/247.25*	Jeb Allen 6.17/232.54 John Collins 6.52/223.88	Dennis Baca 5.84*/243.24 Gordie Bonin 6.04/241.93
1978	Don Garlits 5.84/246.57 Tom McEwen 5.97/245.23	Dennis Baca 5.85/243.90 Denny Savage 6.00/239.36	Kelly Brown 5.77/250.69 Don Prudhomme 6.15/240.67*	Don Garlits 6.16*/238.72* Don Prudhomme 6.58*/228.42*	Rob Bruins 5.76/250.00 Ray Beadle 6.08/241.93
1979	Kelly Brown 5.81/250.69 Gordie Bonin 5.95/246.57	Don Garlits 5.86*/245.90 Simon Menzies 6.14/234.37	Bob Noice 5.81/255.58 Tom Hoover 6.07/243.90	John Abbott 6.30/231.95 Tom Hoover 6.50/229.59	Don Garlits 5.87/243.90 Gordie Bonin 6.22/236.22
1980	Terry Capp 5.68/248.61 Ed McCulloch 5.96/243.90	Con. Kalitta 5.81*/247.00 Jim Dunn 6.04/233.16	S. Muldowney 5.83*/249.30 Dale Pulde 6.05/245.32	Don Garlits 6.20/241.93* Tom Hoover 6.69/231.36	S. Muldowney 5.86/242.58 Ron Colson 6.06/238.09*

NHRA Gator's	NHRA Le Grand's	IHRA Winter's	IHRA U.S. Open	AHRA Finals
D. Chenevert 6.50/231.95*				
Leonard Hughes 7.10/208.81				
Jimmy King 6.53/228.42	Pat Dakin 6.51/220.58	Don Garlits 6.32/225.00	Jim Nicoll 6.54/222.22	Don Garlits 6.40*/229.00
L. Goldstein 6.71*/220.58*	Sam Miller 7.01*/216.86*	Richard Tharp 6.87/215.28	D. Schumacher 6.87/202.24	Dale Pulde 6.43/218.00
Don Garlits 6.15*/243.90*	Art Marshall 6.36/226.70	Don Garlits 6.32*/227.27	Carl Olson 6.26/221.67	Dennis Baca 6.18/232.00
Ed McCulloch 6.54*/224.43*	D. Schumacher 6.54/220.58	Richard Tharp 6.66*/215.82	Richard Tharp 6.56/218.97	Gene Snow 6.53/222.22
H. Petersen 6.05/243.24	Pat Dakin 6.16/238.09	Jim Walther 6.41/220.00*	Don Garlits 6.06/229.59	Don Garlits 5.95*/234.88*
Pat Foster 6.34/233.76*	Dale Emery 6.62/219.50*	Pat Foster 6.63*/211.00	Pat Foster 6.34/204.08	Jim Dunn 6.33/229.00
Dave Settles 6.01/237.45*	Gary Beck 5.91*/242.58*	No event	M. Love 6.45/231.36	Gary Beck 6.00*/232.55
Don Prudhomme 6.38*/225.00*	Shirl Greer 6.51/215.31		Gene Snow 6.65/202.70	Don Prudhomme 6.36*/222.76
Dale Funk 6.04/239.36	Don Garlits 5.89*/245.90*	Don Garlits 6.00/231.36*	Gary Beck 6.21/239.36	Hank Johnson 6.14*/227.56
Don Prudhomme 6.39*/230.17	Don Prudhomme 6.45/224.43	Ray Beadle 6.36*/219.51*	Chas. Lee 7.27/198.00	Gordie Bonin 6.46/217.90
James Warren 5.92/248.61	Gary Beck 5.81*/243.24*	Richard Tharp 6.26/231.95	Clay. Harris 6.13/217.00	James Warren 6.08/231.95
Don Prudhomme 6.24*/236.84*	Don Prudhomme 6.17*/235.60*	Tom Prock 6.52/219.51	Ray Beadle 6.29/227.84	Bob Pickett 6.44/215.82
Don Garlits 5.94*/245.90	S. Muldowney 5.93*/240.00*	P. Longnecker 6.24/230.76	John Abbott 6.11/232.55	Gary Beck 6.17/227.84
Gordie Bonin 6.19*/238.72*	Tom Hoover 6.17/236.00	Tom McEwen 6.63*/216.82*	Dale Emery 6.45/230.76	Tom Hoover 6.40/215.30
Don Garlits 6.00/243.90	Kelly Brown 5.87/241.93	Clay. Harris 6.27/234.37	P. Longnecker 6.19/226.70	Frank Bradley 6.00*/238.09
Dale Pulde 6.23/241.93*	Don Prudhomme 6.04*/237.46*	Dale Pulde 6.35/229.00	Dale Pulde 6.52/235.60	Ray Beadle 6.28/229.59
Kelly Brown 5.89/249.30	Richard Tharp 5.91*/245.90*	Don Garlits 6.20/232.55	Con. Kalitta 6.00/234.98	Don Garlits 5.91/240.61
Gordie Bonin 6.15/243.90	Don Prudhomme 6.11/240.64	Dale Pulde 6.73/230.17	K. Bernstein 6.72/227.27	Tom McEwen 6.15*/223.88
Jeb Allen 5.71/247.25*	Marvin Graham 5.77*/247.93*	Richard Tharp 6.05/243.90	Con. Kalitta 6.16/228.42	Jerry Ruth 5.70*/251.39
Don Prudhomme 5.93*/241.28	Dale Pulde 6.10/239.36	Ray Beadle 6.06/237.46	Gary Burgin 6.26/237.46	Tom McEwen 6.27/227.80

	NHRA Nationals	Bakersfield	NHRA Winter's	AHRA Winter's	NHRA Finals
1981	John Abbott	S. Muldowney	Jeb Allen	Don Garlits	Gary Beck
	5.63/250.69	5.84*/245.00*	5.78/246.57*	6.28/234.90*	5.57*/247.00
	Raymond Beadle	Dale Pulde	Billy Meyer	K. Bernstein	Jim Dunn
	5.91*/249.30*	6.16/242.58	5.93/245.90	6.54*/223.80	5.89/238.72
1982	S. Muldowney	Lucille Lee	Dick LaHaie	Don Garlits	Jim Barnard
	5.48/254.23	5.65*/247.93	5.69/251.39	5.79*/247.24*	5.58/255.68
	Billy Meyer	Tom Ridings	Al Segrini	Ed Moore	T. Shumake
	5.63/254.23	6.10/239.36*	5.86/246.57	6.21*/232.54*	5.76/244.56
1983	Gary Beck	Danny Dannell	S. Muldowney	Don Garlits	S. Muldowney
	5.50*/254.95	5.66/250.00	5.52/251.39	6.08*/240.64*	5.39/257.14
	K. Bernstein	Mike Dunn	Frank Hawley	Don Prudhomme	J. Lombardo
	5.80/257.87	6.00*/243.34	5.88/249.30*	6.80*/229.00*	5.74/249.30
1984	Don Garlits	Gary Beck	Gary Ormsby	Don Garlits	Don Garlits
	5.46/261.62*	5.41*/262.39*	5.40/255.68	5.80/244.56*	5.42/263.92*
	Jim Head	John Force	Al Segrini	Don Prudhomme	Sherm Gunn
	5.69/260.11	5.72/257.14	5.83/255.68*	6.15/232.55	5.69/261.62
1985	Don Garlits	Gary Beck	Joe Amato		Gary Beck
	5.50/263.00*	5.62/251.39	5.43/264.53*		5.44/268.01
	John Lombardo	Rick Johnson	Al Segrini		K. Bernstein
	5.67/260.56	5.72*/254.33*	5.58/263.62		5.61*/264.86
1986	Don Garlits	Don Garlits	Darrell Gwynn		Darrell Gwynn
	5.34/269.29	5.37*/271.90*	5.40/269.46		5.32*/274.22
	Mike Dunn	John Force	Tim Grose		K. Bernstein
	5.50/271.14	5.76/257.87	5.60/260.86		5.56*/266.82
1987	Joe Amato	Don Garlits	Don Garlits		Darrell Gwynn
	5.25/282.13*	5.47*/270.08	5.29*/270.59*		5.09*/283.91
	K. Bernstein	John Force	K. Bernstein		Billy Meyer
	5.46/278.89	5.65*/264.70*	5.48*/268.65		5.35/280.72
1988	Joe Amato	Butch Blair	Dick LaHaie		Darrell Gwynn
	5.00/283.82	5.40*/267.06	5.13*/284.81		5.01/284.90
	Ed McCulloch	John Martin	Dale Pulde		John Force
	5.32*/274.80*	5.57/259.36	5.39/276.24		5.29/277.00
1989	Darrell Gwynn		Gary Ormsby		Gary Ormsby
	4.98*/287.53		5.04/291.54		4.95/290.32
	Don Prudhomme		Bruce Larson		Bruce Larson
	5.17*/278.37		5.32*/279.85		5.28/283.28
1990	Joe Amato		Lori Johns		Joe Amato
	4.99/287.72		4.98/282.37		4.93*/289.38
	Ed McCulloch		K. C. Spurlock		Ed McCulloch
	5.29/277.60		5.27/279.41		5.29/280.19

NHRA Gator's	NHRA Le Grand's	IHRA Winter's	IHRA U.S. Open	AHRA Finals
S. Muldowney 5.62/250.00*	Marvin Graham 5.74/250.00	Richard Tharp 5.92/250.00	Marvin Graham 5.74/246.00	Jerry Ruth 5.57*/254.29
Gordie Bonin 5.92/247.25	Don Prudhomme 5.89/244.56	K. Bernstein 6.13*/243.00*	John Pott 5.98/242.00	Don Prudhomme 5.97*/239.36*
S. Muldowney 5.72/250.00	Con. Kalitta 5.69/250.00	Jim Barnard 6.04/238.09	Mark Oswald 5.72/254.23*	Don Garlits 5.53/252.10*
Frank Hawley 5.91/243.90	Don Prudhomme 5.91/245.90	Dale Pulde 6.26/238.09	Dale Pulde 6.09/238.09	Tom McEwen 5.75/249.30*
Gary Beck 5.44*/257.87	Joe Amato 5.66/252.10	Con. Kalitta 5.81*/243.90*	Jody Smart 5.60/254.23*	Frank Bradley 5.92/241.93
Frank Hawley 5.78/254.23	Mark Oswald 5.62*/252.10	K. Bernstein 6.14*/228.42	Mark Oswald 5.84/254.23*	Ray Beadle 5.78/247.93
Joe Amato 5.47/262.39*	Gary Beck 5.52/259.36*	—d	—d	C. Karamesines 5.47/260.11
K. Bernstein 5.78/260.11	Billy Meyer 5.72/254.95	Tom McEwen 5.76/257.87	Billy Meyer 5.72/252.80	Tom McEwen 5.84/247.25
Dick LaHaie 5.50/259.96	Don Garlits 5.48*/262.39*	—d	—d	C. Karamesines[e] 5.62*/252.80
K. Bernstein 5.64*/261.78*	R. Johnson 5.66/260.26*	Ed McCulloch 5.76/260.10	Mark Oswald 5.92/256.41	John Force 5.87*/252.10*
Don Garlits 5.40*/272.56*	Darrell Gwynn 5.43/268.97	—d	—d	Earl Whiting 5.60*/252.00*
Ed McCulloch 5.59/263.00	K. Bernstein 5.66*/261.93	Dale Pulde 5.68/254.23	Mark Oswald 5.59*/256.41*	Dale Pulde 5.67/255.00
Joe Amato 5.22/276.41*	Dick LaHaie 5.29*/273.55	Joe Amato 5.46/265.48*	Gene Snow 5.38*/268.65*	
Don Prudhomme 5.47/266.98	John Force 5.64/260.26	K. Bernstein 5.73*/256.41*	Mark Oswald 5.63*/255.68*	
Eddie Hill 5.06*/288.73*	Gene Snow 5.11/280.81	Gene Snow 5.19/281.25	Paul Smith 5.15/271.49*	Terry Capp 5.39*/271.90*
K. Bernstein 5.31*/278.46*	Don Prudhomme 5.44/269.21	T. Shumake 5.40/271.90	Dale Pulde 5.51*/269.62	Jim Dunn 5.62*/263.15*
Darrell Gwynn 5.05/286.98	Gary Ormsby 5.11*/282.39*	Pat Dakin 5.38/269.46	Earl Whiting 5.15/284.11	
Ed McCulloch 5.28/280.19	K. Bernstein 5.44*/270.92*	Brad Tuttle 5.79/251.39	K. Bernstein 5.47/273.53*	
Darrell Gwynn 4.98*/284.81	Gary Ormsby 5.05*/281.60*	R. Holcomb 5.29/265.28	Gene Snow 5.32/264.18	
Ed McCulloch 5.28/278.03	John Force 5.38/271.08	C. Etchells 5.61/267.67	J. Caminito 5.82/248.96	

	NHRA Nationals	NHRA Winter's	NHRA Finals
1991	K. Bernstein	Frank Bradley	Pat Austin
	4.95/287.72*	4.93/289.76	4.93/291.45
	Jim White	John Force	Al Hofman
	5.21*/287.81*	5.28/284.00	5.24/291.82
1992	Ed McCulloch	K. Bernstein	C. McClenathan
	4.82/301.20	4.88/292.20	4.77/299.90
	Cruz Pedregon	Jim Epler	C. Etchells
	5.12/289.57*	5.18/287.08	5.11/288.00

NHRA Gator's	NHRA Le Grand's	IHRA Winter's
Joe Amato	K. Bernstein	Gene Snow
4.89*/289.57	5.02*/284.00*	5.16*/276.79
Mark Oswald	Jim White	—[d]
5.17/279.93	5.34*/274.72*	
Eddie Hill	K. Bernstein	Doug Herbert
4.80*/301.70	4.95/292.58*	5.41*/278.63
John Force	John Force	
5.15*/289.01*	5.34/278.81*	

[a]Competitors at NHRA Nationals from 1957 through 1963, and NHRA Winternationals from 1957 through 1962, were restricted to pump gasoline.

[b]Best elapsed time was recorded by winner Cyr, but not in the same car with which he won the event.

[c]Because of darkness the two finalists, Ongais and Bob Sullivan, split the winner's purse; on a coin flip, the trophy went to Ongais and car owner Jim Nelson.

[d]The IHRA dropped fuel dragsters from its fare in 1984, 1985, and 1986, and fuel funny cars in 1991.

[e]American Drag Racing Association after 1984.

Appendix 2

Elapsed Time (ET) and Speed (MPH) Records, 1950–1992

	Team/Driver	ET	MPH	Location	Notes
1950	Don/Harold Nicholson		120.00	Santa Ana, Calif.	Ford roadster powered by Ford/Merc engine
	Bob Ward		120.96	Saugus, Calif.	No ET clocks initially at Saugus or Santa Ana
	Chet Herbert/ Al Keys		121.62	Santa Ana	"The Beast"—Harley-Davidson motorcycle
	Joe Leblanc/ Al Keys		122.44	Santa Ana	"Beauty"—Harley-Davidson motorcycle
1951	Don Zable		126.00	Saugus	
	Bob Fisher		127.00	Santa Ana	
	Chet Herbert/ Ted Irio		129.49	Santa Ana	Herbert "Beast"—top Santa Ana MPH many times, 1951–52
	Paul Leon		131.78	Santa Ana	First to top 130, June 9
1952	Louis Castro		132.81	Santa Ana	Motorcycle, 3 MPH faster than fastest 4-wheelers
1953	Lloyd/Art Chrisman		140.08	Santa Ana	First to top 140, Feb. 7
	Bean Bandits/ Joaquin Arnett		142.98	Santa Ana	6 MPH faster than best 1953 cycle time at Santa Ana
	Bean Bandits	11.08		Pomona, Calif.	First reliable ETs
	Lloyd Krant	10.93		Pomona	Last ET record set by cycle
1954	Bean Bandits	10.86		Madera, Calif.	
	Don Yates		144.85	Santa Ana	Supercharged Ford/Merc
1955	Ollie Morris		144.97	Santa Ana	Mid-engine Ford/Merc
	Lakewood Auto/ Bob Alsenz		147.05	Santa Ana	Merc with Ardun OHV heads
	Bustle Bomb/ Lloyd Scott		151.00	Great Bend, Kans.	Olds + Cadillac engine; first to top 150

	Team/Driver	ET	MPH	Location	Notes
	Ed Losinski		151.77	Long Beach, Calif.	Chrysler Hemi engine
	Hartelt/Dodd/ Calvin Rice	10.30		Phoenix, Ariz.	Slingshot dragster
	Bustle Bomb/ Lloyd Scott	9.44		San Fernando, Calif.	
1956	Fenn/Baney/ Kenny Arnold		152.23	Santa Ana	Chassis Research TE-440 slingshot
	Ernie Hashim/ Bill Replogle		154.00	Bakersfield, Calif.	Slingshot with super-charged Chrysler Hemi
	Red Henslee/ Emery Cook		157.06	Long Beach	Chrysler Hemi-powered mid-engine modified roadster
	Ken Lindley/ Bob Alsenz		159.01	Kansas City, Mo.	Slingshot with super-charged Chrysler Hemi
	Jim Nelson	9.10		San Fernando	Merc-powered Fiat coupe; time questionable
1957	Cliff Bedwell/ Emery Cook		168.85	Colton, Calif.	Chrysler-powered TE-440 slingshot; first to top 160
	Lyle Fisher/ Red Greth		169.00	Tucson, Ariz.	Chrysler Hemi-powered mid-engine modified roadster
	Don Garlits	8.79	176.40	Brooksville, Fla.	First to top 170
1958	Don Garlits		180.00	Brooksville	Last record with naturally aspirated engine
1959	Don Garlits		183.66	Sanford, Maine	
	Robert Johnson/ Jim Nelson	8.35		Riverside, Calif.	Mid-engine chassis with full envelope body
	Don Garlits	8.23		Great Bend	
1960	Don Garlits		189.48	Bainbridge, Ga.	
1961	Don Garlits	7.88		Columbia, S.C.	
1962	Zeuschel/Fuller/ Don Moody		191.96	San Gabriel, Calif.	Kent Fuller chassis
	Masters/Richter/ Bob Haines		192.30	Half Moon Bay, Calif.	
1963	Greer/Black/ Don Prudhomme	7.77		San Gabriel	
	Masters/Richter/ Bob Haines		196.92	San Gabriel	

	Team/Driver	ET	MPH	Location	Notes
1964	Masters/Richter/ Bob Haines		198.66	Fremont, Calif.	From 1964 on, marks listed here were generally backed up (matched within 1%)
	Don Garlits		201.34	Atco, N.J.	
	Jim Brissette/ Bill Alexander		205.94	San Fernando	Woody Gilmore chassis
1965	Broussard/Davis/ Danny Ongais	7.59		Carlsbad, Calif.	Kent Fuller chassis
	Carroll/Oxman/ Buddy Cortines	7.54		Dallas, Tex.	Garlits chassis
	Jimmy Nix		208.32	Richmond, Va.	Kent Fuller chassis
	Ramchargers/ Don Westerdale	7.47		York, Pa.	Chrysler 426 Hemi
1966	Roger Coburn/ James Warren	7.38		Irwindale, Calif.	Woody Gilmore chassis
	Larry Huff/Tommy Allen		213.76	Carlsbad	Frank Huszar chassis
	Skinner/Jobe/ Mike Sorokin	7.34		Bakersfield	Frank Huszar chassis
	Bob Creitz/ Vic Brown	7.26		Bristol, Tenn.	
	Jerry Ruth		218.44	Arlington, Wa.	Woody Gilmore chassis
1967	Pete Robinson	7.08		Phenix City, Ala.	SOHC Ford power; last record-setter not using Chrysler or Chrysler-type engine
	Tony Waters/ John Edmunds		226.12	Carlsbad	
	Adams/Wayre/ John Mulligan	6.98		Pomona	
	Don Johnson	6.97		Carlsbad	
	Jim Lee/Alison Lee/H. Westmoreland	6.88		York	Don Long chassis
	Roger Coburn/ James Warren		227.85	Irwindale	
1968	Tim Beebe/ John Mulligan		229.59	East Irvine, Calif.	
1969	Jerry Ruth	6.68		Bremerton, Wash.	
	Jim Lee/Alison Lee/Tom Raley	6.64		Atco	

	Team/Driver	ET	MPH	Location	Notes
	Tim Beebe/ John Mulligan	6.43		Indianapolis, Ind.	
1971	Bill Schultz/ Gerry Glenn	6.41		Long Beach	
	L. Hendrickson		232.55	Vancouver, B.C.	First NHRA record set outside U.S.
	Don Garlits	6.26		Gainesville, Fla.	First record by mid-engine machine since 1959
1972	Tony Nancy		233.60	Long Beach	Last record by sling-shot
	Gaines Markley		234.43	Seattle, Wash.	
	Jack McKay/ Clayton Harris	6.15		Long Beach	
	Lisa/Rossi/ Bill Tidwell		239.64	Long Beach	Tom Hanna "Super Wedge" body
	Jim Annin/Mike Snively	5.97		Ontario, Calif.	First NHRA-certified run under 6 seconds
	Cerny/Lins/Don Moody	5.91		Ontario	Same day as Snively's 5.97
	Lisa/Rossi/ Danny Ongais		243.34	Ontario	
1973	Gary Beck		243.90	Indianapolis	
	Don Garlits	5.78	247.25	Ontario	
1975	Don Garlits	5.63	250.69	Ontario	ET mark stood until 1982
1977	S. Muldowney		253.52	East Irvine	All records between 1975 and 1984 set with KB blocks
	Jerry Ruth		255.68	Englishtown, N.J.	Car owned by Don Garlits
1982	Candies/Hughes Mark Oswald		256.41	Brainerd, Minn.	Al Swindahl chassis
	Larry Minor/ Gary Beck	5.54		Indianapolis	Al Swindahl chassis
1983	Frank Taylor/ Rocky Epperly		257.14	East Irvine	Don Long chassis
	Larry Minor/ Gary Beck	5.39		East Irvine	
1984	Joe Amato		264.70	Englishtown	Al Swindahl chassis
1985	Don Garlits		268.01	Pomona	First record using JP-1 engine block

	Team/Driver	ET	MPH	Location	Notes
1986	Don Garlits		272.56	Gainesville	SR XXX with canopy and faired nose
	Jerry/Darrell Gwynn	5.25	278.55	Dallas	Dave Uyehara chassis
1987	Joe Amato		287.92	Dallas	
	Jerry/Darrell Gwynn	5.08		Dallas	
1988	Jerry/Darrell Gwynn	5.05		Kent, Wash.	
	Gene Snow	4.99		Dallas	Gene Gaddy chassis
	Eddie Hill	4.93	288.54	Dallas	Car originally owned by Frank Taylor, 1983 record
1989	Con. Kalitta		291.54	Pomona	
	M. Brotherton		294.88	Dallas	
1990	Jerry/Darrell Gwynn	4.90		Houston, Tex.	
	Gary Ormsby	4.88	296.05	Topeka, Kans.	
1992	K. Bernstein		301.70	Gainesville	First to top 300, March 20
	C. McClenathan	4.79		Reading, Pa.	

Appendix 3

Cumulative Points Championships, 1970–1992

	AHRA Grand American Series	IHRA Winston Championship	NHRA Winston Championship
	Top Fuel / Funny Car	Top Fuel / Funny Car	Top Fuel / Funny Car
1970	John Wiebe Gene Snow		
1971	Don Garlits Gene Snow		
1972	Don Garlits Leroy Goldstein		
1973	Don Garlits Don Schumacher		
1974	Don Garlits Don Prudhomme	Dale Funk Ron Colson	
1975	John Wiebe Tom McEwen	Don Garlits Dale Pulde	Don Garlits Don Prudhomme
1976	John Wiebe Tom Hoover	Don Garlits Raymond Beadle	Richard Tharp Don Prudhomme
1977	Jeb Allen Tom Hoover	Don Garlits Dale Pulde	Shirley Muldowney Don Prudhomme
1978	Don Garlits Gene Snow	Clayton Harris Denny Savage	Kelly Brown Don Prudhomme
1979	Don Garlits Tom McEwen	Connie Kalitta Kenny Bernstein	Rob Bruins Raymond Beadle
1980	Don Garlits Don Prudhomme	Jeb Allen Billy Meyer	Shirley Muldowney Raymond Beadle
1981	Shirley Muldowney Don Prudhomme	Richard Tharp Raymond Beadle	Jeb Allen Raymond Beadle
1982	Don Garlits Don Prudhomme	Connie Kalitta Dale Pulde	Shirley Muldowney Frank Hawley
1983	Don Garlits Tom McEwen	Richard Tharp Mark Oswald	Gary Beck Frank Hawley
1984	Don Garlits John Force	Not contested[a] Mark Oswald	Joe Amato Mark Oswald

	ADRA Series[b]		IHRA Championship	NHRA Winston Championship
	Top Fuel		*Top Fuel*	*Top Fuel*
	Funny Car		*Funny Car*	*Funny Car*
1985	Frank Bradley		Not contested	Don Garlits
	John Force		Dale Pulde	Kenny Bernstein
1986	Frank Bradley		Not contested	Don Garlits
	Doc Halladay		Mark Oswald	Kenny Bernstein
1987			Gene Snow	Dick LaHaie
			Mark Oswald	Kenny Bernstein
1988			Gene Snow	Joe Amato
			Ed McCulloch	Kenny Bernstein
1989			Pat Dakin	Gary Ormsby
			R. C. Sherman	Bruce Larson
1990			Mike Brotherton	Joe Amato
			Chuck Etchells	John Force
1991			Gene Snow	Joe Amato
			Not contested	John Force
1992			Doug Herbert	Joe Amato
			Del Worsham	Cruz Pedregon

[a]Last year for RJR/Winston sponsorship of this series.

[b]AHRA ceased operations in 1984, with a few of its tracks being reorganized as the American Drag Racing Association.

Appendix 4

"High Performance Clubs"

The 5-Second Club
The first 16 drivers to clock elapsed times under six seconds, 1972–1974

1. Tommy Ivo	Oct. 22, 1972	5.97	New Alexandria, Pa.[a]
2. Mike Snively	Nov. 17, 1972	5.97	Ontario, Calif.
3. Don Moody	Nov. 17, 1972	5.91	Ontario, Calif.
4. Don Garlits	July 7, 1973	5.95	Portland, Ore.
5. Gary Beck	Sept. 3, 1973	5.96	Indianapolis, Ind.
6. James Warren	Oct. 13, 1973	5.97	Fremont, Calif.
7. Larry Dixon	Nov. 16, 1973	5.94	Ontario, Calif.
8. Dan Richins	Nov. 16, 1973	5.93	Ontario, Calif.
9. John Stewart	Nov. 16, 1973	5.92	Ontario, Calif.
10. Pete Kalb	Jan. 26, 1974	5.96	Phoenix, Ariz.
11. Jerry Ruth	Jan. 27, 1974	5.95	Phoenix, Ariz.
12. Dwight Salsbury	Feb. 2, 1974	5.97	Pomona, Calif.
13. Dwight Hughes	Feb. 2, 1974	5.97	Pomona, Calif.
14. Carl Olson	Mar. 10, 1974	5.94	Famoso, Calif.
15. Gary Ritter	Mar. 23, 1974	5.84	Sacramento, Calif.
16. Frank Bradley	June 29, 1974	5.96	East Irvine, Calif.[b]

[a] *National Dragster* terms this time "controversial and disputed" but still lists it as the "first 5."

[b] Cragar Industries, which sponsored both the 5-Second Club and the 4-Second Club, also established an eight-car 5-Second Club for funny cars. The first driver to gain entry was Don Prudhomme, with a 5.98 on Oct. 12, 1975, at Ontario, Calif.; the eighth slot was not filled until 1981.

The 4-Second Club
The first 16 drivers to clock elapsed times under five seconds, 1988–1991

1. Eddie Hill	Apr. 9, 1988	4.990	Dallas, Tex.
2. Gene Snow	Oct. 6, 1988	4.997	Houston, Tex.
3. Richard Holcomb	Oct. 15, 1988	4.998	Dallas, Tex.
4. Joe Amato	March 3, 1989	4.996	Houston, Tex.
5. Gary Ormsby	June 10, 1989	4.991	Columbus, Ohio
6. Dick LaHaie	Sept. 3, 1989	4.983	Indianapolis, Ind.
7. Darrell Gwynn	Sept. 3, 1989	4.981	Indianapolis, Ind.
8. Shirley Muldowney	Sept. 15, 1989	4.974	Reading, Pa.
9. Mike Brotherton	Oct. 7, 1989	4.996	Dallas, Tex.
10. Frank Bradley[a]	Oct. 25, 1989	4.998	Pomona, Calif.
11. Jimmy Nix	Feb. 16, 1990	4.960	Phoenix, Ariz.
12. Lori Johns	Feb. 17, 1990	4.975	Phoenix, Ariz.
13. Frank Hawley	Aug. 4, 1990	4.982	Kent, Wash.

14. Tommy Johnson, Jr.	Sept. 29, 1990	4.964	Topeka, Kans.
15. Don Prudhomme	Feb. 2, 1991	4.980	Pomona, Calif.
16. Kenny Bernstein	Feb. 24, 1991	4.996	Phoenix, Ariz.

[a]Bradley was the only member of the 5-Second Club also to run one of the first 16 4-second elapsed times.

The NHRA 250-MPH Club

The first 16 drivers to clock speeds faster than 250 m.p.h., 1975–1982

1. Don Garlits	Oct. 11, 1975	250.69	Ontario, Calif.
2. Shirley Muldowney	May 5, 1977	253.52	East Irvine, Calif.
3. Jerry Ruth[a]	July 9, 1977	255.68	Englishtown, N.J.
4. Richard Tharp	Sept. 3, 1977	250.69	Indianapolis, Ind.
5. Gary Beck	Oct. 7, 1979	250.00	Ontario, Calif.
6. Dave Uyehara	Oct. 7, 1979	250.00	Ontario, Calif.
7. Jeb Allen	Mar. 12, 1981	250.69	Gainesville, Fla.
8. Gary Ormsby	Sept. 5, 1981	250.69	Indianapolis, Ind.
9. Terry Capp	Sept. 6, 1981	250.00	Indianapolis, Ind.
10. Johnny Abbott	Sept. 7, 1981	250.69	Indianapolis, Ind.
11. David Pace	Sept. 7, 1981	250.00	Indianapolis, Ind.
12. Marvin Graham	Oct. 4, 1981	250.00	Fremont, Calif.
13. Mark Oswald	Feb. 3, 1982	250.00	Pomona, Calif.
14. Gene Snow	Mar. 12, 1982	250.00	Gainesville, Fla.
15. Don Prudhomme[b]	May 29, 1982	250.00	Baton Rouge, La.
16. Joe Amato[c]	July 16, 1982	250.69	Englishtown, N.J.

[a]Ruth made this run in a car belonging to Don Garlits.

[b]Prudhomme gained entry in a funny car rather than a dragster.

[c]For official records, NHRA requires a "backup"—that is, a driver must clock another time (at the same event) no more than 1 percent slower than the record run. For example, for a speed of 250 to stand as a record, a driver would have to turn another time of at least 247.50. At Englishtown on July 17, 1982, Billy Meyer ran 254.95 in his funny car, and on the 18th he ran 250.69. The 250.69 was not sufficient to back up the faster speed, but the faster speed did back up the 250.69; hence, Meyer may be considered the 17th member of the club. Because the 15th was Prudhomme, when Amato gained entry there were only 15 dragsters. In August 1982, at the NHRA North Star Nationals in Brainerd, Minn., Jody Smart became the 16th with a speed of 252.80.

NOTES

PREFACE

1. James P. Viken, "The Sport of Drag Racing and the Search for Satisfaction, Meaning, and Self: Work and the Mastery of Accrued Skill in Suitable Challenge Situations" (Ph.D. diss., University of Minnesota, 1978).

2. David Pye, *The Nature and Aesthetics of Design* (London: Barrie & Jenkins, 1978), 18; the quote in the epigraph is from p. 12.

3. Brooke Hindle, *Technology in Early America: Needs and Opportunities* (Chapel Hill: University of North Carolina Press, 1966), 10; Eugene Ferguson, "Toward a Discipline of the History of Technology," *Technology and Culture* 15 (Jan. 1974): 13–30; Thomas P. Hughes, "Technological Momentum in History: Hydrogenation in Germany," *Past and Present* 441 (Aug. 1969): 106–32; John B. Rae, "Why Michigan?" in *The Automobile in American Culture,* ed. David L. Lewis and Laurence Goldstein (Ann Arbor: University of Michigan Press, 1983), 6. The quote regarding the aircraft industry is from Bill Trimble's review of Jacob Vander Meulen's *The Politics of Aircraft: Building an American Military Industry* (Lawrence: University Press of Kansas, 1991), in *Technology and Culture* 34 (Apr. 1993): 443; and the quote from Schumpeter's *The Theory of Economic Development* (1934) is from Donald Findlay Davis, *Conspicuous Production: Automobiles and Elites in Detroit, 1899–1933* (Philadelphia: Temple University Press, 1988), 2. The literature on technological enthusiasm is closely related to another that addresses the symbolic and nonrational; for some suggestive remarks, see " 'The Frailties and Beauties of Technological Creativity': An Interview with John M. Staudenmaier by Robert C. Post," *American Heritage of Invention & Technology* 8 (Spring 1993): 16–24.

4. Christopher Jensen, "The All-American Soap Box Derby," *Car and Driver,* Dec. 1987, 153; see also Ralph Iula, *Downhill Heroes* (Princeton, N.J.: Automobile Quarterly Publications, 1986); and Frank Deford, "Real Boys," *Sports Illustrated,* Aug. 1, 1988, 68–82.

5. Ignoring evidence of the hazards of secondary smoke, the men who controlled drag racing would claim that attempts to sunder the commercial relationship between sports and cigarettes subverted "freedom of choice" (Dallas Gardner, president, National Hot Rod Association, to NHRA membership, Feb. 26, 1990). Although I do not intend to dwell upon this relationship, I challenge anyone with benign views of smoking and its glamorization through sporting events to read John A. Meyer's *Lung Cancer Chronicles* (New Brunswick, N.J.: Rutgers University Press, 1990) without changing those views. Less dramatic but equally devastating is the U.S. Public Health Service's *The Health Consequences of Smoking* (1968 et seq. to 1990, with various titles).

6. Tom Wharton, "Salt Flats Produce Speed, Ingenuity," *Salt Lake Tribune,* Aug. 23, 1989.

7. Ellen Goodman, "Women and the Driver's Seat," *Baltimore Sun,* Apr. 25, 1991.

8. The phrase I have quoted is Leo Marx's: "Comment and Response on the Review of *In Context,*" *Technology and Culture* 33 (Apr. 1992): 407.

9. Virginia Anne Bonito, Yale University, to Roberta Rubinoff, Office of Fellowships and Grants, Smithsonian Institution, Dec. 1, 1988. Bonito's sentiments are remarkably similar to those expressed by Akron "Ak" Miller, a dry lakes and drag racing pioneer, nearly forty years earlier: "Hot Rods, I Love 'em," *Hot Rod,* Mar. 1951, 10–11ff.

10. John M. Staudenmaier, S.J., "What SHOT Hath Wrought and What SHOT Hath Not: Reflections of Twenty-Five Years of the History of Technology," *Technology and Culture* 25 (Oct. 1984): 711.

Introduction *A NEW THEATER OF MACHINES*

1. Howard N. Rabinowitz, "The Academic as Sportswriter," *OAH Newsletter* (Organization of American Historians) (May 1991): 3.

2. Edwards Park, "Around the Mall and Beyond," *Smithsonian,* Jan. 1988, 26.

3. Daniel Boorstin, *The Americans: The National Experience* (New York: Vintage Books, 1967), 97; Boorstin quotes the epigraph from von Gerstner's *Die innern Communicationen der Vereinigen Staaten von Nordamerica* on 98. I have developed some of my introductory themes more fully in "Strip, Salt, and Other Straightaway Dreams," in *Possible Dreams: Technological Enthusiasm in Twentieth Century America,* ed. John L. Wright (Dearborn, Mich.: Henry Ford Museum, 1992), 98–109.

4. Ronald Reagan said this in 1971, when governor of California. He added that it was also essential "to the future of thousands of California aircraft industry workers" (*Los Angeles Times,* May 20, 1971, pt. 1, 8).

5. Walt Woron, "Timing the Hot Rods," *Speed Age,* Feb. 1948, 12ff; Wally Parks, "The History of the Hot Rod Sport," *Hot Rods* (Los Angeles: Trend, Inc., 1951), 3–20; "A History of S.C.T.A.," *Drag Racer,* July 11, 1959, 17–20; Don Francisco, "Desert Stronghold," *Hot Rod,* Jan. 1961, 32–33ff; Ed Sarkisian, "El Mirage: Stepping-Stone to Bonneville," *Cars,* Nov. 1961, 44–46; Doc Jeffries, "Rufi and the Flying Chicken Coop," *Bonneville Racing News,* Mar. 1991, 12–13; Wendy Jeffries and Bob Rufi, "Bob Rufi's Photo Album," *Bonneville Racing News,* Nov.–Dec. 1992, 5–6.

6. See H. F. Moorhouse, "Racing for a Sign: Defining the 'Hot Rod,' 1945–1960," *Journal of Popular Culture* 20 (Fall 1986): 83–96. Moorhouse has published several other articles on hot rodding as well as a significant monograph, *Driving Ambitions: A Social Analysis of the American Hot Rod Enthusiasm* (Manchester: Manchester University Press, 1991). It strikes me as likewise phenomenal that such a penetrating academic analyst of this subject could be a Cambridge-educated professor of sociology at the

University of Glasgow, self-portrayed as a "middle-aged, middle-class, non-driving Englishman."

7. Dan Roulston, "The Hot Rod Story," chap. 2, *Car Craft,* Oct. 1966. This three-part series (running from Sept. through Nov.), by a former managing editor of each of drag racing's two leading weekly newspapers, remains the best summary history of hot rodding in print.

8. Half-mile races called "sprints," the Brighton Trials were staged in England long before anyone had coined the term *drag racing;* see Wally Parks interview, Oct. 16, 1989, 6, Drag Racing Oral History Archive, National Museum of American History (hereafter DROHA, NMAH). A "Quarter-Mile Acceleration Knock-Out Competition" took place in Sydney, Australia, as early as 1930; see "The Day They Shook Bondi," *Australian Rodsports Annual '70,* 30–33. There are many opinions about the origins of the term *drag racing,* and I cannot choose among them. One holds that it derives from racing on the main drag, another that it stems from the phenomenon of a car under hard acceleration dragging something in the rear, perhaps a hanging plaque emblematic of club membership. There were "drags" at county fairs involving teams of horses and heavy weights. As a driving technique, dragging meant holding a car in low gear and winding it out.

9. Richard Coniff, "Made in America: An Interview with John Kouwenhoven," *American Heritage of Invention & Technology* 2 (Summer 1986): 30.

10. Michael B. Schiffer, University of Arizona, to the author, Feb. 9, 1993. Copies of all letters to the author cited in the notes are with the *Technology and Culture* correspondence, National Museum of American History.

1 *WARMING UP*

1. On the business firms mentioned, see Les Nehamkin, "The Story of Fuel Injection," *Honk!* Oct. 1953, 18–22; Karen Scott, "Meet the Manufacturer [Hilborn]," *Drag News,* May 10, 1969, 8; "Meet the Manufacturer [Edelbrock]," ibid., Apr. 6, 1968, 16; Dave Wallace, Jr., "Jack Engle Walks Softly, Carries a Big Bumpstick," ibid., Nov. 22, 1975, 16–17; Ed Iskenderian interview, Feb. 6, 1989, DROHA, NMAH; "Interview: Ed Iskenderian, Mickey Thompson, Ray Lavely Discuss Rodding," *Cars,* Dec. 1964, 58–61ff; "Iskenderian Racing Cams," *Drag News,* Mar. 3, 1964, 24–25; Karen Scott, "Meet the Manufacturer [Iskenderian]," ibid., Apr. 12, 1969, 8; Charles Hillinger, "Self-Made Millionaire [Iskenderian] Can't Resist a Bargain," *Los Angeles Times,* July 7, 1971, pt. 2, 1, 3; and "California's Big Wheels [Baney and Edelbrock]," *Best Hot Rods* 1 (1952): 30–33.

2. A brief account of the day's events appeared in Dan Roulston's "Hot Rod Story," 2:22–23. Mike Doherty, editor of *Drag Racing,* gave it the title and published an embellished version in his April 1969 issue (42–45). It stirred so much response that he printed it again in July 1970, and in October 1973 he told another tale, "The First Top Eliminator," which added embroidery. "The Day Drag Racing Began" also appeared a few years later in *Drag News.*

3. Dave Wallace, Jr., "C.J. Pappy Hart," in *Petersen's History of Drag Racing,* ed. Dave Wallace, Jr. (Los Angeles: Petersen Publishing Co. [PPC], 1981), 20–21; see also Wallace, "The First Dragstrip," ibid., 4–11. In the Wallace Family Archives in Mokelumne Hill, California, there is a file pertinent to Hart's career which includes clippings, correspondence, and a fragment of a family biography begun by Peggy Hart before her death in 1980.

4. H. F. Moorhouse, "The 'Work' Ethic and 'Leisure' Activity: The Hot Rod in Post-War America," in *The Historical Meaning of Work,* ed. Patrick Joyce (Cambridge: Cambridge University Press, 1987), 243.

5. Gene Balsley, "The Hot Rod Culture," *American Quarterly* 2 (Winter 1950): 353.

6. One needs to keep this in mind when considering that long afterward, with sanctioned drag strips from coast to coast, there was still a thriving street racing subculture around most big cities. For suggestive material on the persistence of street racing in the

Los Angeles area in the 1960s, see Rob Ross, "The Subterranean World of Los Angeles Street Racing," *UCLA Daily Bruin,* Dec. 15, 1965, 7, 10; and "Big Willie . . . King of the Street," *Drag Racing,* Dec. 1968, 42–47.

7. Wallace, "First Dragstrip," 5.

8. On the early days at Santa Ana, see: "Santa Ana: A History of Drag Racing, 1950–1959," *Drag News,* Mar. 28, 1959, 8–9ff; "Meet the Manufacturer: Chet Herbert Cams," ibid., Apr. 16, 1960; Dick Day, "Our Point of View," *Car Craft,* Oct. 1966, 10; W. A. Huggins, "Hot Shots for Safety," *Best Hot Rods* 1 (1952): 62–63ff; Don Montgomery, "Drag It Out," *Street Rodder,* Feb. 1987, 78–83; *Hot Rods as They Were* (Fallbrook, Calif.: Author, 1989), 146–57; Wally Parks, Editor's Column, *Hot Rod,* June 1959, 3. In July and August 1965 *Drag Sport* ran a series titled "Santa Ana Relived," which included photos from the collection of Don Tuttle, who had been the Santa Ana announcer. In the archival holdings of the National Hot Rod Association in Glendora, California, there is a typescript history, apparently derived from the same source as the *Drag Sport* series, which discusses a nearby military field at which racing had previously been permitted for a short time.

9. "Yeakel 426 Club, or Lou Baney Rides Again," *Drag News,* July 6, 1963, 12. Ostensibly an interview by *DN*'s Al Caldwell, this is actually a lengthy (and extremely informative) monologue by Baney.

10. See Rob Kling, Spencer Olin, and Mark Poster, eds., *Postsuburban California: The Transformation of Orange County since World War II* (Berkeley: University of California Press, 1991); see also Greg Klerkx, "Drag Strips to Runways: JWA's Seen It All," *Orange Coast Daily Pilot,* Jan. 29, 1989, A1–2.

11. Lou Baney, "Drag Racing: Bean Fields to Big Bucks," *Hot Rod,* Jan. 1973, 132; see also Louis Kimzey, "Saugus Drags," *Hop Up,* Oct. 1951, 12–15.

12. Baney, "Drag Racing," 132. Yeakel, who later ran for mayor of Los Angeles and still later died when his plane crashed on an L.A. freeway, had made local news in 1947 when he mounted a "jet engine" (it may actually have been some kind of rocket) on a '36 Ford chassis and tested it at Rosamond, a dry lake near Muroc (Los Angeles *Daily News,* Jan. 4, 1947, qtd. in Spencer Murray, "The First Jet Car," *Rod and Custom,* June 1953, 18–21ff).

13. Eugene S. Ferguson, "The American-ness of American Technology," *Technology and Culture* 20 (Jan. 1979): 3.

14. Ferguson, "Toward a Discipline of the History of Technology," 21.

15. George Basalla, *The Evolution of Technology* (Cambridge: Cambridge University Press, 1988), 14.

16. Klaus Maurice and Otto Mayr, eds., *The Clockwork Universe: German Clocks and Automata, 1550–1650* (New York: Neale Watson Academic Publications, 1980), vii.

2 STAGING

1. Bob Cress interview, Mar. 1990, DROHA, NMAH, 39. This interview was conducted by Don Jensen, whose epigraphic quote is from a letter to the author, dated Sept. 14, 1989.

2. A top-notch survey is Richard P. Hallion, *Test Pilots: The Frontiersmen of Flight* (Washington, D.C.: Smithsonian Institution Press, 1981), but see also Laurence K. Loftin, Jr., *Quest for Performance* (Washington, D.C.: NASA, 1985).

3. "Northern California Speed Sprints," *Hot Rod,* Sept. 1951, 16.

4. Montgomery, "Drag It Out," 80; see also "The Early Drag Racers," *Hot Rods as They Were,* 147–58.

5. Griff Borgeson, "Salinas Drag Races," *Hot Rod,* Feb. 1951, 13.

6. "Northern California Speed Sprints," 16.

7. Roger Huntington, "Lightening Holes," *Rod and Custom,* Mar. 1955, 34–37ff.

8. Most knowledge about the use of nitro was transmitted orally, although the literature on fuel became voluminous, even in popular periodicals. See, for example, Keith

Thorpe, "Secrets of the Fuel Burners," *1964 Drag Racing Annual* (New York: Argus Publishers, 1964), 6–11; "Racing Fuels," *Drag Racing,* Dec. 1968, 54–57; Bill Sievers, "Hot Fuels for Street and Strip," ibid., May 1971, 42–44ff, and June 1971, 19–22ff; John Hogan, "Behind the Scenes: Hot Topics," ibid., Feb. 1972, 12ff; Golden West Fuels, "Nitro Safety Procedures," reprinted in *National Dragster,* Feb. 11, 1983, 8ff; Andy Carl, "Liquid Horsepower," *Drag Racing,* Nov. 1989, 68–71. A full-scale technical treatise is Carl's *Nitromethane-based Racing Fuels and Detonation Additives* (Fullerton, Calif.: ADC Performance Engineering, n.d.). Two interviews in the DROHA, NMAH, with Don Montgomery and with Keith Black, provide insight into the early experimentation with exotic chemicals.

9. Manny Clinnick, "Tracy Throttle Mashers," *Hot Rod,* May 1951, 36. Clinnick designed the timing equipment used by the Cal-Neva Timing Association at Tracy, Kingdon, and elsewhere, which was considered more accurate than most; see Cress interview, Mar. 1990, DROHA, NMAH, 39.

10. Robert Devereux, "Dago Drags," *Hop Up,* Sept. 1951, 38.

11. "Early Lakesters," *Hop Up,* Apr. 1952, 9–13; "0 to 140 MPH in 9 seconds!" *Hot Rod,* May 1953, 34–37. For many years Chrisman's "25" was on display in the lobby of the NHRA's sumptuous headquarters in Glendora, California.

12. "Drag Racers' Paradise," *Hot Rod,* Aug. 1951, 9; see also "San Diego Drag Races," *Hop Up,* July 1952, 12–15.

13. Mike Nagem, "The History of Drags in San Diego," *Bonneville Racing News,* Oct. 1990, 4; Wendy Jeffries, "First NHRA Champion," ibid., Nov. 1990, 13; Lee Swanson, "Bean Bandits Are Back," ibid., Dec. 1990, 19; Roger M. Showley, "Local Dragsters Have Rich History Here," *San Diego Union,* Apr. 21, 1991.

14. "The Bean Bandit," *Hot Rod,* Feb. 1953, 36.

15. "Kern County Drags," *Hop Up,* Aug. 1952, 32–35.

16. "City Sponsored Drags," *Hot Rod,* Dec. 1952, 14–17; see also Ralph Parker, "Harnessing the Hot Rods," ibid., June 1955, 44–45.

17. "Motorama 1952," *Hot Rod,* Jan. 1953, 14–15.

18. "Draggin' in Paradise," ibid., Feb. 1953, 27.

19. Pye, *Nature and Aesthetics of Design,* 12.

20. Ray Brock, "The Ardun: What Makes It Run," *Hot Rod,* July 1956, 30–33.

21. Barney Navarro, "144 MPH Dragster," *Rod and Custom,* June 1954, 37–39; "How Fast Did He Go?" ibid., Sept. 1954, 51–54ff; "The Keeper of the Stable," ibid., Apr. 1955, 16–21ff. See also Mark Dees, "A Technical History of the Racing Flathead," ibid., Apr. 1973, 42–49.

22. Jack Baldwin, "The Bear: Buick Powered Dragster," *Car Craft,* Jan. 1954, 38–41.

23. "Southern California Championship Drags," *Hot Rod,* June 1953, 26.

24. Ibid., 77.

25. Ibid.

26. The quote is from Dave Wallace, Jr., "Wally Parks," in *Petersen's History of Drag Racing,* 23. A 1985 Diamond P Sports Video production, *Gathering Speed,* aimed to show that (as Chris Martin put it in *National Dragster* [Feb. 7, 1986]), NHRA was not merely "important . . . in the infancy of drag racing. It *was* drag racing."

3 GATHERING SPEED

1. Walter G. Vincenti, *What Engineers Know and How They Know It* (Baltimore: Johns Hopkins University Press, 1990). Cf. Donald Campbell, "Blind Variation and Selective Retention in Creative Thought as in Other Knowledge Processes," *Psychological Review* 67 (Nov. 1960): 380–400.

2. On design paradigms and normal technologies the work of Edward W. Constant is fundamental: *The Origins of the Turbojet Revolution* (Baltimore: Johns Hopkins University Press, 1980). See also Rachel Laudan, ed., *The Nature of Technological Knowledge:*

Are Models of Scientific Change Relevant? (Dordrecht, Neth.: D. Reidel, 1984). Peter Hugill has suggested the three-phase periodization of automotive technology.

3. Lee Swanson, "Otto Ryssman's Drag Days," *Bonneville Racing News,* Feb. 1991, 4; C.J. Hart interview, Jan. 30, 1990.

4. John Durbin, "Marvin Rifchin," *March Meet Souvenir Program* (Bakersfield, Calif.: Kern County Racing Association, 1977), 21–22; Suzy Beebe, "Spotlight: Marvin Rifchin," *Drag News,* Apr. 20, 1968; "The Racing Tire Story," *Drag World,* Nov. 19, 1965, 14; Marvin Rifchin interview, Oct. 20, 1987, DROHA, NMAH. See also Bob Behme, "Racing Rubber," *Car Craft,* Aug. 1954, 10–15ff; Bob Pick, "Answers to Your Racing Tire Problems," *Hot Rod,* Feb. 1958, 62–65; and Scotty Fenn, "Meet the Manufacturer: Inglewood Tire Service," *Drag News,* Nov. 14, 1959, 4–5.

5. "Not Your Grandfather's Oldsmobile," *Team Oldsmobile in Action,* Apr. 1989, 2.

6. Bill Martin to Don Garlits, Mar. 23, 1990; Wally Parks to Martin, Mar. 30, 1954; clippings from the Palatka (Fla.) *Daily News,* 1953–54: copies in the Archives of the Museum of Drag Racing, Ocala, Florida (hereafter AMDR).

7. Mickey Thompson with Griffith Borgeson, *Challenger: Mickey Thompson's Own Story of His Life of Speed* (New York: Signet Key Books, 1964), 80–83; see also Gray Baskerville, "Mickey Thompson," in *Petersen's History of Drag Racing,* 26–29. Despite the posting of huge cash rewards, nobody was brought to justice for the murder of Thompson and his wife, Trudy, in March 1988. See *New York Times,* Mar. 17, 1988; see also Shav Glick, "Mickey Thompson's World: Speed, Ideas," *Los Angeles Times,* Mar. 17, 1988; and Jim Murray, "The Shocking, Sudden End for Mickey Mach I," ibid., Mar. 24, 1988.

8. On the LeBlanc machine, see Lynn Wineland, "Wrinkled Rails," *Rod and Custom,* May 1954, 42–43. Jones told Don Jensen that this slingshot "actually preceded Mickey by . . . three or four months." But, he added, "us guys up north [Jones was from Niles, Calif.] were always forgotten" (Red Jones interview, June 1990, DROHA, NMAH, 6). And in this case so were the guys from down south. In a letter to the author (Nov. 4, 1990) Jensen sketched a five-step model for the evolution of dragster design, from a track or lakes roadster of 1950 to a slingshot with no drive shaft and no rear suspension in 1955.

9. Thompson with Borgeson, *Challenger,* 81.

10. Ibid.

11. Mickey Thompson, "Keep It Straight!" *Hot Rod,* Oct. 1955, 40–41ff.

12. In reconstructing what happened to Gendian, I appreciate the help of his friend Frank Carey, an eyewitness. Not long after Thompson left Lions to take over another strip at Fontana in San Bernardino County, NHRA canceled the Lions sanction, claiming that in a five-month period 37.6 percent of all accidents in competition nationwide occurred on that one strip and that NHRA's insurance program had been put in serious jeopardy (*National Dragster,* June 12, 1964). According to C.J. Hart, Thompson's successor, the actual reason was Hart's failure to suspend operations, as demanded, while NHRA and *Hot Rod* staged an event at Riverside.

13. Don Jensen calls attention to another consideration: "As the horsepower increased and the tires began to spin quite a bit [it helped] for the driver to see the tires out of his peripheral vision, and to control the amount of wheelspin by watching the color of the tires turn from grey to black . . . then go to white in the center." Jones interview, June 1990, DROHA, NMAH, 28.

14. Dick Day, "Santa Ana-versary Waltz," *Car Craft,* Oct. 1954, 50; Bob Greene, "Smokin' White Owl," *Hot Rod,* Nov. 1954, 32–35.

15. Duane DePuy, "An Experiment in Acceleration: The Problems Besetting a Rear-Engine Dragster," *Hot Rod,* Oct. 1954, 44.

16. Gray Baskerville, "Breakthrough: Henslee-Cook's Modified Roadster May Be Drag Racing's Missing Link," *Drag Racing,* Sept. 1989, 34–36; Dean Brown, "Special for Speed," *Drag News,* Dec. 14, 1957, 4ff.

17. David Pye notes that "the visible shape of a streamlined craft depends as much

upon what you have chosen to streamline, i.e., to put inside it, as on the laws of nature." *Nature and Aesthetics of Design,* 30.

18. Chuck Kurzawa to "The Roving Reporter," *Drag News,* Apr. 7, 1967, 9.

19. Bob Greene, "One-Five-0 Comin' Up," *Hot Rod,* Oct. 1955, 34–37ff.

20. Roger Huntington, "Getting a Bite," ibid., Dec. 1955, 39.

21. Karen Nelson, "Personality Profile: Ernie Hashim," *Drag Racing,* Sept. 1967, 33–37ff.

22. Roulston, "The Hot Rod Story," 2:70.

23. Dean Brown, "Master of the Quarter," *Drag News,* May 11, 1956, 11–12; "Torrid Topolino," *Car Craft,* Jan. 1955, 28–31. Nelson's flathead did have some unusual (and expensive) components, including a crankshaft with a 4⅜ stroke that had been machined from a steel billet.

24. "A 'Right' Set of Rails" (Collins), *Hot Rod,* July 1956, 48–49; "Cornelius Man Builds Dragster" (Hall), *Forest Grove News Times,* Mar. 3, 1955; "He's Nothing but a Dragster" (Hall), *Fort Ord Diamond Dust Panorama,* Feb. 15, 1957. I am thankful to Carolyn Hall-Salvestrin for the references on her father.

25. Eric Rickman, "Dragsters Walk the Straight and Narrow," *Hot Rod,* Sept. 1956, 44–45; see also Terry Cook, "On the Carpet: Scotty Fenn," *Drag World,* May 28, 1965, 3ff. Nelson's Fiat also had differential caster.

26. Rickman, "Dragsters Walk."

27. Dean Brown, "Slide Rule Rebel," *Drag News,* Aug. 10, 1957, 8–9.

28. "Santa Ana Strip Places Immediate Ban on Any Fuel but Straight Gasoline," *Drag News,* Feb. 9, 1957, 2; see also "Automotive Pump Gas Only Operation Adopted by Growing Drag Strip Organization March 1," ibid., Feb. 23, 1957, 1; the quote in the epigraph is from Dean Brown's editorial on p. 2 of the Feb. 23 issue.

29. See, for example, Bill Likes, "Engineering vs. Chemistry," *Hot Rod,* Dec. 1952, 21; see also Don Francisco, "Nitro! The Poor Man's Supercharger," ibid., Sept. 1951, 9ff; Barney Navarro, "More Horses through Chemistry," *Hop Up,* Dec. 1951, 28–29; California Bill, "How to Use Hot Fuels," ibid., Feb. 1953, 8ff; W. G. Brown, "Danger—High Explosives," *Car Craft,* July 1954, 20–25ff; Brown, "The Nitro Engine," ibid., Sept. 1954, 18–23ff.

30. Wally Parks, "The Editor Says," *Hot Rod,* June 1957, 5; see also "Gas Only for Nationals," 56.

31. In early 1961 this was one of the questions Fenn circulated as part of an "Open Letter to Wally Parks." Terming this a "vicious, unwarranted attack," NHRA's *National Dragster* published the questions along with Parks's answers in its issues of March 10 and 24, 1961.

32. *Hot Rod* began pretty much ignoring the fuel racers, though their exploits could still be followed in the ads. Parks finally did run an article on the Cook-Bedwell machine but called its marks unofficial (Miller Mack, "Top Rail," Dec. 1957, 43–47).

33. Bob Pendergast, "The Championship Money Can't Buy," *Hot Rod,* July 1957, 62.

34. This is Don Garlits's description of his onetime competitor and longtime friend in *"Big Daddy": The Autobiography of Don Garlits* (Ocala, Fla.: Museum of Drag Racing, 1990), 25.

35. "The Draggin' Machines," *Rod and Custom,* July 1953, 46.

4 POWER

1. Rick Voegelin, "The Ten Most Powerful Men in Drag Racing," *Car Craft,* May 1975, 40–43.

2. Wendy Jeffries, "Wally Parks, a Racing Original," *Bonneville Racing News,* Sept. 1990, 1. See also Wallace, "Wally Parks," 22–25; "Wally Parks," in *The Encyclopedia of Auto Racing Greats,* ed. Robert Cutter and Bob Fendell (Englewood Cliffs, N.J.: Prentice-Hall, 1973), 458–59; and the Wally Parks interview in Tom Medley, *Hot Rod History,*

Book One: The Beginnings (Osceola, Wisc.: Motorbooks International, 1990), 146–48.

3. Wallace, "Wally Parks," 24. This exposition was the forerunner of a long-running annual event called the Motorama.

4. Lee O. Ryan, "The Hot Rod Story," *Hot Rod,* Mar. 1952, 62. (The epigraph is taken from this same article, a reprint from an NHRA pamphlet.) Parks admitted: "We hesitated before using the name [hot rod] for our purposes. . . . Hot rodders had a terrible reputation" (qtd. in Ted Hilgenstuhler, "Genuine Hot Rodder Is Right Guy," *Los Angeles Herald Express,* May 16, 1959, B1).

5. *How to Form a Hot Rod Club* (Los Angeles: NHRA, 1961), 8; "Draggin' Around," *Hot Rod,* Mar. 1956, 40.

6. "Your NHRA Bulletin Board," *Hot Rod,* Jan. 1958, 71. The extent of NHRA insurance coverage in the mid-1960s is detailed by Dean Brown in "No Elephants," *National Dragster,* Apr. 10, 1964, 2. The exploits of the Drag Safari (or, sometimes, Safety Safari) are central to NHRA lore; Coons was Wally Parks's very first choice for induction into Don Garlits's Drag Racing Hall of Fame, established in 1990. Hence, a fundamental source is Parks's own account, "Safety Safaris—Spreading the Word," in *Drag Racing: Yesterday and Today* (New York: Trident Press, 1966), 43–104; but see also Karen L. Dunbar, "The Original Hunt for Drag Racing Safety," *Drag Racing,* Nov. 1987, 142–47. The Wally Parks interview (Oct. 16, 1989, DROHA, NMAH) is richly evocative of Parks's personal style.

7. Parks interview, Oct. 16, 1989, 42. The AAA sanctioned competitive events, including the Indianapolis 500, from 1909 until 1955, when it withdrew from this activity in the face of attacks on auto racing led by Oregon senator Richard L. Neuberger.

8. NHRA's eastern regional director, Ed Eaton, made these remarks in the course of presenting Parks with a plaque at the 1959 NHRA advisors' meeting in Detroit (see minutes of the meeting, AMDR). Formerly employed in the tool and die department of Grumman Aircraft in Bethpage, New York, Eaton had also headed the Long Island Hot Rod Association, which, said *Hot Rod,* "was a pretty nebulous affair" until Parks began lending a hand (Fred Horsley, "Draggin' in the East," *Hot Rod,* July 1953, 52–53ff). In the latter 1950s Eaton rose quickly through the ranks of the NHRA leadership.

9. Wally Parks, "The Editor Says," *Hot Rod,* Aug. 1956, 5.

10. *Operation Drag Strip* (Los Angeles: NHRA, ca. 1958), 3.

11. "Will Fuel Stay Legal?" *Rod Builder and Customizer,* Sept. 1957, 18–19ff. Moorhouse, *Driving Ambitions,* 101ff, presents an astute analysis of the "coded language" Parks's *Hot Rod* used to impugn the motives of those who remained involved in fuel racing and whose ultimate aim was to topple the amateur ideal.

12. "A.T.A.A. Merger with N.H.R.A. Announced, Wally Parks to Head," *Drag News,* Mar. 22, 1958, 1; Jim Lamona, "Hot Rodding's Big Merger," *Rod Builder and Customizer,* Aug. 1958, 8–9; "A.T.A.A. Series Date Aug. 20–24, Fuel-Gas-Stocks," *Drag News,* Aug. 9, 1958, 8.

13. Wally Parks, "The Editor Says," *Hot Rod,* June 1958, 5. Iskenderian's criticism appeared, in his customary full-page ad, in May.

14. Wally Parks, "The Editor Says," *Hot Rod,* Aug. 1957, 5; May 1958, 5.

15. *Drag News,* Apr. 6, 1960, 5.

16. There were subtle ways of getting even. Even though Fenn's Chassis Research was thriving by 1958, *Hot Rod* called the rather obscure Joe Itow "Southern California's foremost chassis builder" (July 1958, 31). Later it was widely believed that NHRA deliberately "legislated" (i.e., changed its technical rules) to exclude Fenn's standard frontend design.

17. Sy Mogel to "Open Forum: Three More Thoughts on NHRA Profit Picture," *Drag News,* Oct. 16, 1976, 4; Dick Carol to "Open Forum: NHRA Dollars, Pt. 2: Where They Go, and Why," *Drag News,* Oct. 30, 1976, 2.

18. Doris Herbert, "Editor's Exhaust," *Drag News,* Oct. 1, 1960, 1ff. Gibbs, who signed

himself "In the Dark" and asked rhetorically, "Did the NHRA ban your reporters and photographers?" later became NHRA's vice president for competition.

19. Ernie Schorb, "Forward Message to Speed Week Drag Races," *Official Program: Speed Week Drag Races* (NHRA/NASCAR, 1960), 3.

20. Cutter and Fendell, *Encyclopedia of Auto Racing Greats,* 459.

21. Don Garlits and Brock Yates, *King of the Dragsters* (Radnor, Pa.: Chilton Book Co., 1975), 126–28.

22. Shav Glick, "NASCAR Is Still All in Family," *Los Angeles Times,* Nov. 2, 1989, C1ff. See also "William H. G. France," in Cutter and Fendell, *Encyclopedia of Auto Racing Greats,* 225–28.

23. Mark Kram, "Fame and Terror at 200 mph," *Sports Illustrated* (offprint, AMDR), 28–29; Garlits and Yates, *King of the Dragsters,* 127. When Parks successfully defended NHRA against Garlits's Professional Racers Organization in the 1970s he must surely have taken inspiration from France's 1969 defeat of the Professional Drivers Association, an organization headed by Richard Petty, who was to stock car racing what Garlits was to drag racing.

24. Roger Huntington, "Powerhouse Unlimited," *Hot Rod,* July 1959, 44–46.

25. Wally Parks, "Editor's Column," *Hot Rod,* Nov. 1959, 5; see also "Detroit Bid Wins Nationals," *Rod Journal,* June 1959, 1.

26. Bob Pendergast, "Motor City Spectacular," *Hot Rod,* Nov. 1959, 34.

27. "Let's *All* Play '20 Questions,'" *National Dragster,* Mar. 10, 1961, 3. In their glossary Cutter and Fendell (*Encyclopedia of Auto Racing Greats*) identified the NHRA as "National Hot Rod Association, sometimes also known as Wally Parks."

28. Parks, *Drag Racing,* 207.

29. Dave Wallace, Jr., "Allison Wonderland," *Drag Racing,* July 1985, 88; see also "Art Arfons," in Motorsports Hall of Fame, *Third Annual Induction Ceremony* (Novi, Mich.: Motorsports Hall of Fame of America, 1991), 10; and "The Arfons Brothers," in Cutter and Fendell, *Encyclopedia of Auto Racing Greats,* 22–23.

30. See, for example, Bob Truby, "To the Members of the Drag Fraternity," *Drag News,* Nov. 1, 1958, 13.

31. Don Francisco, "The Sidewinder," *Hot Rod,* Oct. 1959, 36.

32. See "The Magwinder," *Hot Rod,* Aug. 1961, 51–53.

33. "Meet the Manufacturer: Howards Cams," *Drag News,* Mar. 21, 1959, 8–9; "Pages from the Scrapbook of Howard Johansen," *Drag Racing,* July 1967, 66–71; "The First Top Eliminator," *Drag Racing,* Oct. 1973, 38–39; Pamela Carnline, "Motor Sports World Mourns Johansen Passing," *Sun City News* (Arizona), Nov. 3, 1988.

34. Bob Behme, "Drag Experts Predict," *Car Craft,* Aug. 1960, 14–19.

35. "The Big Go West," *Hot Rod,* May 1961, 26–37ff.

36. Quoted in Al Caldwell, "Northern Briefs," *Drag News,* May 6, 1961, 3ff.

37. LeRoi Smith, "Second Annual Winternationals Championship Drags," *Hot Rod,* May 1962, 92.

38. Terry Cook, "On the Carpet: Pete Robinson," *Drag World,* Feb. 11, 1966, 3.

39. LeRoi Smith, "1961 National Drags," *Hot Rod,* Nov. 1961, 43.

40. Smith, "Second Annual Winternationals," 92, 38.

41. Garlits and Yates, *King of the Dragsters,* 169. Garlits noted that a factor in NHRA's decision was "the development of aluminized fire suits that protected the drivers almost completely from burns" (163). Fiery explosions were certainly a much greater danger with nitro, but the first firesuits did not afford anything like complete protection, and it would be only a short time before Chuck Branham, a regional technical director for NHRA, would die of burns suffered at Pomona in the Pomona Valley Timing Association's "Starlight" dragster.

42. *National Dragster,* Dec. 14, 1962.

43. [LeRoi] Tex Smith, "Pop Rod Debate: Is Hot Rodding Too Organized?" *Popular Hot Rodding,* July 1964, 53.

44. Al Caldwell, "Northern Briefs," *Drag News,* Aug. 25, 1962, 3. Caldwell was obviously ambivalent about this matter, for a couple of years later he would decry national organizations as "depressing dead weight" ("Pop Rod Debate," 510).

5 *FAME*

1. Speed equipment (or, more palatably to the general public, "specialty" equipment) is a subject deserving of its own monograph. There is one full-scale biography of a key figure (Art Bagnall, *Roy Richter: Striving for Excellence* [Los Alamitos, Calif.: Author, 1990]) and several excellent interviews in the Drag Racing Oral History Archive (Ed Iskenderian, Keith Black, and Marvin Rifchin), but the opportunities for further work are enormous. For an overview, see Don Montgomery, "Speed Equipment and Shops," *Hot Rod Memories Relived Again* (Fallbrook, Calif.: Author, 1991), 89ff. For a few particulars: "Schiefer Manufacturing," *Drag Sport,* Apr. 23, 1966, 10; "Meet the Manufacturer [Schiefer]," *Drag News,* Apr. 19, 1969, 8; Ray Brock, "Paul Schiefer Eulogy," ibid., Sept. 5, 1970, 2; "Meet the Manufacturer: Edelbrock," ibid., Apr. 6, 1969, 16; Dave Wallace, Jr., "An Afternoon on the [Edelbrock] Dynamometer," ibid., July 12, 1975, 18; "Bell Auto . . . World's First Speed Shop," *Drag Digest,* Nov. 11, 1966, 27; Karen Scott, "Meet the Manufacturer [Crower]," *Drag News,* May 24, 1969, 23; Dave Wallace, Jr., "And Now, from the Folks Who Brought You the 'U-Fab' Intake Manifold," ibid., Aug. 2, 1975, 16–17; Wallace, "Jack Engle Walks Softly, Carries a Big Bumpstick," ibid., Nov. 22, 1975, 16–17; Terry Cook, "Ed Winfield: The Father of Hot Rodding," *Hot Rod,* Jan. 1973, 107; "Meet the Manufacturer: Venolia," *Drag News,* Aug. 22, 1970, 18–19; Al Caldwell, "Go with Moon!" ibid., Sept. 28, 1963, 8–9; Karen Scott, "Meet the Manufacturer: Hays Clutches and Flywheels," ibid., May 31, 1969, 18; Scott, "Meet the Manufacturer [Jardine]," ibid., May 17, 1969, 8; Scott, "Meet the Manufacturer [Donovan]," ibid., June 28, 1969, 23; Al Caldwell, "Meet the Manufacturer: Ed Pink Racing Engines," ibid., Aug. 18, 1967, 26–27; "A Look at the Old Master's Shop," *Drag Sport,* Sept. 6, 1965, 3–4ff; "Tips from Ed Pink," *Drag Racing,* Mar. 1967, 38–39; Jerry Brandt, "Ed Pink," ibid., May 1975, 61–62; Karen Scott, "Meet the Manufacturer: B&M," *Drag News,* Apr. 5, 1969, 6; Scott, "Meet the Manufacturer: Mickey Thompson," ibid., May 3, 1969, 8; "Meet the Manufacturer [Crane Cams]," ibid., Dec. 29, 1973, 16; "Meet the Manufacturer: Lakewood Industries," ibid., Nov. 20, 1971, 18–19; June 1, 1974, 22–23.

2. "Ready-made Dragster Parts," *Hot Rod 1961* (Los Angeles: Trend, Inc., 1961), 72–83. My thanks to Rudi Volti for this reference.

3. The quote in the epigraph is from Garlits and Yates, *King of the Dragsters,* 67.

4. "Bakersfield: The Heritage of the Racer's Race," *Drag News,* Mar. 7, 1970, 5.

5. Garlits and Yates, *King of the Dragsters,* 7, 12.

6. Ibid., 19, 28.

7. Ibid., 41.

8. Ibid., 46.

9. Ibid., 47–55.

10. Ibid., 56–67. Though Postoian lived in Detroit, he was a longtime friend of Californians such as Iskenderian and Joe Pisano (who worked for Frank Venolia), having street raced around Los Angeles in the very early 1950s. Postoian interview, Mar. 20, 1992.

11. Ibid., 75–86; see also Dan Roulston, "Romeo and Go!" *Drag News,* Nov. 30, 1957, 8–9; and Dean Brown, "Garlits Stops Big California Fuel Challenges," ibid., July 25, 1958, 1ff.

12. "Who Is Pete Ogden," *Drag Sport,* Nov. 15, 1965, 9.

13. Setto brought along Connie Kalitta, a neighbor of his from Mt. Clemens, who would soon become a star on the dragster circuit himself.

14. Ernie Schorb, "This 'Slingshot' Is Loaded," *Hot Rod,* Feb. 1959, 56–57; John Durbin, "An Open Letter to Don 'Big Daddy' Garlits," *March Meet '76 Program* (Bakersfield, Calif.: Kern County Racing Assn., 1976). There was a lot of controversy about

how Garlits's work compared. Contemporary photos appear to show a machine that was about average by California standards; see, for example, "Don Garlits—East Coast 'King,'" *Drag News*, Feb. 21, 1959, 10–11.

15. Quoted in Bob Behme, "Controversial 'King of the Drags,'" *Car Craft*, Oct. 1959, 56.

16. Garlits and Yates, *King of the Dragsters*, 104.

17. "Garlits-Howard Cam Meet in Anniversary Race Final," *Drag News*, Mar. 28, 1959, 1ff.

18. Quoted in Jeff Burk's editorial in *Super Stock*, Aug. 1990, 6.

19. Don Elliott, "Chrysler Dragster Wins First DNI Meet," *Drag News*, June 20, 1959, 1.

20. Iskenderian ad, *Drag News*, June 20, 1959, 13.

21. Garlits and Yates, *King of the Dragsters*, 111. Garlits presents an even more graphic account of this incident in Darryl E. Hicks, *Close Calls* (Shreveport, La.: Huntington House, 1984), 89–90.

22. Garlits and Yates, *King of the Dragsters*, 113–15.

23. Ibid., 116. As Malone recalls, it was Garlits who asked *him* if he would drive: Dale Wilson, "Art Malone, the Sport's 'Other' Old Man," *Super Stock*, May 1991, 93. See also Ken Weddle, "Inside the World of Art Malone," *Drag News*, Feb. 25, 1966, 5ff.

24. Garlits to *Drag News*, Aug. 22, 1959, 1ff.

25. Garlits and Yates, *King of the Dragsters*, 121; Kram, "Fame and Terror," 27. See also Joe Deggendorf, Jr., to *Drag News*, May 16, 1960, 14; Ben Brown, "Al's Speed Shop Stable," ibid., Dec. 31, 1960, 5; "Pop Rod X-Ray: The Chizler," *Popular Hot Rodding* (offprint, AMDR), 70–73; Ken Weddle, "Behind That Mustache . . . Is a Real Drag Racer," *Drag News*, June 16, 1967, 9; and "Legend of Al's Speed Shop," *Drag Racing*, Sept. 1967, 55–61.

26. Bob Creitz, qtd. in Jeff Burk, "Guest Editorial: The First 200," *Drag Racing*, Aug. 1987, 4.

27. Ibid.

28. Quoted in Garlits and Yates, *King of the Dragsters*, 122–23.

29. Palamides "killed some of his drivers," Don Jensen remarked (Dec. 1989 tapes, DROHA, NMAH).

30. Bill Friend to *Drag News*, Nov. 28, 1959, 14.

31. Friend to *Drag News*, Dec. 12, 1959, 14.

32. Bruce Bieker to Bill Friend, *Drag News*, ibid.; Donald A. Noyes to *Drag News*, ibid.

33. Garlits to *Drag News*, Dec. 5, 1959, 14.

34. Williams had gained considerable notoriety with a blown Chrysler fueler that weighed only 1,170 pounds, partly by virtue of his use of rectangular aluminum for the entire frame. See Don Francisco, "Quarter-Mile Pioneers," *Hot Rod*, Sept. 1959, 78–81.

35. Garlits and Yates, *King of the Dragsters*, 139.

36. Ibid., 141.

37. Ibid., 141–45. See also "On the Carpet: Connie Swingle with Terry Cook," *Drag World*, June 4, 1965, 5–6ff.

38. Qtd. in Kram, "Fame and Terror," 26. This is a perceptive article, even though Garlits was supremely disillusioned with the author for portraying him, in the wake of his Chester fire, as "a reformed junkie who went clammy with fear every time [he] climbed into the race car" (Garlits and Yates, *King of the Dragsters*, 118). The articles that have been written about Garlits are literally countless, but special favorites of mine include: Tom Madigan, "Pop Rod Drag Star: Don 'Big Daddy' Garlits," *Popular Hot Rodding*, Mar. 1971, 18–21ff; Richard Conniff, "In Florida: Old-Fashioned Ingenuity on Wheels," *Time*, Apr. 12, 1986, 16–17; Sam Moses, "The Man Who Would Be Greyhound," *Sports Illustrated*, Sept. 29, 1986, 46–49ff; Dave Wallace, Jr., "Mr. Garlits Goes to Washington," *Drag Racing*, Mar. 1988, 68–74; and J. Edwin Smith, "'Big' Is Back," *Drag Racing Today*, June 16, 1989, 10–11ff (from the Gwinnitt, Georgia, *Daily News*). I have had many conversations with Garlits, never once coming away without a head full of new

ideas, but the only transcript in the NMAH Drag Racing Oral History Archive (Feb. 1988) is from an interview with a deliberately narrow focus—on dragster materials and fabrication—the purpose having been to garner material for a video produced by the museum.

39. Jeff Burk, "Interview: Wally Parks," *Super Stock,* Nov. 1992, 35. Moses talks about "how thoroughly Garlits has intimidated NHRA," but that may be stretching a point ("The Man Who Would Be Greyhound," 50).

40. "On the Carpet: Connie Kalitta with Terry Cook," *Drag World,* Jan. 21, 1966, 3. Kalitta was a resourceful racer, and he would enjoy a long career, but he usually took his technological and theatrical cues from others. When he first began to attract attention he called himself "Crazy Connie" (*Drag News,* Jan. 7, 1960, 10). Later, with more originality, he dubbed himself the "Bounty Hunter," listing the names of racers he had bested on the side of his cockpit (see "Personality Profile: Connie Kalitta," *Drag Racing,* July 1966, 10–13). But for many years Garlits remained the only racer who could be depended on to draw "on the strength of [his] name alone" (Bob Ramsay, "Exhibition Evolution," *Drag World,* Apr. 25, 1965, 9).

41. As late as the spring of 1964, in noting how nearly a dozen fuelers converged on a Pennsylvania track to compete for a $300 purse, NHRA's *National Dragster* lamely explained that "big money fuel meets are still in the experimental stage in the East" ("Big Fuelers Invade York," May 1, 1964, 18). See also Bob Ramsay, "Who's Who: Eastern Fuelers," *Drag World,* June 4, 1965, 10; and "Fuelers Abounding in East," ibid., Apr. 15, 1966, 10.

42. LeRoi Smith, "Dragsters on the Prowl," *Hot Rod,* Jan. 1964, 50.

6 FORTUNE

1. Dave Densmore, "It's Not Just the Money, It's the Principle," *Super Stock,* Aug. 1991, 11.

2. By one rather improbable estimate the total value of these manufactures was $57 billion annually. Jay Storer, "Racing behind the Scenes: The Big Trade Shows," *Race Time,* Nov. 1991, 5.

3. David Jernigan (Marin Institute for the Prevention of Alcohol and Other Drug Problems), qtd. in Densmore, "It's Not Just the Money"; "Potholes Ahead on Tobacco Road," *Newsweek,* June 18, 1990, 16–17; Richard Harwood, "No More Tobacco Ads," *Washington Post,* Apr. 21, 1991.

4. Allen E. Brown, publisher of the annual *National Speedway Directory,* estimated in 1992 that there had been a total of some 550 drag strips in North America. This estimate was no doubt conservative, but in any event the economics were typically marginal. For example, when the owner of the strip at Amarillo put it on the market for $500,000 in 1992, he reported a 1991 gross of $275,000 and recast earnings (before interest, taxes, depreciation, and owner compensation) of only $95,000 (see Alessandra Bianchi, "Business for Sale—Texas Drag Strip," *Inc.* 14 [July 1992]: 146). Although Amarillo had once been the site of the NHRA World Finals, no "national" events had been staged there in many years. There were, however, three other strips within a day's drive that did host such events, at Denver, Dallas, and Topeka. My thanks to Robert Staples for the *Inc.* reference.

5. In 1990 only about a quarter of all the top-fuelers in the United States were from California, but eight out of the Winston top ten were either from California (two), Michigan (two), or Texas (four).

6. Gray Baskerville, "Tommy Ivo," in Wallace, ed., *Petersen's History of Drag Racing,* 71.

7. Ivo subsequently fostered the notion that all his machines were showstoppers, but actually his first roadster looked crude enough to be a deliberate joke.

8. "And Four to Go," *Hot Rod,* Dec. 1961, 62–65.

9. A friend of Ivo's named Tom McCourry later bought the Showboat and clothed it in a body that mimicked a Buick station wagon.

10. Steve Gibbs, "Ivo-Zeuschel Wins Match Race with Untouchable," *Drag News,* Apr. 6, 1963, 3; "Untouchable Beats Ivo 2 Straight in Re-Match," ibid., Apr. 20, 1963, 3. Ivo wound up his career in the 1980s in a jet car himself.

11. Iskenderian ad, *Drag News,* Nov. 25, 1961, 11.

12. "The Lions Drag Strip History," *Tach* (AHRA), Sept. 1965, 26; "History of Lions Drag Strip," *The Last Drag Race: Lions* (Costa Mesa, Calif.: Newport Productions, 1972), 1ff; "Drag Racing Profile: Larry Sutton," *National Dragster,* Feb. 2, 1979, 5; Kay Presto, "Behind the Line: The Life and Times of Starter Larry Sutton," *Super Stock,* May 1986, 26–29ff. The first night racing in southern California actually took place at the old Saugus strip a year or two before Lions. See Mike Doherty, "Under the Lights," *Drag Racing,* July 1969, 48–51.

13. Doris Herbert, "Editors Exhaust," *Drag News,* Jan. 13, 1963, 2.

14. Cash purses noted in this and the following paragraph are from *advertised* amounts in *Drag News* in 1963–64 and probably do not always have an exact correspondence with reality. For one thing there is no way of knowing when contestants agreed in advance to a split.

15. Terry Cook to the author, Apr. 2, 1965. On the economics of racing a top-fueler, see Cook, "The Grass Is Always Greener," *Drag World,* Mar. 19, 1965, 6ff; see also Eric Dahlquist, "There's No Substitute for 'Cubic Money,'" *Hot Rod Magazine Championship Drag Races* (Los Angeles: Petersen Publishing Co., 1965), 16–17. The newspapers might play up an exceptionally large purse (see, e.g., Bob Thomas, "Drag Racing: A Big Payoff for a Short Duel," *Los Angeles Times,* Aug. 9, 1968, pt. 3, 7), but, as Cook pointed out, "the average racer spends a fortune to go out and butt heads with dozens of guys like himself, and a few big buck teams who can afford spare engines."

16. Mike Doherty, "The Leading Money Winner," *Drag World,* Apr. 2, 1965, 2.

17. See Rob Burt, *Surf City Drag City* (Poole, Dorset: Blandsford Press, 1986), 14–15. This book addresses the popular media's effort to link surfing, hot rodding, and drag racing as quintessential "California" activities.

18. Terry Cook, "Meet the 'Surfers,'" *Drag World,* Mar. 11, 1966, 3ff.

19. Ibid., 3.

20. Ibid., 16.

21. On the dragster Skinner and Jobe never campaigned, see Don Prieto, "It's What's Happening," *Drag Racing,* May 1967, 33–35, an article that sought to link drag racing with another California pop culture icon, the hippie.

22. Cook, "Meet the 'Surfers,'" 3; see also Chris Martin, "Ten Best Top-Fuelers of All Time," *Drag Racing,* Nov. 1989, 26–27. Skinner and Jobe later became involved in other forms of racing, however, including a stint in the early 1970s as mechanics on a Formula A McLaren driven by John Cannon.

23. "Personality Profile: Mike Sorokin," *Drag Racing,* Aug. 1967, 49–53ff.

24. JoAnn Haines to Doris Herbert, *Drag News,* Sept. 24, 1960, 14.

25. Terry Cook, "Jackson Taken," *Drag World,* Feb. 11, 1966, 5; letter from D. C. to Art Irwin, Feb. 8, 1966, "New England Notes," *Drag News,* Mar. 18, 1966, 2.

26. Terry Cook, "So You Want to Go on Tour," *Drag World,* Apr. 8, 1966, 6. Californian Tony Nancy tells about getting bullied out of his appearance money at a small southwestern strip, in Tom Madigan, *The Loner: The Story of a Drag Racer* (Englewood Cliffs, N.J.: Prentice-Hall, 1974), 66–70.

27. Cook, "So You Want to Go on Tour"; Dave Wallace, Jr., "Out of the Gate," *Drag News,* Sept. 4, 1976, 4.

28. Don Garlits, "Ten Tips for Cutting Costs," *Drag News,* Nov. 6, 1976, 16. Garlits added, only slightly immodestly, that "only about five people in the whole damn world have the financial backing and 'staying power'" to run a fueler consistently for a year. He had made this same point about "five people" for many years. In an article in the June

1965 issue of *Popular Mechanics* titled "Here's How I Win Drag Races," he wrote, "I'd guess there are more than 500 unlimited fuel dragsters in the United States, but most of the drivers earn little more than 'food and fuel.' . . . about 5 earn a good living" (206). See also Dave Wallace, Jr., "A Candid Conversation with Our 1975 Drag News Driver of the Year," *Drag News*, Dec. 6, 1975, 14–15.

29. Mike Lewis, "Open Forum: Track Operator's Views on NHRA Dollars," *Drag News*, Dec. 26, 1976, 4.

30. Terry Cook, "Baney Steps Down," *Drag World*, Nov. 12, 1965, 7. See also "On the Carpet: Lou Baney with Terry Cook," ibid., Apr. 22, 1966, 3ff; April 29, 1966, 3ff. This chapter's epigraph is from the first page of the second part of this interview.

31. Terry Cook, "Motorman," *Car Craft*, Feb. 1967, 14. See also Ralph Guldahl, Jr., "Gene Adams Speaks Out," *Popular Hot Rodding*, Oct. 1964, 58–61ff; "King of Garden Grove," *Drag Racing*, Nov. 1966, 28–31ff; and "Adam[s] and Wayre AFD," *Drag Sport*, Aug. 16, 1965, 3–4 (which, not at all coincidentally, appeared in conjunction with an ad for Wayre's car lot).

32. Cook, "Motorman," 71; "Our Point of View," *Car Craft*, Mar. 1967, 13. In the latter, Cook was responding to Dale Ham, an NHRA official, who had decried the practice of splitting purses by drawing on a fatuous analogy with the Vietnam War (ibid., 12).

33. Hindle, *Technology in Early America*, 24.

34. Ralph Guldahl, Jr., "Why 200 MPH?" *Popular Hot Rodding*, Jan. 1965, 30–33ff.

35. John Durbin, "Zoom Headers," *Drag Racing*, Mar. 1965, 36–39; see also "Frank Cannon's 'Hustler V,'" *Hotrod Parts Illustrated*, Oct. 1964, 10–12; Garlits and Yates, *King of the Dragsters*, 175. This story is a bit tangled, for Gilmore claimed that he sold Garlits "the third set we made" (Al Yates, "The Woody Chassis Story," *Drag Racing*, Aug. 1965, 55).

36. Guldahl, "Why 200 MPH?" 72.

37. The information in the next eight paragraphs is drawn primarily from Dave Scott, "The Surfers—World's Quickest Dragster," *Popular Hot Rodding*, June 1966, 28–33ff.

7 COMPETITION

1. Montgomery, *Hot Rods as They Were*, 158; see also *Hot Rods in the Forties* (Fallbrook, Calif.: Author, 1987), 143.

2. Dan Roulston, "Star Coupe . . . a Chrysler for a Quarter," *Best Hot Rods* 3 (1957): 38–39. Montgomery designed the injector himself. Rather than a GMC, the blower was a SCOT, a unit designed specifically for automobile engines; Montgomery recalled that "the castings were awful." Montgomery interview, Feb. 8, 1989, DROHA, NMAH, 33.

3. "San Pedro [Jack Ewell], Russ Palmer Posts Torrid Class Wins in Invitational," *Drag News*, Oct. 19, 1956, 2; Montgomery to the author, Apr. 10, 1991.

4. L. C. Taylor, "When #554 Was King," *Drag Racing*, July, 1974, 44–45.

5. Roger Huntington, "150 [*sic*] MPH Ford," *Rod and Custom*, Dec. 1964, 30–32.

6. Dave Scott, "Record Roadster," *Hot Rod*, Sept. 1960, 42–45.

7. Mike Doherty, "A History of Altereds," *Drag Racing*, Sept. 1970, 38–45ff (essentially this same article appears in Lyle Kenyon Engel, *The Complete Book of Fuel and Gas Dragsters* [New York: Four Winds Press, 1968], chap. 4). See also Terry Cook, "Fiat," *Car Craft*, Apr. 1967, 40–42. On the persistence of fuel altereds in the 1980s, see Sky Wallace, "Topless Dancers," *Drag Racing*, May 1987, 49–59.

8. The sanctioning bodies also allowed for roadsters that were nominally street legal, but these were not as numerous as gas coupes. When NHRA merged the two classes in 1968 technical restrictions rendered the roadsters noncompetitive (see Mike Doherty, "A History of Street Roadsters," *Drag Racing*, Oct. 1970, 46–51), but machines with fiberglass bodies which vaguely resembled Model-T roadsters later became a mainstay of a class called "Super Gas." As suggested throughout this chapter, a history of drag racing could be written solely in terms of the endless revamping of its classification systems.

9. Don Montgomery, *Supercharged Gas Coupes* (Fallbrook, Calif.: Author, 1993),

provides a comprehensive technological analysis of these machines. See also Mike Doherty, "A History of Gas Coupes," *Drag Racing,* Aug. 1970, 44–51ff.

10. An organization called the Street Racers of Los Angeles County claimed a membership of several thousand in 1968. The leader was Willie Andrew Robinson III, a towering weightlifter who worked at the Harvey Aluminum plant in Torrance. See "Big Willie," *Drag Racing,* Dec. 1968, 42–47. Only once would there ever be formal pari-mutuel betting (in the 1970s at Budds Creek, a track in southern Maryland), but open informal gambling was commonplace.

11. Montgomery interview, Feb. 8, 1989, 13.

12. See Editors of Consumer Guide, *Cars of the 30s* (New York: Beekman House, 1980), 91–93.

13. *Drag News,* Sept. 24, 1960; Oct. 1, 1960.

14. "Doug Cook: Personality Profile," *Drag Racing,* Feb. 1967, 20–23ff; "On the Carpet: 'Big John' with Terry Cook," *Drag World,* Apr. 23, 1965, 3ff; Terry Cook, "On the Carpet: Tim Woods," ibid., Apr. 30, 1965, 3ff; Cook, "On the Carpet: Big John vs. Tim Woods," ibid., May 7, 1965, 3ff; "Ohio George and His '33 Willys," *Drag Racing,* July 1966, 30–32ff; Gray Baskerville, "Gasser Wars," *Hot Rod,* Feb. 1988, 38.

15. Baskerville, "Gasser Wars."

16. Terry Cook, "Drag World Interview of Junior Thompson," *Drag World,* Apr. 16, 1966, 3ff; "Tandem Tangle," *Drag Racing,* Nov. 1966, 37–39ff.

17. Baskerville, "Gasser Wars."

18. Ibid.

19. Dorothy Bakity et al. to *Drag News,* Sept. 1, 1962.

20. "Person to Person with 'Dyno Don' Nicholson," *Drag Racing,* Mar. 1966, 13. The first quote in the epigraph is from this same interview.

21. Dean Brown, "Factory Participation," *National Dragster,* May 15, 1964, 2. Brown also mentioned the need to protect the "little guy," a matter of express concern for some time; as early as 1956, NHRA had classified ten "factory-produced 'hot models'" as gas coupes rather than stockers, presuming that this system of rules would "provide the ultimate in fair competition" and "remain in effect for a long time to come" ("What to Do with Stockers?" *Hot Rod,* June 1956, 42). The class system—that is, the political context of technological change—would, however, undergo constant reconstruction in the years to come. As for concern about the little guy, in 1965 Terry Cook estimated that, without stock competition, "95% of our nation's drag strips . . . would operate in the red" ("Blessed Are the Meek . . . ," *Drag World,* Apr. 9, 1965, 6).

22. Dave Wallace, Jr., "Out of the Gate," *Drag News,* Oct. 16, 1976, 2.

23. Frank Wylie, former chief of Dodge public relations, qtd. in Dave Wallace, Jr., "Factory Warfare," *Super Stock,* June 1990, 83.

24. Engel, *Complete Book of Fuel and Gas Dragsters,* 90; see also Norm Mayersohn, "Those Fabulous Sixties," *Car Craft,* Oct. 1973, 58–64; and Al Kirschenbaum, "Galactic Journey," *Hot Rod,* July 1990, 84–89.

25. See Dave Wallace, Jr., "Revolutionary Iron," *Drag Racing,* Dec. 1988, 26–27; and Wallace, "'64 Funny Cars! They Revolutionized a Motorsport," *Super Stock,* Jan. 1990, 16–19.

26. Eric Rickman, "Elapsed-Time Bomb," *Hot Rod,* July 1965, 77. See also Eric Dahlquist, "Super Fuel Xterminator," ibid., Sept. 1964, 26–29; "Drag Racing's First Lady," *Drag Racing,* May 1965, 28–30; and "Personality Profile: Jack Chrisman," ibid., Aug. 1965, 14–17ff.

27. "First Funny on Fuel," *Drag Racing,* Jan. 1989, 18–20; Dan Roulston, "Chrisman's New Match Maker," *Car Craft,* Apr. 1967, 28–33; "Pure Pandemonium," *Drag Racing,* July 1965, 24–27.

28. Al Kirschenbaum, "Altered States," *Hot Rod,* Sept. 1990, 84–87. The Engel book attributes the designation "funny car" to railbirds at Bee Line (*Complete Book of Fuel and Gas Dragsters,* 94); others credit it specifically to an announcer named Jon Lundberg, but

in fact the term was already commonplace in auto racing circles: The first mid-engine machines that showed up for the Indy 500 were dubbed funny cars.

29. In 1965 NASCAR announced that it would not permit using the new 426 hemi (or the Ford SOHC 427) until it was an actual production option, and Chrysler staged a Grand National boycott. See Phil Hall, "Catch the Competitors Sleeping," *Old Cars*, Feb. 7, 1991, 8–9; see also "DRM Interviews Richard Petty," *Drag Racing*, July 1965, 50–53. Petty's car was called "Outlawed."

30. Wallace, "Out of the Gate," Oct. 16, 1976, 2.

31. Engel, *Complete Book of Fuel and Gas Dragsters*, 99.

32. "Strip Blazer IV," *Drag Racing*, Sept. 1966, 40–43.

33. Don Prieto, "Stone-Woods-Cook's Funny Mustang," *Drag Racing*, Mar. 1967, 11–14; "Stone-Woods-Cook Go the 'Funny' Route," *Car Craft*, Mar. 1967, 51–52ff.

34. Don Prieto, "Breaking Loose," *Drag Racing*, Dec. 1966, 5.

35. Forrest Bond, "For Your Second Magical Trick," *Drag Digest*, Oct. 28, 1966, 5; "Garlits's Dart," *Drag Racing*, Feb. 1967, 48–51; "Garlits Withdrawing Dart: Too Successful—No Competition," *Drag News*, Apr. 21, 1967, 11.

36. Dave Wallace, Jr., "Out of the Gate," *Drag News*, July 26, 1975, 4; "The Three-Ring Circus and Traveling Side Show," *Car Craft*, July 1976, 38–39ff.

37. Terry Cook, "Bench Racing Championships," *Drag World*, Apr. 16, 1965, 7.

38. "Drag Races, Inc.," *Rod and Custom*, June 1954, 22–27; Peter Lisa, "Drag Races, Inc.," ibid., July 1954, 6; "Draggin' at Bakersfield," ibid., Mar. 1955, 39–43; "Drag Racers Inc.," *Best Hot Rods* 3 (1957): 24–27; "D.R.I. Opens Big Membership Drive," *Drag News*, Nov. 1, 1958, 3; "Big Meet for DRI," ibid., Nov. 3, 1959, 2.

39. See Dave Wallace, Jr., "Out of the Gate," *Drag News*, Feb. 7, 1976, 2; see also Rick Voegelin, "Lou Baney," *Car Craft*, Nov. 1972, 84–85.

40. Terry Cook, "Our Group Wants 40% of the Gate," *Drag World*, Nov. 26, 1965, 4ff. The idea was hardly new. Doris Herbert had advocated a percentage plan several years before ("Editors Exhaust," *Drag News*, Aug. 20, 1960, 2) and received seconds from her readership (e.g., A. E. McGown to *Drag News*, Oct. 15, 1960, 14). Rick Stewart was in favor of a guarantee plus a percentage (Ralph Guldahl, Jr., "Gene Adams Speaks Out," *Popular Hot Rodding*, Oct. 1964, 85). Forty percenters had been around other racing constellations at least as far back as the 1940s; in 1946 an organization called the American Society of Professional Automobile Racing (ASPAR) threatened to strike the Indy 500 (see Terry Reed, *Indy Race and Ritual* [San Rafael, Calif.: Presidio Press, 1980], 139).

41. "On the Carpet: Roy 'Goob' Tuller with Terry Cook," *Drag World*, Oct. 15, 1965, 3; see also "Personality Profile: Roy 'Goober' Tuller," *Drag Racing*, Apr. 1967, 55–59.

42. In between Santa Ana and Lions, Hart had done stints at Taft and Riverside (on Taft, see "Pappy's Back in Action," *Hot Rod*, Feb. 1960, 75). The last strip Hart managed was Orange County International Raceway in Irvine, only a few miles from his first in Santa Ana (see Dave Distel, "C.J. Hart Returns to the Sport He Helped Found—Drag Racing," *Los Angeles Times*, Aug. 23, 1973, pt. 3, 18).

43. "On the Carpet: Harry Hibler with Terry Cook," *Drag World*, Oct. 22, 1965, 3, 15.

44. Terry Cook, "The '40% Group' Speaks Out," *Drag World*, Nov. 26, 1965, 11.

45. Ibid. See also Karen Nelson, "Personality Profile: Bob Tapia," *Drag Racing*, Nov. 1967, 57–62.

46. Terry Cook wondered both publicly and privately about McEwen; see "1966—A Crucial Year for UDRA," *Drag World*, Dec. 24, 1966, 6ff; and Cook to the author, Dec. 15, 1965: "Why does McEwen get a match race of some sort . . . every other week at Lions when there are so many other cars more in need of, or deserving of, opportunity?" See also "The Mount of the Mongoose," *Drag Racing*, Mar. 1966, 24–27ff; and Don Prieto, "Personality Profile: Tom Mongoose McEwen," ibid., Sept. 1966, 12–14ff.

47. Terry Cook, "Wildcat Boycott Fails," *Drag World*, Dec. 10, 1954, 4ff.

48. Ibid.; Wallace, "Three Ring Circus," 39; Gary Cagle (paid ad), *Drag News*, Mar. 18, 1966, 28; "On the Carpet: Don Garlits with Terry Cook," *Drag World*, Aug. 13, 1965, 22.

49. Cook, "1966"; "Lions Boycott Fails," *Drag World,* Dec. 10, 1966, 1.

50. "The History of Wheelie Cars," *Drag Racing,* Mar. 1974, 52–58ff; Terry Cook, "The Christ Formula—Godsend or Gargoyle?" *Drag World,* Jan. 14, 1966, 6ff; Cook, "FX Foothold," ibid., Oct. 29, 1965, 4.

51. Cook, "FX Foothold."

52. Landy quoted in "Who Is . . . Dick Landy?" *Drag Sport,* Nov. 8, 1965, 5ff. See also "UDRA Elections: A Profile of the Hopefuls," *Drag World,* Nov. 19, 1965, 5; "Dick Landy—a Profile," ibid., May 21, 1971, 2.

53. Bob Ramsay, "The Answer for Dragsters," *Drag World,* Feb. 25, 1966, 15; see also John Jodauga, "Funny Car Evolution," *Car Craft,* Aug. 1972, 69.

54. Mike Doherty, "I'm Sorry Fellas, but . . . ," *Drag World,* Aug. 12, 1966, 2.

55. Gibbs quoted in *Drag News,* Jan. 13, 1967, 12 (the same source as the Gibbs quote in the epigraph); "Funny Cars Replaced the Fuelers at Irwindale," ibid., Mar. 17, 1967, 7. The editor of *Drag Racing,* Ralph Guldahl, Jr., spotlighted Smyser (who had formerly raced a top-fueler) as part of a campaign against funny cars. See John Smyser, "Why I Quit Funny Car Racing," *Drag Racing,* May–June 1968, 16–19; and July–Aug. 1968, 49–51; see also Bob Cruse, "Are Funny Cars Ruining Drag Racing?" ibid., Sept. 1967, 12–15ff; and Dave Scott, "The Flip Side of Funny Cars," ibid., Nov. 1967, 29–32. These authors and others such as Terry Cook ("No Longer Funny," *Drag World,* May 13, 1966, 4) and Jim McFarland ("Not-so-Funny Cars," *Car Craft,* Sept. 1969, 43ff) emphasized the inherent dangers of funny cars, but the toll was in fact considerably less than in fuel dragsters. Arguing in favor of the funnies was Rick Lynch ("Positive Statement by a DRM Reader on the Future of Funny Cars," *Drag Racing,* Mar. 1968, 54–56), and even the avidly pro-dragster Guldahl had to concede the key point that "today's youthful spectator . . . cannot *identify* or *involve* himself with fuelers as he can with Funny Cars" (ibid., Apr. 1967, 5).

56. Ralph Guldahl, Jr., "Bakersfield '67: Wake Year for the Big One?" *Drag Racing,* June 1967, 16–25ff.

57. Don Prieto, "The Pro Race," *Drag Racing,* Nov. 1967, 16–25ff (Kruse quote on 17); [Martin Kasindorf], "The Drag Bag," *Newsweek,* Aug. 5, 1968, 64; see also Ralph Guldahl, Jr., "Diggers!" *Drag Racing,* Dec. 1968, 16–19ff. Kruse staged this event for eight years, at Orange County after Lions closed in 1972.

58. Joanne Peters, "Zookeeper Wins Pomona," *Drag News,* Feb. 8, 1969, 4; George Houraney, "Garlits Beats Ramchargers at 11th Annual Smokers," ibid., July 5, 1969, 8; Joe Buysse to *Drag News,* Jan. 18, 1969, 4.

59. Gilmore qtd. in Fred M. H. Gregory, "Is the Dragster Dead? Did the Funny Car Kill It?" *Car Craft,* Mar. 1971, 16; "Are 200-MPH Dragsters Washed Up?" *Super Stock,* Feb. 1968, 25. When McEwen joined an exodus to funny cars *Drag Racing* could not ignore the irony, terming him a "dragster diehard who formerly served as president of . . . an organization which scorned FX cars during its existence" ("The Wild New Funny Cars," Apr. 1969, 33).

60. See "Constant Funny Cars," *Drag Racing,* Mar. 1967, 16–24 (photo on 20).

61. Nicholson qtd. in Terry Cook, "Sneak Preview: 1967 Funny Cars," *Car Craft,* Feb. 1967, 27; see also Nicholson, "Attention Funny Car Drivers," *Drag News,* Jan. 12, 1968, 10. For various reasons Nicholson and others sought to compete in "identifiable stock-bodied race cars," and their efforts eventually led to creation of the pro-stock class (Rick Voegelin, "Remembrance of Things Past," *Car Craft,* May 1973, 140). See also chap. 11.

8 HUSTLING

1. Qtd. in Cook, "The '40% Group,'" 11.

2. Keith Thorpe, "Jet Dragsters . . . the Grounded Missiles," *1964 Drag Racing Annual* (Los Angeles: Argus Publishers, 1964), 22–29; "All about the Wienie Roasters," *Modern Rod,* 1964 (offprint, AMDR), 9–13ff; "Palamides—Father of Jet Cars," *Drag News,*

July 8, 1963, 22–23; " 'The Untouchable' Real Crowd Pleaser at V.V.," ibid., Apr. 21, 1962, 3; Steve Gibbs, "Palamides Jet Dragster Appears at San Gabriel," ibid., 1; Don Elliot, "Arfons Turns 217 at Kansas City," ibid., 13; "Arfons Appears as Crowd Pleaser at Long Beach," ibid., Apr. 14, 1962, 4.

3. Moody qtd. in John Richardson, "Zeuschel-Moody-Fuller A/FD," *Drag News,* Oct. 26, 1963, 21; Stan Adams, in ibid., Apr. 21, 1962, 6.

4. LeRoi Smith, "The Jet: A Short Fused Bomb?" *Hot Rod,* Aug. 1963, 84–85ff.

5. "What Ever Happened to 'Jet Car' Bob Smith?" *Drag World,* Mar. 19, 1965, 3; "A Complete History of Jet Cars," *Drag Racing,* May 1972, 51–59ff; Robert C. Post to ibid., "Jet Accident Record," Aug. 1972, 68. Palamides turned jet cars into a minor industry; in 1966, when there were ten of them all told, he owned four. (Endorsement to Continental National American Group Policy RDS 9499111, AHRA, Apr. 1, 1966, copy in AMDR.)

6. "Petty 'Cuda into Crowd," *Drag World,* Mar. 5, 1965, 1ff. Four years later a funny car accident at another unsanctioned Georgia track, Yellow River, had a far more gruesome outcome ("Eleven Killed, 50 Hurt as Drag Racer Rips into Crowd near Covington," *Atlanta Journal,* Mar. 3, 1969, 1ff; Bill Robinson, "A Day of Terror at Yellow River Brings back Tragic Memories," *Atlanta Journal and Constitution,* Mar. 26, 1989, C1ff). While the drag racing press said almost nothing about such occurrences, they were naturally played up nationwide in the daily papers (e.g., "Dragster Rams Crowd, Kills 11," *Los Angeles Times,* Mar. 3, 1969, pt. 1, 12).

7. The quote is from Roger Gustin, a Palamides protégé who was instrumental in persuading NHRA to rescind its ban (Gustin to *Drag News,* Jan. 15, 1977, 4); Gustin praised NHRA for having previously sought to protect racers "from themselves" ("Speed World" [ESPN], July 2, 1990). When *Drag Racing* published its "complete history" in May 1972 it estimated that there were only eight jet cars then performing, but the number increased steadily after the NHRA lifted its ban, to more than fifty by 1990 (see *The Drag Racing List* 5 [Fall 1990]: 30–31). Many promoters headlined jets rather than fuelers of any kind (see Jeff Burk, "Under the Big Top: But Don't Call It a Circus," *Drag Racing,* Aug. 1988, 94–99; and Chris Martin, "Staging Light," *National Dragster,* Oct. 18, 1991, 6).

8. See Jon Asher, "Circus Acts and Freak Shows," *Drag Racing,* Jan. 1991, 66–71.

9. Nathan Cobb, "The Speedway Spiel," *Boston Globe,* Apr. 16, 1986.

10. *Drag News,* Dec. 19, 1970, 2. By the late 1960s Knievel "had a daredevil image equaled by no other person in the twentieth century" (Benjamin G. Rader, *In Its Own Image: How Television Has Transformed Sports* [New York: Free Press, 1984], 110). His first drag strip appearance was said to have drawn the largest single-day gross in Lions's fifteen-year history ("Evel Knievel Flies at Lions," *Drag Racing,* Apr. 1971, 29–30).

11. Dave Wallace, Jr., "Out of the Gate," *Drag News,* June 14, 1975, 4.

12. Brock Yates, "The Editorial Side," *Car and Driver,* Aug. 1966.

13. For a driver's personal recollection, see "Drag Racing Profile: Jim Herbert," *National Dragster,* Oct. 26, 1979, 5.

14. Terry Cook, "Progress . . . but to Where?" *Drag World,* July 8, 1966, 4. Cook had parodied the extremes to which exhibitionism could be stretched in "So You Want a Gimmick," ibid., Mar. 18, 1966, 4.

15. See Mike Civelli, "Sammy Miller Spending Less, Enjoying More with 'Spirit,' " *Drag News,* Dec. 18, 1976, 25. This is part of a three-part article on jets, wheelstanders, and rockets, "The Exhibitionists," 19ff.

16. See Dave Wallace, Jr., "Factory Warfare," *Super Stock,* June 1990, 82–87. An article on the "Ten Greatest Super Stockers in Drag Racing History" (*Drag Racing,* Nov. 1969, 50–55ff) included Golden's 1963 Dodge.

17. *Drag News,* May 10, 1975, 2, and May 31, 1975, 2; Nick Paciulli to ibid., June 14, 1975, 24. On Doner, see Wallace, "Three-Ring Circus," 38–39ff. Evans, who started out helping his father, Jay, Irwindale's concessionaire, would eventually become drag racing's most visible TV personality. The Orange County track, which opened in 1967 and closed in 1983, had various swings of fortune; see "Orange County International

Raceway," *Car Craft,* Mar. 7, 1967; Bud DeBoer, "Drag Racing U.S.A.," *Drag News,* Dec. 20, 1969, 11ff (an interview with the Orange County raceway's initial manager, Mike Jones); and especially Dave Wallace, Jr., "The County: Tracing the Rise and Fall of Drag Racing's Original 'Supertrack,'" *Drag Racing,* Nov. 1984, 24–31.

18. Qtd. in Tim Berry, "Fuel for Thought," *Drag Racing Today,* Apr. 7, 1989, 10. Even twenty-five years before it was clear to one journalist that "neither Black nor Prudhomme are in this for Christian charity" ("Who Is Roland Leong?" *Drag Sport,* Dec. 13, 1965, 17).

19. Keith Black interview, Feb. 6, 1989, DROHA, NMAH, 13–14.

20. Ibid., 21, 26.

21. Ibid., 25.

22. Ibid.

23. *Drag News,* June 10, 1961; Black interview, Feb. 6, 1989, 25.

24. Black interview, Feb. 6, 1989, 26.

25. Ibid., 27.

26. Chris Martin, "Black on Black," *National Dragster,* Apr. 26, 1985, 15.

27. Black interview, Feb. 6, 1989, 28, 30. On the theory that "spinning tires can conceivably produce *more* than 100% efficiency while those that do not spin have a lesser coefficient of friction," see "The Draggin' Machines," *Rod and Custom,* July 1954, 49; see also Tom Daniel and Dick Dean, "200 MPH Drags," ibid., Dec. 1964, 37.

28. Ibid., 30.

29. Stuckey to Woody Gilmore, in *Drag World,* Apr. 23, 1965, 18. Dave Scott credits three men—Fuller, Garlits, and Stuckey—for independently designing the type of dragster prevalent by 1960, with the engine way back but tilted down to counter wheelstands. See "Dragster Chassis Design," *Popular Hot Rodding,* Aug. 1965, 29–31ff.

30. Black interview, Feb. 6, 1989, 33.

31. Ibid., 36, 37; "Black on Black," 16.

32. Hal Higdon, *Six Seconds to Glory* (New York: G. P. Putnam's Sons, 1975), 29. During an interview at a race long afterward Prudhomme told a reporter, "I was a loser in school but I'm a winner out here" (Al Martinez, "A Kiss and a Cup," *Los Angeles Times,* Feb. 5, 1991, B2).

33. "Personality Profile: Dave Zeuschel," *Drag Racing,* June 1966, 10–13ff; Doris Herbert, "Editors Exhaust," *Drag News,* Mar. 10, 1962, 2.

34. Howard Pennington, "Quick Quarter Miler," *National Dragster,* Feb. 21, 1964, 20.

35. For descriptions of this machine at two stages, before and after the wheelbase was lengthened to 136 inches, cf. Don Francisco, "8.09 Special," *Hot Rod,* Dec. 1962, 30–33ff; and Dave Scott, "Greer-Black-Prudhomme," *Popular Hot Rodding,* Oct. 1964, 36–41ff.

36. Dave Wallace, Jr., "The Man in Black," *Super Stock,* July 1991, 83; Prudhomme qtd. in Higdon, *Six Seconds,* 28.

37. Scott, "Greer-Black-Prudhomme," 37; Black interview, Feb. 6, 1989, 38–39.

38. Black interview, Feb. 6, 1989, 39.

39. Higdon, *Six Seconds,* 29.

40. Ibid.

41. Qtd. in "Who Is Roland Leong?" 17–18; see also Kay Presto, "Roland Leong," *Super Stock,* Mar. 1987, 42–46ff.

42. "Who Is Roland Leong?" 18.

43. Higdon, *Six Seconds,* 31.

44. Kay Presto, "The Snake Awakes," *Drag Racing,* May 1987, 74.

45. Higdon, *Six Seconds,* 31, 37.

46. "Black on Black," 17.

47. Black interview, Feb. 6, 1989, 44, 43. See also "Keith Black—Engines Unlimited," *Drag World,* Oct. 20, 1967, 22.

48. "Black on Black," 17.

49. Black interview, Feb. 6, 1989, 45. Black also developed many other engines for Chrysler, including wedge-head V-8s for USAC and NASCAR racing.

50. Karen Nelson, "Sid Waterman," *Drag Racing*, Jan. 1968, 43–47; Peter Fong, "Sid Waterman," ibid., May 1975, 53–55ff.

51. Holly Hedrich, "Dual Drives for Dragsters," *Drag Racing*, June 1966, 22, 23, and July 1966, 22–25ff.

52. Bob Pendergast, "Competition Clutches and Flywheels," *Hot Rod*, April 1959, 22–27ff; Jerry Mallicoat, "The Reverend Mr. Black and His 'Super Slipper,'" *Car Craft*, Oct. 1968, 37.

53. Hedrich, "Dual Drives," 23; Mallicoat, "Reverend," 37; Bob Shafer, "A Tragedy of Our Times," *Santa Monica Evening Outlook*, Jan. 3, 1968.

54. Mallicoat, "Reverend," 82–83.

55. "Black on Black," 18; see also "The Great Aluminum Hemi Race," *Drag Racing*, Sept. 1972, 21–24ff. On Black's venture into another major engine component, crankshafts machined from alloy-steel billets, see John Hogan, "The Kryptonite Krank," *Drag Racing*, July 1973, 45–47ff. Black wrote a monthly series on fuel engine technology for *Drag Racing* magazine, "The Latest from Keith Black."

56. Black interview, Feb. 6, 1989, 12.

57. Qtd. in Steve Alexander, "Keith Black," *Drag Racing*, May 1975, 59. Black was notoriously "businesslike," as was Garlits (C.J. Hart recalled that Garlits was the only racer who invariably demanded *formal* contracts). People loved Black's competitor Joe Pisano for his liberal credit policies, but Black did not put anyone on "the deal"—his prices were the same to all customers—and in certain instances he insisted on cash. See "On the Carpet: Keith Black with Terry Cook," *Drag World*, May 21, 1965, 3ff; Jeff Burk, "Winners: Keith Black," *Super Stock*, Mar. 1989, 15. Even beyond what I have cited, the periodical literature on Black is extensive; when *Drag Racing* launched a series on "individuals who are responsible for making drag racing what it is today," it commenced with Black (May 1965, 15–17).

58. "On the Carpet: Don 'The Snake' Prudhomme with Terry Cook," *Drag World*, Mar. 25, 1966, 21.

59. Quotes from *Don Prudhomme: Professional Racer* (Diamond P Sports Video, Inc., 1988); and "Don Prudhomme," *National Dragster*, Feb. 6, 1993, 38–39ff. Shelby, who had dabbled in just about every form of racing, including Formula 1, got heavily involved in the drags in the late 1960s, even taking a seat on the NHRA board.

60. "Don Prudhomme."

61. Ibid.

62. Qtd. in John Ernesto, "Snake," *Reading Eagle* (Pa.), Sept. 11, 1988, supp., 6.

63. Mike Doherty, "Wildlife Enterprises: The Richest Act in Racing," *Drag Racing*, June 1972, 36–39ff; see also Bud DeBoer, "Person to Person" (interview with McEwen and Prudhomme), *Drag News*, Apr. 4, 1970, 2ff.

64. "Mattel's Million Dollar Team," *Drag Racing*, May 1971, 35.

65. "Don Prudhomme." See also Chris Martin, "'The Snake' and 'The Mongoose': Model Performers," *National Dragster*, June 14, 1991, 21–23; as well as "The Day 'The Snake' Met 'The Mongoose,'" ibid., June 7, 1991, 27ff.

66. Ibid.

67. Dave Wallace, Jr., "The Life and Times of Tom McEwen, World Champion," *Drag Racing*, May 1984, 39.

68. Doner qtd. in Earl Gustkey, "Expenses a Drag to the Mongoose," *Los Angeles Times*, Mar. 18, 1976, pt. 3, 4. The PRA race paid thirty-eight thousand dollars, but rumor had it that McEwen and the three other finalists at Tulsa agreed in advance to divide the purse money evenly.

69. Larry Carrier to *Drag News*, Apr. 3, 1976, 2; Jake Trent, "Don't Tread on Me," *Drag Review*, June 1977, 9; Dave Wallace, Jr., "A Candid Conversation with Our 1975 Drag News Driver of the Year," *Drag News*, Nov. 29, 1975, 10–11.

70. "Don Prudhomme."

71. Shav Glick, "A $50,000 Car Is Not That Funny," *Los Angeles Times*, Jan. 27, 1977,

pt. 2, 6; Wallace, "Life and Times of Tom McEwen," 42; *Don Prudhomme: Professional Racer*; Shav Glick, "Mongoose Is at a Loss for Sponsors, Not Words, as Pomona Streak Ends," *Los Angeles Times,* Feb. 6, 1988, pt. 3, 6.

9 *REVOLUTION*

1. Garlits, "*Big Daddy*," 128.

2. See, for example, *Drag News,* Sept. 6, 1969, 16. See also John Brasseaux, "Drag Racing's Mechanical Milestones," *Drag Racing,* Sept. 1989, 31.

3. "Allison & Crow," *Drag Racing,* Mar. 1967, 49–53ff. This machine had an aluminum-block Chevy engine acquired in the wake of one of Mickey Thompson's Indy 500 ventures. Driver Ed Allison later suffered paralyzing injuries when it crashed.

4. Garlits, "*Big Daddy*," 130; see also *Los Angeles Times,* Feb. 3, Sept. 2, 18, 25, 1969. Mulligan, who had made his name driving for Gene Adams, was twenty-six. In July he had won an award for Top-Fuel Driver of the Year, named in honor of Mike Sorokin.

5. Garlits, "*Big Daddy*," 131.

6. Rick Lynch (AHRA director of public relations) to Doris Herbert, Feb. 9, 1970, AMDR.

7. Qtd. in Shav Glick, "Tharp Is on the Road Again, This Time Chasing Shirley," *Los Angeles Times,* Jan. 29, 1983.

8. Garlits, "*Big Daddy*," 132.

9. Eric Rickman, "Life Preservers Called 'Scattershields,'" *Hot Rod,* Apr. 1958, 57–59; "It's in the Can," *Drag Racing,* June 1969, 30–31; "Clutch Explosions: Housing and Blanket Safety," ibid., Mar. 1971, 44–47ff; Bill Holland, "'71 Rules and What They Mean to You, the Racer," ibid., Apr. 1971, 21. Don Jensen recalled that the first custom-fabricated steel scattershield he ever saw was on the car in which Red Case lost his life in 1958 (July 1990 tapes, DROHA, NMAH, 7).

10. Garlits qtd. in Bud DeBoer, "Person to Person," *Drag News,* Mar. 28, 1970, 2.

11. Qtd. by Don Rackemann in *Drag News,* June 13, 1970, 2; see also "Personality Profile: Connie Swingle," *Drag Racing,* Mar. 1967, 55–59ff.

12. Glick, "Tharp Is on the Road."

13. Garlits, "*Big Daddy*," 136, 138. The quote from Garlits in the epigraph is from p. 137; the quote from Mike Sorokin is from *Drag World,* Nov. 26, 1965, 13.

14. Jensen, July 1990 tapes, 2.

15. These are the recollections of Vaughn Bowen, Jr., and Peter Massett, with whom the author had a racing partnership in the mid-1950s.

16. Jensen, July 1990 tapes, 7.

17. Weber ad, *Drag News,* Oct. 18, 1958; "Editors Exhaust," ibid., Jan. 16, 1960, 2; see also Bruce Tawson to ibid., Feb. 15, 1960, 14.

18. Larry Morgan, "Lace Killed," *Drag News,* Sept. 21, 1963, 12; Dave Scully, "John Wenderski Fatally Injured at Ramona," ibid., Feb. 24, 1964, 10; see also Steve Scott, "The Wenderski-Winkel Dragster," *Hot Rod,* July 1964, 52–54.

19. Don Carsten to *Drag News,* Dec. 26, 1959, 10.

20. The procedure for obtaining a license in the late 1960s, with seasoned drivers required to attest to an applicant's capabilities, was depicted quite nicely in the 1983 movie about Shirley Muldowney, *Heart like a Wheel.* Ultimately, would-be drivers could be licensed in conjunction with formal schooling, but that was still a long way off.

21. "Big Daddy Says, 'Too Dangerous,'" *Drag World,* July 2, 1965, 1; "Garlits-Goodsell Collide," ibid., 2; Mike Doherty, "Drag Racing's Black Weekend," ibid.; "On the Carpet: Don Garlits with Terry Cook," ibid., July 9, 1965, 3ff; "Garlits Quits Driving," *Drag Sport,* July 5, 1965, 3.

22. ". . . Accident Takes Novice Fueler," *Drag World,* July 16, 1965, 6; Kaye Trapp, "The Prince of Wales," *Hot Rod,* Oct. 1964, 66–67; "'Q-Ball' Louisiana Fatality," *Drag World,* July 23, 1954, 2; "Tex Randall Victim of Wheel," ibid., 1; "Cangelose Chute Failure

Fatal," ibid., July 2, 1965, 3. The manufacturer of the parachute on Cangelose's machine addressed specific and general problems in "On the Carpet: Jack Carter with Mike Doherty," ibid., July 16, 1965, 3ff.

23. See, for example, John Bowen, "Drag Notes," *Drag Digest,* Oct. 28, 1966, 3.

24. Mike Doherty, "Saving Your Life from Flames," *Drag Racing,* Jan. 1971, 38–43ff; Terry Cook, "You Bet Your Life," *Car Craft,* July 1970, 52; John Raffa, "Perspective," ibid., 6; "An Interview with Gas Ronda," *Drag Racing,* July 1970, 31; Larry Schreib, "SEMA's Lou Baney Looks at Funny Car Safety," *Car Craft,* Aug. 1972, 82–85ff.

25. All such rules, of course, addressed only effects, though there were scattered attempts to address causes; see, for example, Joe Panek to George Phillips, in *National Dragster,* Feb. 7, 1986, 24ff.

26. John Law, "Theory and Narrative in the History of Technology: Response," *Technology and Culture* 32 (Apr. 1991): 383.

27. What the racers called "rear-engine" machines are properly termed "mid-engine." The distinction is not trivial. Of the first two commercially produced automobiles, one, the Daimler, was mid-engine, the other, the Benz, was rear-engine. The first Fords and the curved-dash Oldsmobile had mid-engines; the Volkswagen and the Corvair were true rear-engine machines. With dragsters Garlits preferred "front driver," accurate terminology, but the misnomer rear-engine took hold anyway.

28. Mike Doherty, "A History of Competition Coupes and Roadsters," *Drag Racing,* Nov. 1970, 54–59; Don Prieto, "The Passing of the Speed Sport Roadster," ibid., Feb. 1968, 30–33; "Firepowered Simplicity," *Hot Rod,* Feb. 1960, 50–51; "Bad Daddy from Memphis," ibid., June 1962, 54–55; "Handler Up Front," ibid., 64–65; "Dragging the Shoehorn," ibid., Sept. 1960, 76–78; Jack Simondson, "Shoehorn Day," *Drag News,* June 23, 1962, 1; John O'Brien, "Israeli Rocket," *Super Stock,* June 1988, 8; Bob Greene, "The Wedge," *Hot Rod,* Aug. 1964, 90–91; Madigan, *The Loner,* 25–29.

29. Qtd. in Garlits, *"Big Daddy,"* 139. Mid-engine modified roadsters got such a bad reputation at El Mirage and Bonneville that the SCTA eventually outlawed the configuration.

30. Dave Scott, "Back-Rigger Eldorado," *Drag Racing,* Oct. 1967, 10–15ff.

31. Vincenti, *What Engineers Know;* Todd Barker, "Has the Slingshot Dragster Had It?" (offprint, ca. 1965, AMDR).

32. "Drag Wedge," *Car Craft,* Sept. 1969, 54–55; "Drag Racing Profile: Junior Kaiser," *National Dragster,* Feb. 29, 1980, 5; "From the Front Desk," *Drag News,* Dec. 20, 1969, 16.

33. Law, "Theory and Narrative," 383.

34. "Garlits Leads a Rear-Engine Revival," *Drag Racing,* May 1971, 19. Swingle displayed aggressive pride in his empiricism. Asked how he had learned to weld so well, he replied, "Born knowing it" ("Personality Profile: Connie Swingle," 55).

35. Robert M. Vogel, "Draughting the Steam Engine," *Railroad History* 152 (Spring 1985): 17.

36. Ralph Guldahl, Jr., "Garlits Up Front," *Hot Rod,* May 1971, 55.

37. Garlits, *"Big Daddy,"* 140–41; "Garlits Unveils New Rear-Engine Dragster at Sunshine," *Drag News,* Jan. 10, 1971, 15.

38. Guldahl, "Garlits Up Front," 55.

39. Frank Hawley with Mark Smith, *Drag Racing: Drive to Win* (Osceola, Wisc.: Motorbooks International, 1989), 41.

40. And the existence of that piece of the puzzle had been known for at least sixteen years; Duane Depuy and Joe Scarpelli realized that their "Cleveland Clipper" required slower steering (Depuy, "An Experiment in Acceleration"). For an admirably nonpresentist contrast of Garlits's machine to one from 1954, see "Rear-Engine Revolution: From 'White Owl' to 'Swamp Rat,'" *Hot Rod Yearbook* 11 (1971): 323–27.

41. Garlits, *"Big Daddy,"* 141; Guldahl, "Garlits Up Front," 55.

42. Shav Glick, "Garlits's Rear-Engine Car Could Cause Revolution in Drag Racing," *Los Angeles Times,* Jan. 26, 1971, pt. 3, 7; Guldahl, "Garlits Up Front," 53.

43. Guldahl, "Garlits Up Front," 57.

44. Qtd. by Glick, "Garlits's Rear-Engine Car."

45. See, for example, "Wings Are the Thing," *Drag Racing*, Dec. 1971, 46–47.

46. Pete Robinson, "Use Chutes Only in Emergency," *Drag World*, July 30, 1965, 19 (this piece was published in both *Drag News* and *Drag Sport* too). Pete claimed that the 1964 crash of his superlight Woody Gilmore machine resulted from a crosswind catching the (conventional) chute as it opened ("The Life and Death of Pete Robinson's 'Tinkertoy,'" *Hotrod Parts Illustrated*, Oct. 1964, 13–15).

47. Garlits to Fellow Racers, *Drag News*, Nov. 30, 1963, 16.

48. Pete Robinson, "Reasons for Use of 'Jacks,'" ibid.

49. Lee Pendleton to Fellow Racers, ibid., Jan. 25, 1964, 15; Ron Colson to ibid., Jan. 11, 1964, 17; Brian McGoff to ibid., Feb. 8, 1964, 26. Also see "When 'The Jacks' Got to Don Garlits," *Drag Racing*, Nov. 1973, 60–61ff.

50. Terry Cook, "Vacuum Cleaner," *Car Craft*, Oct. 1969, 27.

51. "Racer Critical after Crash at 199 m.p.h.," *Los Angeles Times*, Feb. 7, 1971, D2; Terry Cook, "Point of View," *Car Craft*, May 1971, 6.

52. Shav Glick, "Prudhomme Feels Flying Wedge Dragster Can Hit 235–240 m.p.h.," *Los Angeles Times*, July 20, 1971, pt. 3, 7; "AA/Piece'a Pie," *Drag Racing*, Sept. 1971, 17–19; "Mid-Engine Mania," ibid., Dec. 1971, 32–39; Ray Marquette, "Garlits' Five-Second Surprize: A Fully Enclosed Streamliner for '72," *Drag Racing*, Apr. 1972, 21.

53. For a vivid account of a high-speed crash at Pomona which Tommy Ivo survived unhurt, see Jim Kelly, "I Knew I'd Bought the Farm," *Drag Racing*, May 1974, 28–31ff. In the decade after 1967, throughout all of drag racing's many competitive classes, the number of fatalities remained quite steady at seven or eight per year (see "Fatalities in Motor Racing," *Drag Review*, Mar. 1977, 33).

54. Doherty, "Saving Your Life," 42.

10 *FINESSE*

1. "New All Time E.T. Record of 8.35 Sec. Set at Riverside," *Drag News*, June 6, 1959, 1; "Streamlined for Draggin'," *Car Craft*, Nov. 1959, 34–37; Terry Cooke [sic], "Streamlining—Dragster Dilemma?" *Hot Rod*, Aug. 1965, 94–101ff; Norm Porter to Mike Doherty, *Drag Racing*, Jan. 1971, 4–6ff. The blower and carburetors stuck up too far to use the fairing over the engine, but otherwise Jocko's liner looked the same as when new.

2. Fred M. H. Gregory, "Garlits Goes for 275/5.60," *Car Craft*, July 1972, 73.

3. Ibid., 74.

4. Roger Huntington, "Aerodynamics on the Quarter Mile," pt. 2, *Drag News*, Dec. 29, 1956, 6 (also quote in epigraph).

5. Karl Ludvigsen, qtd. in James J. Flink, "The Aerodynamic Automobile," MS, 3. A condensed version of Flink's stimulating essay was published as "The Path of Least Resistance," *American Heritage of Invention & Technology* 5 (Fall 1989): 34–44.

6. Huntington, "Aerodynamics."

7. Don Francisco, "King of the Kilo," *Hot Rod*, Sept. 1961, 85.

8. Flink, "Aerodynamic Automobile," 6.

9. "Spirit II," *Hot Rod*, Feb. 1964, 76–77; "The Spirit Moves," ibid., June 1964, 62–67.

10. "Spirit II," 76.

11. Flink, "Aerodynamic Automobile," 7.

12. "In the Beginning a Backyard Venture," *Hot Rod*, Oct. 1963, 32. A perceptive biographical sketch of Breedlove appears in Cutter and Fendell, *Encyclopedia of Auto Racing Greats* (84–86); but see also Breedlove's own *Spirit of America* (Chicago: Henry Regnery Co., 1971); and Irwin Stambler, *The Supercars and the Men Who Race Them* (New York: G. P. Putnam's Sons, 1975).

13. See Ewart Thomas, "Hot-Rod Derby on the Salt Flats," *Popular Mechanics*, Aug. 1950, 65–68ff.

14. On LSR attempts at Bonneville in the early 1960s, the material in *Hot Rod* magazine alone is voluminous: see, for example, Griffith Borgeson, "Bonneville Encore," Feb. 1960, 34–35ff; Borgeson, "A Jet Car for the Big Records," Apr. 1960, 42–45; Ray Brock, "The Flying Caduceus," Oct. 1960, 24–29ff; "'60 Bonneville National Speed Trials," Nov. 1960, 30–39ff; "Bonneville Preview," Aug. 1961, 76–79; Don Francisco, "The Spirit of America," Oct. 1963, 26–31ff; Francisco, "World Land Speed Record Cars—1964," Jan. 1965, 38–43ff; Roger Huntington, "Looking through the Sound Barrier," July 1965, 42–44ff; Eric Rickman, "Spirit of America: Sonic I," Oct. 1965, 46–51. See also William A. Moore and David Burke, "Jets at Bonneville," *Popular Hot Rodding,* Jan. 1965, 16–19ff; Thompson with Borgeson, *Challenger;* Frederic Katz, *Art Arfons: Fastest Man on Wheels* (London: Routledge, 1965); Deke Houlgate, *The Fastest Men in the World—On Wheels* (New York: World Publishing, 1971); Jeff Scott, "The Last American Hero," *Cars,* Mar. 1980, 68–74; Peter J. R. Holthusen, *The Land Speed Record* (Newbury Park, Calif.: Haynes Publications, 1986). The fastest speed ever recorded at Bonneville was 622.407, by the rocket-powered "Blue Flame" on Oct. 23, 1970. The driver was Gary Gabelich, who had honed his skills in the cockpit of numerous dragsters, including jets; see Karen Nelson, "Personality Profile: Gary Gabelich," *Drag Racing,* May 1967, 25–31; and "Gary Gabelich," in Cutter and Fendell, *Encyclopedia of Auto Racing Greats,* 232–33.

15. The quotes are from "Bonneville Encore," 24; see also Thompson with Borgeson, *Challenger,* 167–69; and Holthusen, *Land Speed Record,* 32.

16. Eric Rickman, "Four in a Row Gotta Go!" *Hot Rod,* Mar. 1965, 45–47; Rickman, "Three-Ton Rod," ibid., June 1965, 66–69; "Summers on Standby," ibid., Nov. 1965, 86–87; "Summers Brothers," in Cutter and Fendell, *Encyclopedia of Auto Racing Greats,* 481. For the sake of comparison, a conventional dragster of the 1970s or 1980s had a Cd of about 1.0, and commercial automakers in the latter 1980s were striving for .30.

17. See Robert C. Post, "The Machines of Nowhere," *American Heritage of Invention & Technology* 8 (Spring 1992): 28–35.

18. Bob Greene, "New Approach for Dragster Design," *Hot Rod,* Oct. 1965, 71.

19. Qtd. in E. K. von Delden, "New Concepts for Fuelers," *Hot Rod,* Aug. 1971, 40. One mid-1960s streamliner was the creation of Ed Roth—"Big Daddy" before Don Garlits was—a hero of Tom Wolfe's *Esquire* epic on hot rods and custom cars. See *The Kandy-Kolored Tangerine-Flake Streamline Baby* (New York: Noonday Press, 1963), 98–107; see also Dennis McLellan, "Flamboyant 'Big Daddy' Just a Father Now," *Los Angeles Times,* Jan. 27, 1981, pt. 5, 1–2.

20. Qtd. in Cooke [*sic*], "Streamlining," 100.

21. Mike Doherty, "The Ten Big Mistakes in Drag Racing," *Drag Racing,* Oct. 1970, 22.

22. Greene, "New Approach," 70. Even body panels that only covered the frame were costly; Garlits charged $800–$1,000 for adding a set to a $1,500 rolling chassis.

23. Quotes from Terry Cook, "Dragster of Tomorrow," *Car Craft,* Apr. 1967, 82, 36. Flink emphasizes that wind tunnel work did not *inevitably* yield better results than eyeball aero, noting that John Tjaarda, the designer of Ford's prewar Lincoln Zephyr—which had a much better Cd than Chrysler's 1934 Airflow—was a proponent of what he called "guessamatics."

24. Terry Cook, "Let's Talk about Streamlining," *Drag World,* Apr. 23, 1965, 14, 17; "New Jersey News," *Drag News,* Dec. 4, 1964, 17.

25. Terry Cook, "The Rail Renaissance," *Car Craft,* May 1971, 64–67ff. Cook's upbeat article followed by only a few months one by his former editor at *Drag World,* Mike Doherty, which called streamlining one of the "ten big mistakes."

26. Marquette, "Garlits' Five-Second Surprise"; Gregory, "Garlits Goes for 275/5.60," 72–74; "Don Garlits' 'Statue of Liberty' Option Play," *Drag Racing,* Oct. 1972, 16–18; Garlits, "*Big Daddy,*" 150–52.

27. Garlits, "*Big Daddy,*" 159.

28. *Drag News,* Apr. 3, 1971, 2. See also Karen Nelson, "Personality Profile: Tom

Hanna," *Drag Racing*, Feb. 1968, 55–60ff; among Hanna's employees at this time were onetime "Surfers" Tom Jobe and Bob Skinner.

29. Arthur Irwin III, "New England Notes," *Drag News*, Sept. 16, 1972, 2. See also "A Patriotic Pusher," *Drag Racing*, Apr. 1973, 24–27. Lehman reports that "the body was going to take one month and $2,500; instead, it took $6,000 and four months."

30. John Fuchs, "Dragster of the Future," *Hot Rod*, Sept. 1972, 71–74; "Barry Setzer Bursts into Top Fuel," *Drag Racing*, Oct. 1972, 40–44. A machine similar to Buttera's was built by S&W Race Cars, a shop in Pennsylvania.

31. See, for example, Rick Voegelin, "Formula Fuel," *Car Craft*, May 1975, 64–68.

32. See "Funny Car Body Language," *National Dragster*, Jan. 21, 1983, 32. Air dams and spoilers had first appeared on Ferraris in the early 1960s; see Flink, "Path of Least Resistance," 43.

33. Phil Burgess, "Setting the Tempo in Funny Car," *National Dragster*, Feb. 10, 1984, 10–11; "Aerodynamics: The 263 MPH Solution," *United Racer*, Mar. 25, 1985, 24. When the Detroit factories stepped up their wind tunnel research in the 1970s the main site was the Lockheed-Georgia facility; see Flink, "Aerodynamic Automobile," 48.

34. Ed Shaver, "Guest Columnist," *Drag Racing*, Apr. 1976, 20; "The Sleekest Funny Car," ibid., Feb. 1973, 32–33; "Dunn and Reath: Out Front with a Trick Mid-Engine Funny Car," ibid., May 1972, 48–50.

35. See, for example, Cliff Morgan to Dave Wallace, *Drag Racing*, Jan. 1985, 7: "Today's Funny Cars are really streamlined front-motored Top Fuelers." See also Morgan to *National Dragster*, Apr. 13, 1984, 4.

36. Qtd. in John Brasseaux and Phil Burgess, "Streamliners: An Endangered Species?" *National Dragster*, May 22, 1987, 21.

37. "Specialty Control and Mfg.," *Drag News*, Oct. 26, 1963, 20; see also ibid., Dec. 21, 1963, 13.

38. See Walter Vincenti, "The Davis Wing and the Problem of Airfoil Design: Uncertainty and Growth in Engineering Knowledge," *Technology and Culture* 27 (Oct. 1986): 725. In 1986 Bob Jinkins of Jackson, New Jersey, employed a technique similar to Davis's to test a dragster nosepiece he had designed for Joe Amato (Bob Doerrer, "Poor Man's Wind Tunnel," *Drag Racing*, May 1987, 68–71). On the Harrell-Borsch roadster, see George Roe, "A Dragster in Roadster's Clothing," *Drag Racing*, Nov. 1964, 30–32ff.

39. John Brasseaux, "The Basics of Aerodynamics, Part II," *National Dragster*, Apr. 25, 1986, 9. See also "History Made in Frantic Florida Finale," ibid., Mar. 30, 1984, 1ff; ibid., May 4, 1984, 2–3; "World's Fastest," *Drag Racing*, Jan. 1985, 14–18. Down-force inhibited the phenomenon called "pneumoplaning."

40. "Behind the Scenes with John Hogan," *Drag Racing*, Sept. 1972, 14.

41. Dave Wallace, Jr., "Lola's Touch," *Drag Racing*, Nov. 1985, 49.

42. Phil Burgess, "Out the Back Door," *National Dragster*, Jan. 17, 1986; Eloisa Garza interview, Jan. 1988, DROHA, NMAH.

43. Huntington, "Aerodynamics of the Quarter Mile."

44. Qtd. in Castrol GTX/NHRA Keystone Nationals press release, Sept. 1987, 2.

45. Qtd. in Brasseaux and Burgess, "Streamliners," 21.

46. Qtd. in Dale Wilson, "Top Fuel Streamliners: Taking the Drag Out of Drag Racing," *Super Stock*, July 1986, 55. On Swamp Rat XXX, see also Phil Burgess, "Rat under Glass," *National Dragster*, Apr. 11, 1986, 6–7ff; Ron Colson, "A Talk with 'Big Daddy' Don Garlits," *United Racer*, April–May 1986, 10–11; Sky Wallace, "Thunderliners," *Drag Racing*, July 1986, 64–66; Phil Elliott, "Up in Smoke," *Super Stock*, Aug. 1986, 10ff; and Robert D. Friedel, "A Materials Showcase," *A Material World* (Washington, D.C.: National Museum of American History, 1988), 51.

47. Patrick Hale to the author, Aug. 1, 1986. Locating the fuel tank entirely ahead of the front axle had been tried at least once before, in 1983, and had long been standard placement in funny cars.

48. Brasseaux and Burgess, "Streamliners," 20, 26.

49. Garlits immediately promised Ron Colson "I'll Be Back" (*United Racer,* Sept. 14, 1987, 25, 29), and he did make a few runs in the next few years, but a posterior vitreous detachment in his left eye forced him out of the cockpit permanently in 1992.

50. Frank qtd. in an interview with Lou Baney, in *Hot Rod,* July 1972, 46; Robert C. Post, "The Cars Won't Fly," *Air and Space* 1 (Aug.–Sept. 1986): 76–84.

51. I must express appreciation for all the coaching on aerodynamics which I have received from Harry Lehman. Much of his insight is distilled in an eight-page letter to the author dated Nov. 30, 1992.

52. Certain historians have attempted to analyze such processes through "techno-morphology"; see "The Langen Suspended Monorail & the Railplane," *Newcomen Bulletin* 135 (Aug. 1986): 12.

53. Qtd. in Rich Carlson and Dick LaFayette, "Caution! 300 mph Speed Zone: A Look at the Top Fuel Chassis of Today and the Future," *Super Stock,* Apr. 1989, 45, 51. See also "Blowover," ibid., June 1991, 71–73. In 1989 NHRA limited the size of wings to 1,600 square inches as well as restricting their location vis-à-vis the rear axle, but everyone remained apprehensive and blowovers continued to occur.

54. Richard Conniff, "In Florida," 16–17.

11 *TELEVISION*

1. "Behind the Scenes with Jerry & Darrell Gwynn," *National Dragster,* Nov. 28, 1980, 5; Bob Abdellah, "Darrell Gwynn Puts the Squeeze on Top Fuel," *Drag Racing,* Sept. 1986, 60–69; "The Old Man and the Kid," *Super Stock,* Nov. 1986, 64–65ff; Norm Froscher, "Gwynn's Mark: Lighter, Faster," *Gainesville Sun,* Mar. 22, 1987, 1Fff; Terry Bickhart, "Gwynn Still Young and Restless," *Reading Eagle* (Pa.), Sept. 11, 1988, suppl., 3ff; Thomas Pope, "Darrell Gwynn: The *Real* Miami Sound Machine," *Drag Racing Today,* May 5, 1989, 61–65; Andy Willsheer, "Innerview: Darrell Gwynn," *Drag Racing,* Aug. 1989, 78–79ff; Tim Berry, "Darrell Gwynn's Hoping 1990 Is His Year," *Drag Racing Today,* Dec. 1989, 12–13; Maya Bell, "Fiery Crash Revs Racer's Will, Spirit," *Orlando Sentinel,* July 29, 1990, 1ff; Jon Asher, "Temporarily Down, But on the Comeback Trail," *Drag Racing,* Aug. 1990, 84–86; Ercie Hill, "Postcard from the Pits," *Super Stock,* Nov. 1990, 78–79.

2. "Consumer Legislation," in *Congress and the Nation* (1969–72) (Washington, D.C.: Congressional Quarterly, 1973), 3:660, 662–65, 671–72.

3. Ibid., 3:665.

4. The epigraph is quoted from a letter that NHRA sent to all members on Feb. 26, 1990, along with a stamped postcard addressed to Secretary Louis W. Sullivan of the Department of Health and Human Services (HHS). The message on the card read, in part: "I do not agree with your efforts to eliminate tobacco advertising and support of sporting events. . . . Tobacco is a legal product in this country and should have the right to be advertised. The growth and health of sporting events and the general sports industry, which provides tremendous economic impact and jobs in America, would suffer significantly from your irresponsible position." HHS must have received thousands of these cards, although several drag racing stars, including Don Garlits, told me they had penned in an indication that they did support Sullivan.

5. See, for example, Ed Sarkisian, "Better Think Twice about Politics in Drag Racing," *Drag Sport,* Nov. 21, 1964, 5; see also Don Elliott, "Midwest Outlook," *Drag News,* Jan. 7, 1960, 5; Bob Ramsay, "Sanctioning: It's What's Happening," *Drag World,* Sept. 10, 1965, 8; Terry Cook, "Point of View," *Car Craft,* July 1970, 8.

6. Reno Zavagnon, "UDRA: The Best Kept Secret in Drag Racing," *Super Stock,* Aug. 1985, 28–31ff; "NASCAR Goes Drag Racing," *Drag World,* Apr. 16, 1965, 1ff; NASCAR, Drag Race Division, *Official 1966 Rule Book* (Carnegie, Pa.: NASCAR, ca.1965).

7. Fred M. H. Gregory, "AHRA vs. NHRA: A Comparison," *Car Craft,* July 1970, 62.

See also American Hot Rod Association, *1971 Drag Racing Rules and Technical Manual* (Overland Park, Kans.: AHRA, 1970).

8. "Carrier to Head Up New Sanctioning Body," *Drag News,* Nov. 21, 1970, 17–18; ibid., Aug. 28, 1971, 24; Gregory, "AHRA vs. NHRA," 62; "New Racing Associations," *Drag Racing,* May 1971, 14; "The Association Battle," ibid., Jan. 1972, 59–64.

9. Bud DeBoer, "Funny Car Cavalcade of Stars" (Litchey interview), *Drag News,* June 20, 1970, 2; "Gold Agency Sets the Pace," ibid., Dec. 19, 1970, 9ff; "Coke Cavalcade Cars," *Super Stock,* Sept. 1970, 26–30ff; Dave Wallace, Jr., "Out of the Gate," *Drag News,* July 12, 1975, 4; "Chicago Gold: It All Belongs to Ben," ibid., Apr. 23, 1977, 5; Mike Doherty, "The Ten Most Important Men in Drag Racing," *Drag Racing,* June 1969, 29. Doherty's other nine were: Bob Cahill, Chrysler's director of Special Performance Events; Jacque Passino, manager of Ford's Special Vehicle Department; Andy Granatelli, president of STP; George Hurst of Hurst Performance, Inc.; Wally Parks, Jim Tice, Gil Kohn, Marvin Rifchin, and Don Garlits. A distinct minority had even token experience in actual quarter-mile competition.

10. Dave Hederich, "Don Garlits on the Carpet," *Motorsports Weekly,* Sept. 25[?], 1973, 3 (clipping, AMDR).

11. [Don Garlits], "An Open Letter to Owners and Drivers," *Drag News,* July 5, 1972, 27. See also "Drag Racing Comes of Age," ibid., Aug. 19, 1972, 21; Doris Herbert, "From the Front Desk," ibid., Sept. 23, 1972, 2; "Garlits and Tice Buck NHRA Nationals," *Drag Racing,* Aug. 1972, 14; John Hogan, "More on the Tice-Garlits Race," ibid., Sept. 1972, 14; Ro McGonegal, "Point of View," *Car Craft,* Aug. 1972, 10; and Garlits, *"Big Daddy,"* 150. Ultimately, the pro-stock purse was cut back, and racers in other classes were encouraged to participate.

12. "Tulsa Challenge/Indy Nationals," *Car Craft,* Nov. 1972, 56–79; "96 Make History," *Drag Racing,* Dec. 1972, 26–30ff.

13. Steve Alexander, "Where Angels Fear to Tread," *Super Stock,* Dec. 1974, 20; see also "Garlits-PRA Doings," *Drag Racing,* Jan. 1974, 14.

14. "Of Trauma and Tragedy," *Super Stock,* Dec. 1974, 19.

15. Ed Eaton, "The Rightest and the Rest," ibid., 21. In his autobiography Garlits did not omit mention of the 1974 PRO race altogether (as Wally Parks ignored the 1960 NASCAR/NHRA Winternationals in his own book), but he gave it only one short paragraph (166), placing the blame squarely on Eaton. Later he suggested that, deep down, Eaton wanted the event to fail, that his long-standing loyalties to Parks were that strong. For his part Eaton suggested that key members of PRO sought "to undermine the event" ("Rightest and the Rest," 76), presumably because of smoldering animosities of their own.

16. Jon Asher, "The Short and Chaotic History of the Professional Racers Organization," *Car Craft,* Mar. 1977, 68.

17. Ed Sarkisian to the author, Mar. 20, 1992.

18. Terry Cook, "Who Is Larry Carrier?" *Car Craft,* Dec. 1971, 66.

19. Qtd. in ibid., 68.

20. Gardner qtd. by Phil Elliott in *Super Stock,* Feb. 1988, 6. If racers thought that NHRA officials could be high-handed, they had not seen anything until Meyer took over the IHRA. The son of the millionaire founder of the Success Motivation Institute, "Waco Willie" had begun driving a funny car as a teenager. With IHRA he vowed to "take drag racing to the next plateau," but his brief tenure ended "in total and complete failure"—in the words of Densmore, who, since leaving NHRA, would have been favorably disposed to any rival organization ("Dave Densmore Speaks," *Super Stock,* Apr. 1989, 23). See also Jeff Burk, "Billy Ball," *Drag Racing,* Mar. 1988, 23–33; Jon Asher, "Back in Bristol Again," ibid., May 1989, 68–73; and Meyer's series "On the Positive Side" in *Drag Racing Today* during 1988.

21. Jim Kelly, "Behind the Scenes," *Drag Racing,* Nov. 1974, 16.

22. Mike Doherty, "Can Drag Racing Survive?" *Drag Racing,* Mar. 1974, 36.

23. "The Last Good-Bye," *Super Stock,* Dec. 1974, 61. The politics of pro stock could easily be accorded an extended treatment; even Wally Parks would admit, long afterward, that he regretted the measures NHRA took in the early 1970s. For further background, see: Bud DeBoer, "Drag Racing U.S.A." (Nicholson interview), *Drag News,* Jan. 24, 1970, 14; Gray Baskerville, "Dyno Don Nicholson," in *Petersen's History of Drag Racing,* 75–77; "Sox & Martin," *Drag Racing,* Apr. 1971, 32–37ff, and May 1974, 56–64; Ronnie Sox and Buddy Martin, *The Sox and Martin Book of Drag Racing* (Chicago: Henry Regnery Co., 1974); "Sox & Martin Split with Chrysler," *Drag Racing,* Sept. 1974, 16; Thomas Pope, "Where Are They Now? Buddy Martin," *Drag Racing Today,* Dec. 1989, 55; "Dick Landy, Pro Innovator," *Super Stock,* Sept. 1970, 48–51ff; " 'Grumpy,' 'Dyno,' and Sox Reveal New Pro Stock Trends for 1972," *National Dragster,* Mar. 3, 1972, 2; "Bill Jenkins' Big Year," *Drag Racing,* Jan. 1973, 36–39ff; "Fords Take Over Pro/Stock," ibid., Jan. 1974, 28–31; "Gapp & Roush 'Breakaway' Mustang Pro/S," ibid., July 1974, 32–35; Norm Mayerson, "Pro Stocker to Go," *Car Craft,* May 1973, 48–49; and Jon Asher, "It Wasn't Always Pro and It Was Never Stock," *Drag Racing,* Feb. 1991, 18–27. Pro-stock racers were the first to traffic in glitzy press kits, and these also provide historical background.

24. Billy Meyer, "On the Positive Side," *Drag Racing Today,* Mar. 24, 1989, 2.

25. Dave Wallace, Jr., "Out of the Gate," *Drag News,* July 3, 1976, 2.

26. *Don Prudhomme: Professional Racer.*

27. Dave Wallace, Jr., "Tobacco Road," *Drag Racing,* Nov. 1987, 105–6. See also Wallace, "Out of the Gate," *Drag News,* May 3, 1975, 4; and "Winston, NHRA Announce $100,000 WCS Program," ibid., 8.

28. Qtd. in "Consumer Legislation," 663.

29. Wallace, "Tobacco Road," 106. See also John Lawlor, "Reynolds Spends Good . . . ," *Popular Hot Rodding,* Aug. 1981, 66–69; and Dave Densmore, "Taken for Granted," *Drag Racing Today,* Dec. 1989, 58–61.

30. "Hot Rodders Gain Nationwide Audience," *NHRA Tie Rod,* Sept. 1, 1956, 1; Erik Barnouw, *Tube of Plenty: The Evolution of American Television,* rev. ed. (New York: Oxford University Press, 1982), 350–53; Sky Wallace, "Show Time!" *Drag Racing,* July 1987, 80–88; Randall McCarthy, "Film at Eleven," *Super Stock,* June 1989, 46–48.

31. Qtd. in Rader, *In Its Own Image,* 136.

32. Jones qtd. in Eddie Perkins, "The TV Effect, Part II: The Networks," *Drag Racing Today,* May 19, 1989, 60. While it remained doubtful whether TV would ever be a paying proposition for the sanctioning bodies, hope could spring eternal: In the early 1950s both NBC and CBS ignored pro football, and, when the Chicago Bears improvised a small network in 1951, George Halas actually had to pay two of the stations to carry the games. See Rader, *In Its Own Image,* 86–87.

33. Qtd. in Perkins, "The TV Effect, Part I: The Sanctioning Bodies," *Drag Racing Today,* May 5, 1989, 66. Tracy had formerly been involved with Dallas Gardner at Ontario Motor Speedway, as was Don Krashaur, NHRA's comptroller.

34. Tracy indicated that the focus was on the machinery rather than on personalities as a matter of "the overall philosophy the NHRA has always subscribed to" (Mike McGovern, "NHRA Enjoying Big Season on TV," *Reading Eagle* [Pa.], Sept. 7, 1986, 15ff).

35. Carl Steward, "An Outside Look," reprinted from the *Hayward Daily Review* (Calif.), Oct. 3, 1983, in *National Dragster,* Oct. 21, 1983, 21.

36. In that regard Diamond P did not help, for it also produced monster truck and "pulling" shows. For typical reaction of the big-city press to such spectacles, see Barry Meier, "The Hot Rods of the 90's Go Crash! Crunch! Crush!" *New York Times,* Aug. 7, 1991, C1ff; Dana Milbank, "Tractor-Pull Sport Fights Muddy Image to Win More Fans," *Wall Street Journal,* Sept. 25, 1991, A1–2.

37. Qtd. in Jon Asher, "Top Fuel 1985: The Quickest & Fastest, But Will It Survive?" *Hot Rod,* Apr. 1985, 36.

38. Bud DeBoer, "Person to Person," *Drag News,* Aug. 15, 1970, 2.

39. Qtd. in Budweiser King Racing news release, 1988.

40. Rick Voegelin, "Decal Money," *Car Craft,* Oct. 1972, 37.

41. Steve Earwood, NHRA media relations director, qtd. in Kay Presto, "Where Have All the Sponsors Gone?" *Super Stock,* Aug. 1986, 87.

42. "Profile: Kenny Bernstein," *National Dragster,* Jan. 26, 1979, 5.

43. Leonard Emanuelson, "Kenny Bernstein's Chelsea King Funny Car," *Popular Hot Rodding Yearbook 1980* (Los Angeles: Argus Publishers, 1980), 42–45; Norm Froscher, "Froscher's Forum," *National Dragster,* Aug. 3, 1979, 8; Dave Densmore, "King Kenny" (offprint, AMDR).

44. Presto, "Where Have All the Sponsors Gone," 74; Michael Lufty, "Innerview: Kenny Bernstein," *Drag Racing,* June 1989, 70–74; July 1989, 98–101 (quote on 99).

45. "Kenny Bernstein 'Bud' Bound for 1981 Season," *National Dragster,* Dec. 5, 1980, 1; see also "Budweiser's Beasts," *Car Craft,* July 1980, 82–84.

46. Quoted in Dave Wallace, Jr., "1975 Drag News Driver of the Year: Dale Armstrong, Pro Comp Eliminator," *Drag News,* Nov. 15, 1975, 6; see also Wallace, "'AA/Dale' Repeats as Driver of the Year," ibid., Dec. 4, 1976, 6–7ff; Randy Black, "AA/Dale: Driver of the Year Again!" ibid., Dec. 3, 1977, 15–16.

47. Dave Wallace, Jr., "The Nice Guy Everyone Loves to Hate," *Drag Racing,* July 1987, 4.

48. Lufty, "Innerview," July 1989, 101. See also Gray Baskerville, "Long-Arm Inspection," *Drag Racing,* Oct. 1987, 73–82; and John Bowen, "Killer Coupler," ibid., July 1987, 70–78.

49. Steve Collison, "Wouldn't You Really Rather Design a Buick?" *Super Stock,* June 1987, 40ff; see also Ray and Shirley Strasser to *Drag Racing,* Aug. 1987, 8–9; and Tate Calvert to ibid., Oct. 1987, 10–11.

50. Keith Black interview, Feb. 6, 1989, DROHA, NMAH, 10. The one scholarly address to this activity (Frederick L. Honhart, "Speed and Spray: Technology, Its Implication, and Unlimited Hydroplane Racing, 1946–1988," *Journal of American Culture* 13 [Spring 1991]: 19–29) is steadfastly innocent of the political implications of Budweiser's switch to turbines—the key, in Black's view, to its control.

51. Qtd. in Nina Murphy, "Public Relations," *Drag Racing,* Nov. 1989, 88.

52. Don Gillespie, "Will 'The King' Reign in Top Fuel?" *Drag Racing Today,* Oct. 20, 1989, 73; see also Mark Zeske, "Funny-Car Drivers Down-Shift from Danger, Cost," *Dallas Times Herald,* Dec. 26, 1989.

53. *Sponsors Report Almanac: Exposure, Analysis, Trends and Projections* (Ann Arbor: Joyce Julius and Associates, 1991), 107. See also Thomas Amshay and Christina Clement, *Get Ready, Get Set, Get Sponsored* (Cuyahoga Falls, Ohio: RFTS Marketing and Consulting, 1991); Phil Burgess, "Sponsorship Scholarship," *National Dragster,* Aug. 30, 1991, 56–59; Thomas Gregg, "Corporate Sponsorship Marketing in the 1990s," ibid., 6ff; Jon Asher, "Myths and Facts about Sponsorships," *Drag Racing,* Feb. 1991, 5.

54. Art Buchwald, "Cigaret Men Try to Filter Back into Big TV Picture," *Los Angeles Times,* Jan. 10, 1971; Alan Blum, "The Marlboro Grand Prix: Circumvention of the Television Ban on Tobacco Advertising," *New England Journal of Medicine* 324 (Mar. 28, 1991): 915.

55. Jerry Bonkowski, "Kenny Bernstein's Talents Extend Far beyond the Quarter Mile," *Drag Racing Today,* Mar. 1989; "The Corporate Elite: Anheuser Busch," *Business Week,* Oct. 19, 1990, 68.

56. "Dave Densmore Speaks," *Super Stock,* Oct. 1990, 88.

12 *MEN AND WOMEN*

1. Ruth Schwartz Cowan, "From Virginia Dare to Virginia Slims: Women and Technology in American Life," *Technology and Culture* 20 (Jan. 1979): 51.

2. Jack O. Baldwin, "Mother Is a Hot Rodder," *Hot Rod,* Oct. 1953, 59.

3. Ibid.

4. Fran Deggendorf to *Drag News,* May 18, 1963.

5. Marianne Beckstrom to *Drag News,* Feb. 13, 1960.

6. "On the Carpet: Shirley Shahan with Mike Doherty," *Drag World,* Oct. 29, 1965, 3.

7. Debi Shepherd, "Women in Drag Racing, Part II: The Drivers," *Drag Racing Today,* Apr. 7, 1989, 25.

8. "Women Invade Drag Racing," *Drag Sport,* Aug. 9, 1965, 17.

9. Qtd. in "The Staging Lane," ibid., Aug. 20, 1965, 1.

10. See Karen Nelson, "Personality Profile: Paula Murphy," *Drag Racing,* July 1967, 27–31ff; "Paula: Still the Fastest Woman on Wheels," ibid., Mar. 1974, 20–23; and especially Connie Strawbridge, "Paula Murphy: Confessions of the First Female Fuel Racer," *Super Stock,* May 1989, 76–90.

11. Terry Cook, "The Perils of Paula," *Drag World,* Aug. 19, 1966, 6.

12. Qtd. in Jon Asher, "Leading Lady," *Drag Racing,* May 1989, 74–77.

13. "Girl Does 169, Flips New A/FD into Field," *Drag World,* Aug. 5, 1966, 25; "The Digger Lady," *Drag Racing,* Dec. 1967, 15; "Ginger Watson Heads Card at St. Petersburg," *Drag News,* Feb. 24, 1968, 6.

14. "Lady License Rulings," *National Dragster,* Aug. 18, 1967, 3.

15. Barbara Hamilton to Jack Hart, July 18, 1967; also July 20, 1967, and Hart to Hamilton, July 10, 1967 (copies from Hamilton, courtesy of Dave Wallace, Jr.). See also Doris O'Donnell, "Woman Driver Races to Fame as Inductee in Hot-Rodder Hall," *Cleveland Plain Dealer,* July 12, 1992. My thanks to Roger Grant for this reference.

16. Qtd. in Strawbridge, "Paula Murphy," 80.

17. Bud DeBoer, "Drag Racing U.S.A." (Della Woods interview), *Drag News,* Nov. 22, 1969, 18; ibid., Jan. 12, 1968, 14; "Della Woods," press release, Bernella Racing Team, 1968; Shepherd, "Women in Drag Racing," pt. 2, 28–29. Bernie and Della's name was actually Drzewinski, which they changed "for publicity reasons."

18. "Capitol Hosts 'Miss America' of Drag Racing," *Drag News,* Aug. 15, 1970, 13, 22.

19. "Men over Women at Maple Grove," ibid., June 19, 1971, 26.

20. Strawbridge, "Paula Murphy," 90.

21. "Indy: Females out in Force," *Drag News,* Sept. 4, 1976, 9; Jeff Tinsley, "Who Is 'That Girl'" (Shay Nichols), *Drag Racing,* June 1975, 21–24ff; Francis C. Butler, "Quick like a Rabbit" (Carol Burkett), *Drag Racing,* Sept. 1986, 52–54; "BB/Flopper Bucks Long Odds with Chevy Power, Lady Driver" (Carol Henson), *Drag News,* July 17, 1976, 6; Randy McCarthy, "Paula Gage: Second Best Won't Do," *Super Stock,* Jan. 1987, 39–41ff; Thomas Pope, "Amy Faulk: From Turkey Farm to World Champ and Corporate Executive," *Drag Racing Today,* July 14, 1989, 61–63; John Baechtel, "'Who Was That Lady?'" *Car Craft,* Mar. 1977, 45–47; Shepherd, "Women in Drag Racing," pt. 2.

22. Bud Dzamba, "Marsha Smith: Pioneering at Speed," *National Dragster,* Nov. 22, 1985, 13; Dale Wilson, "Aggi Hendriks Is Pushing 290 MPH," *Drag Racing,* Jan. 1988, 34ff.

23. See Dusty Brandel, "Linda Vaughn: First Lady of Racing," *Motorsports Illustrated,* Mar. 1990, 34–46. Hurst said, "Sex sells." NHRA knew it did, too, even though it might make a show of being prepared to revoke the sanction of a strip that was promoting topless dancers along with the racing (see G. R. Hill to *National Dragster,* June 16, 1972, 4; and the reply of *ND*'s editor, Bill Holland).

24. "The Girls of Drag Racing," *Drag Racing,* Apr. 1973, 60–67. See also "Jungle Pam," ibid., Apr. 1974, 37–39; "Nobody Did It Better . . . ," *Super Stock,* Jan. 1978, 42–47; Kay Presto, "Jungle Jim," ibid., Jan. 1987, 51–56ff.

25. Dave Densmore, "Surely Shirley," *Popular Hot Rodding: 1983 Drag Racing Yearbook* (Los Angeles: Argus Publishers, 1983), 79.

26. *Shirley* (Diamond P Video Productions, 1987); *Drag World,* Aug. 26, 1966, 5; *Drag News,* Sept. 15, 1966, 9; ibid., Aug. 15, 1970, 13; "Drag Racing Profile: Shirley Mul-

downey," *National Dragster,* June 16, 1972, 5; "Shirley Muldowney: America's Foremost Female Auto Racer," news release, Maple Grove Dragway, 1975.

27. Hawley with Smith, *Drag Racing,* 41; Muldowney qtd. in Densmore, "Surely Shirley," 81.

28. Hawley with Smith, *Drag Racing,* 19; Densmore, "Surely Shirley," 80.

29. "Drag Racing Profile: Connie Kalitta," *National Dragster,* Mar. 30, 1979, 5.

30. Densmore, "Surely Shirley," 81.

31. Ibid.; Garlits, *"Big Daddy,"* 232; Liberman qtd. in "Drag Racing Profile: Shirley Muldowney."

32. Qtd. in Dave Wallace, Jr., "Shirley Scores Another 'First': Drag News Driver of the Year," *Drag News,* Dec. 25, 1976, 7.

33. Densmore, "Surely Shirley," 81.

34. *Shirley;* Garlits, *"Big Daddy,"* 235.

35. John Hall, "Bye Cha Cha," *Los Angeles Times,* Jan. 27, 1977.

36. Qtd. in Wallace, "Shirley Scores Another 'First,' " 7.

37. "Behind the Scenes With: Rahn Tobler & John Muldowney," *National Dragster,* Apr. 25, 1980, 5.

38. "Drag Racing Profile: Dave Uyehara," *National Dragster,* Oct. 27, 1978, 5; "Drag Racing Profile: John Kimble," ibid., May 16, 1980, 5.

39. Muldowney qtd. in Tim Cline, "The World of Shirley Muldowney," *Echelon* 9 (Sept. 1987): 10–16.

40. George Phillips, "Shirley Muldowney's Banner 1980 Season," *National Dragster,* Nov. 21, 1980, 8–10ff; Phillips, "Shirley Muldowney: Where No Man Has Ever Gone Before," *Drag Racing Annual, 1981* (Los Angeles: NHRA, 1981), 24–25; Phil Elliott, "World Finals," *Drag World,* Sept. 15, 1981, 1ff; Garlits, *"Big Daddy,"* 241; Shepherd, "Women in Drag Racing," 18–23; "Muldowney Wins Titus Award," *National Dragster,* Feb. 4, 1983, 1.

41. Qtd. in *Heart like a Wheel,* Twentieth Century-Fox playbill, 1983.

42. Hawley with Smith, *Drag Racing,* 65.

43. John Baechtel, "It's What I Do," *Hot Rod,* May 1986, 60–64; for a sampling of the journalistic accounts of Muldowney's comeback, see also: "Shirley Muldowney: On the Road to Recovery," *Winston Drag Racing Media Guide* (Winston-Salem, N.C.: R. J. Reynolds Tobacco Co., 1985), 56–57; Steve Earwood, " '1984 Wasn't That Bad': An Interview with Shirley Muldowney," *National Dragster,* Feb. 1, 1985, 5ff; John Bowen, "Shirley," *Drag Racing,* Sept. 1985, 80–84; Michelle Krebs, "Muldowney's Heart behind Wheel Again," *USA Today,* Jan. 15, 1986, 1ff; Shav Glick, "Shirley Muldowney Is Back behind the Wheel," *Los Angeles Times,* Jan. 28, 1986; Phil Burgess and Chris Martin, "The Return of Shirley Muldowney: Few Trials, Lots of Tribulation," *National Dragster,* Feb. 7, 1986, 1ff; Asher, "Leading Lady"; Norm Froscher, "Shirley Muldowney's Come Back the Last Quarter-Mile," *Gainesville Sun,* Mar. 14, 1986, suppl., 2; Sky Wallace, "Comeback of the Year," *Drag Racing,* May 1986, 12–13.

44. Jon Asher, "Reprieve," *Drag Racing,* June 1990, 30.

45. J. E. Vader, "Two Foes Bury the Hatchet, but Not the Competition," *Sports Illustrated* (offprint, AMDR); see also "A Match Made in Heaven: Muldowney, Garlits Join Forces," *Drag Racing Today,* May 5, 1989, 50–51; Shav Glick, "Steaming Back," *Los Angeles Times,* Oct. 25, 1989; Jon Asher, "The Day the Old Man Drove Shirley's Car," *Drag Racing,* Jan. 1990, 76–79; Dave Densmore, "Garlits Returns," *Super Stock,* Jan. 1990, 46–49.

46. Virginia Scharff, *Taking the Wheel: Women and the Coming of the Motor Age* (New York: Free Press, 1991), 13. This is an extraordinarily good book, but, by ignoring Muldowney, Scharff misses an opportunity to underscore her perceptions regarding female race drivers; she chooses instead the less evocative persona of Janet Guthrie. On Guthrie, see Harvey Duck, "How Janet Guthrie Revolutionized Auto Racing," *Auto Racing Digest,* Jan. 1977, 12–19.

47. Rosellen Brown, "Transports of Delight," *Women's Review of Books* 9 (Sept. 1992): 7.

48. Qtd. in Shav Glick, "These Seniors Enjoy Life in the Fast Lane," *Los Angeles Times,* Feb. 4, 1990, C6; see also Jon Asher, "The Greek: Still Golden after All These Years," *Drag Racing,* Oct. 1990, 28–32.

49. There is a wealth of literature on each of the people mentioned here. On the Hoovers, for example, see: Karen Nelson, "Personality Profile: The Hoovers," *Drag Racing,* Dec. 1967, 56–60ff; "Behind the Scenes With: George Hoover," *National Dragster,* Dec. 12, 1980, 5; Joe Sherk, "Challenging Corporate America," *Drag Racing Today,* Sept. 22, 1989, 44–46; Shav Glick, "Elder Hoover's in Fast Lane," *Los Angeles Times,* Feb. 2, 1990, C4.

50. Glidden qtd. in Kay Presto, "What Makes Glidden Run—and Win?" *Super Stock,* Mar. 1989, 39. For Glidden at his most outrageously outspoken, see "Interview: Bob Glidden," *Drag Racing Illustrated,* Jan. 1989, 37–40ff.

51. McCulloch and Muldowney qtd. in Bob Abdellah, "Drawing Power," *Drag Racing,* May 1987, 87, 82.

52. "Behind the Scenes With: Tony Ceraolo," *National Dragster,* Dec. 26, 1980, 5; David M. Leavitt, "Kathy and Tonia Ceraolo: Drag Racing's Daughters," *Super Stock,* Feb. 1982, 19–20ff. In the 1970s one of Ceraolo's crew members had been Virginia Anne Bonito, who is quoted in the preface of this book.

53. Kim LaHaie qtd. in "LaHaie/McCulloch/Minor 1988 Media Guide," Miller High Life Racing, 1987, 36; see also Chris Martin, "Like Father . . . like Daughter," *National Dragster,* Apr. 13, 1984, 12ff; Dan Carpenter, "Small Business Is Alive, Well in Big-Bucks Drag-Racing World," *Indianapolis Star,* Sept. 3, 1985, 23; Martin, "All the Way with . . . Dick LaHaie," *National Dragster,* Sept. 4, 1987, 32ff; Dave Wallace, Jr., "LaHaie & LaHaie," *Drag Racing,* Nov. 1984, 18–22; Shav Glick, "Like Father, like Daughter," *Gainesville Sun,* Mar. 20, 1987, suppl., 30; Dale Wilson, "Family Fuel," *Super Stock,* Oct. 1987, 32–34; Debi Shepherd, "Women in Drag Racing, Part I: Women behind the Scenes," *Drag Racing Today,* Mar. 24, 1989, 74; Jerry Bonkowski, "On the Rebound," ibid., Oct. 20, 1989, 62–65; Kay Presto, "Innerview: Dick & Kim LaHaie," *Drag Racing,* Sept. 1989, 90–93.

54. Scharff, *Taking the Wheel,* 175. The phallus theme has been elaborated by Terry Reed in *Indy: Race and Ritual:* "Vividly painted, handsomely lettered, low slung, and fearsomely aggressive, [the cars] have a way of letting even the most sexually moribund observer know that they mean erotic business" (188ff). Reed is referring to Indy 500 cars, but the sexual suggestiveness of dragsters is, of course, multiplied enormously.

55. Lee qtd. in Thomas Pope, "Where Are They Now?" *Drag Racing Today,* Dec. 1989, 55; see also "Drag Racing Profile," *National Dragster,* May 21, 1971, 5; Neil Britt, "Down Home," *Drag Racing,* Feb. 1975, 40–43ff; Pope, "Alison Lee: Dragdom's Best Lady Mechanic?" *Drag News,* Nov. 26, 1977, 4.

56. Dunn qtd. in Thomas Pope, "Like Father, like Son," *Drag Racing Today,* Dec. 1989, 63; Anderson qtd. in Shav Glick, "Speed Becomes a Family Value," *Los Angeles Times,* Feb. 3, 1993, C10.

57. The second quote in the epigraph is from a video produced by the National Museum of American History for the "A Material World" exhibit; the other quote is from a P.A.W. press kit from Roven Productions, 1988.

13 ENTHUSIASM

1. Quote from "Gotelli—the Man," *Drag News,* Mar. 2, 1963, 27. Don Jensen reported that Gotelli "didn't care what blew up." "If it didn't work, [he] changed it" (transcription entitled "Construction, Materials, Attitudes," Sept. 1990, DROHA, NMAH, 37). See also *Hot Rod,* July 1962, 68–69.

2. Ted Robinson, "1320 Carat Diamond" (Annin), *Drag Racing,* Aug. 1970, 16, 72; "Ben Brown on Funny Cars" (Setzer) *Car Craft,* Apr. 1973, 36.

3. Kay Presto, "A Minor Miracle," *Drag Racing,* Sept. 1984, 16. For more on Minor and the people involved in his drag racing ventures, see: "Minor Takes the Wheel, Hits 6.05," *National Dragster,* Jan. 21, 1983, 40; "The World's Two Quickest Cars . . . Live in the Same House," ibid., July 8, 1983, 8–9; Joe Sherk, "The Spud King," *Super Stock,* July 1986, 42–45ff; Jeff Burk, "The Brew Crew," ibid., May 1989, 24–27; "Larry Minor," *Drag Racing Today,* Sept. 8, 1989, 19–20; Kathy Folk, " 'Ace' McCulloch's Career Standing the Test of Time," *Reading Eagle* (Pa.), Sept. 11, 1988, suppl., 20–21; Jeff Burk, "Innerview" (McCulloch), *Drag Racing,* Jan. 1989, 32–33; "Gary Beck—Top Fuel Driver of the Year," *Drag News,* Dec. 28, 1974, 16–17; Al Carr, "Beck: A Drag Racing Legend," *Los Angeles Times,* Feb. 24, 1977, pt. 3, 12.

4. Amato qtd. in Don Gillespie, "Man in Motion: Joe Amato," *Drag Racing Today,* Feb. 10, 1989, 47. See also "Behind the Scenes With: Joe Amato," *National Dragster,* Dec. 4, 1981, 4–5; "The Business Side of Joe Amato," ibid., Oct. 19, 1984, 16–17; Kay Presto, "Too Fast to Stop Now," *Super Stock,* Mar. 1988, 69–73ff; Jim Hofman, "He Means Business," *Reading Eagle* (Pa.), Sept. 11, 1988, suppl., 36; Thomas Pope, "Following the Golden Rule," *Drag Racing Today,* June 30, 1989, 22–25; Jon Asher, "There's No Stopping Him," *Drag Racing,* May 1990, 32–35.

5. Richards qtd. in "The General," *Super Stock,* Mar. 1988, 90.

6. Candies qtd. in Dave Densmore, "The Odd Couple," *Drag Racing Today,* Aug. 25, 1989, 102; see also "Candies & Hughes: The Odd Couple at 27," *Super Stock,* July 1990, 40–45.

7. "Gary Ormsby," *Drag Racing Today,* Sept. 8, 1989, 21; Chuck Schifsky, "What a Year It Was" (Ormsby), *Drag Racing,* Mar. 1990, 68–77; Gary Falanga, "Return of the Texan" (Hill), *Drag Racing,* July 1987, 108–9ff; Ercie Hill, "The Crash and the Thrash," *Super Stock,* May 1989, 22–23ff. Ormsby drove his first slingshot in 1965; Hill had gained prominence even before that (see, e.g., "Texas Tornado," *Hot Rod,* Aug. 1962, 52–53), but he had spent many years racing motorcycles and boats (see Shav Glick, "On Land or Water, This Texan Is Known as Fastest Eddie," *Los Angeles Times,* Oct. 28, 1988, pt. 3, 6).

8. Qtd. in "The Independents of Top Fuel: John Carey," *Drag Review,* Dec. 10, 1988, 52; see also Randall McCarthy, "Double Vision," *Drag Racing,* Aug. 1987, 19–21.

9. Force qtd. in Jerry Bonkowski, "Life According to Force," *Drag Racing Today,* June 2, 1989, 65; see also, "Drag Racing Profile: John Force," *National Dragster,* Feb. 23, 1979, 5; and Kay Presto, "The Force Is with Us," *Drag Racing,* Mar. 1985, 58–63.

10. See Steve Smith, *The Racer's Tax Guide* (Santa Ana, Calif.: Steve Smith Autosports, 1978), 20, 88–89.

11. Burk, "Innerview," 33. Old-time funny car racers regarded the best years of their lives as those just prior to the mid-1970s recession, when they could book as many matches as they could get to; see Jon Asher, "Circuit Riders," *Drag Racing,* Jan. 1990, 18–35.

12. Wally Parks, "The NHRA Staging Light," *National Dragster,* May 14, 1982, 4.

13. "What's It Like to Be the World's Fastest Teenager?" *Drag Racing,* July 1972, 62; Thomas Pope, "Where Are They Now? Jeb Allen," *Drag Racing Today,* July 28, 1989, 29–30.

14. Qtd. in Kay Presto, "The Snake Awakes," *Drag Racing,* May 1987, 73.

15. Ibid., 67.

16. McEwen qtd. in Don Gillespie, "The Goose's Candor," *Super Stock,* Feb. 1989, 45. For many veteran drivers in this position the waiting was forever. But McEwen's angel actually came along—Jack Clark, the designated hitter for the Boston Red Sox. Under McEwen's guidance Clark assembled a top-fuel operation topped for opulence only by Bernstein's. McEwen was, as his old friend Don Garlits put it, "in hog heaven." Ironically, when Clark filed for bankruptcy in 1992 (halfway through a three-year $8.7 million contract) his lawyer cited "expensive hobbies." His business manager elaborated for the

Orange County Register: Jack Clark Racing was an operation that had cost him "well over $1 million" and from which he had received "virtually nothing."

17. Karamesines qtd. in Bob Cruse, "Chris Karamesines . . . Our Leader," *Drag Racing,* Nov. 1967, 42. The Karamesines quote in the epigraph is also from this article; the Hoover quote is from a letter to *National Dragster,* July 24, 1981, 4.

18. Pastorini qtd. in Jeff Burk, "Quarterback Blitz," *Drag Racing,* Jan. 1987, 31.

19. Susan Bernstein qtd. in Joe Sherk, "The Mammas & the Pappas," *Drag Racing Today,* June 16, 1989, 35; Martin qtd. in "Paula Martin's Risky Business," ibid., May 5, 1989, 60. The personality of the "thrill seeker," and the possibility of it having biochemical origins, is addressed in Charles S. Carver and Michael F. Scheier, *Perspectives on Personality* (New York: Simon & Schuster, 1988).

20. Carolyn Hall-Salvestrin to the author, Sept. 30, 1991.

21. Ormsby ran his last race in May 1991 and died in September.

22. Bud DeBoer, "Person to Person," *Drag News,* Apr. 11, 1970, 3.

23. Fred M. H. Gregory, "Jerry Ruth," *Car Craft,* Mar. 1972, 93.

24. Chris Martin, "Some Reasons Why Ernie Still 'Hauls,'" *National Dragster,* Mar. 8, 1985, 8ff. Hall's daughter Carolyn has kindly shared with me a logbook that her father kept in 1969, just before he switched from gas to fuel. He entered fifteen events and won four of them, making forty-six runs all told and grossing a little less than two thousand dollars. At Woodburn, Oregon, on July 27 he made five runs and won four hundred dollars; at Pomona in February he made five runs, failed to qualify, and came away with nothing.

25. Phil Burgess, "Ray Stutz: Back in the Saddle Again," *National Dragster,* Mar. 8, 1985, 8ff; Taylor qtd. in Burgess, "Dave Braskett and Dennis Taylor: 'We Know We Are Underdogs,'" ibid.

26. Strauss qtd. in *Drag Review,* Dec. 10, 1988, 66.

27. Chris Martin, "Coburn Just Wants to Run Well Again," *National Dragster,* Mar. 8, 1985, 9; see also "Warren & Coburn," *March Meet '76 Program,* 9–11ff.

28. Warren qtd. in Ralph Guldahl, "Personality Profile: James Warren," *Drag Racing,* Dec. 1966, 54.

29. Martin, "Coburn Just Wants to Run"; Jon Asher, "Distant Thunder," *Car Craft,* July 1976, 105.

30. Eddie Flournoy qtd. in Phil Burgess, "Rodney Flournoy: It's a Family Affair," *National Dragster,* Mar. 9, 1984, 7; see also Judy Strawn to ibid., Apr. 6, 1984, 4.

31. Birky qtd. in Chris Martin, "Despite the Ante, Arnold Birky Is Still Playing the Game," ibid., Mar. 8, 1985, 30.

32. "Who Is John Garrison?" *Drag Sport,* Nov. 1, 1965, 15; Robinson, "Use Chutes Only in Emergency."

33. Mike Briganti, "Comeback of the Year: Welcome Back, Doug Kerhulas," *Drag Racing,* May 1985, 16–21; Bret Kepner, "Inner Vision" (Edstrom), *Super Stock,* Apr. 1987, 33–34.

34. Hicks, *Close Calls,* 147; Presto, "Snake Awakes," 77.

35. Qtd. in Martin, "Despite the Ante," 30.

36. Ferguson, "Toward a Discipline," 21. For an analysis of automotive enthusiasm, as well as a compelling refutation of much of the cultural critique set forth by Adorno and Marcuse, see Moorhouse, *Driving Ambitions,* chap. 1, "Theoretical Perspectives."

37. Rader, *In Its Own Image,* 108.

38. Miglizzi interview, Oct. 2, 1987, DROHA, NMAH.

39. Eric Nilsson, "The Legend of the Chi-Town Hustler, Part 2: 1971 to 1984," *Drag Racing,* Sept. 1984, 49. See also Paul Stenquist, "The Legend of the Chi-Town Hustler," *Drag Racing,* May 1975, 46–51ff.

40. Coil and Richards both added another title in 1991.

41. Ormsby qtd. in "On the Run," *National Dragster,* Apr. 18, 1991, 9; Beard qtd. in

Thomas Pope, "The Crew Chief, Part III: Lee Beard," *Drag Racing Today,* Aug. 25, 1989, 37.

42. Madigan, *Loner,* 25.

43. By the 1970s there were apprehensions that the SEMA specs program, managed by Holly Hedrick and by Bob Spar of B&M Automotive, left the organization with a "huge legal vulnerability" due to "product liability problems"—hence, the creation of SFI as a separate and distinct entity. See Kevin C. Osborn, "SEMA: The First 25 Years," *Drag Review,* Nov. 13, 1989, 23 (reprinted from SEMA's *Performance Aftermarket*). Note the name change from the Specialty Equipment Manufacturers Association.

44. Malone qtd. in "Richard Holcomb," *Drag Review,* Dec. 10, 1988, 60.

45. Head qtd. in *National Dragster,* Dec. 13, 1991, 59.

46. Dan Roulston, "Chassis Research Company Builds 'Slingshots,' " *Drag News,* May 3, 1958, 8; Scott Fenn, "Long Wheelbase or Short," in *1964 Drag Racing Annual* (Los Angeles: Argus Publishers, 1964), 66–67ff.

47. Roger Huntington, "Weight Transfer," *Hot Rod,* Apr. 1963, 78; Greg Curtis, "Weight and Power in the Balance," in *1964 Drag Racing Annual,* 54–57; Forrest Bond, "Big Daddy's Latest," *Popular Hot Rodding,* Dec. 1967, 26–27.

48. Most designers prominent in the 1960s were out of business by the 1970s. Woody himself quit making chassis in the wake of a lawsuit by a customer, Herm Peterson, who suffered disfiguring burns when his car upset and he was drenched with fuel from a tank just behind the cockpit. Lawsuits were the touchiest of all subjects, and the touchiest of all suits occurred in 1986 when Corpus Christi businessman Terry Johns sued after another driver collided with his twenty-year-old daughter Lori's car and inflicted injuries as serious as Muldowney's two years before. Every competitor signs a waiver of liability, but, as NHRA's Carl Olsen put it, "If you can get a case like this to trial, a jury is liable to do almost anything." The suit was eventually dropped but not until it acquired dimensions of "The Case That Nearly Killed Racing" (the title of an article by Steve Spence in *Autoweek,* Sept. 25, 1989, 26–28ff).

49. Phil R. Elliott, "The Guru" (Swindahl), *Super Stock,* May 1991, 82; Garlits, "*Big Daddy,*" 230; see also Jeff Morton, "Al Swindahl: The *Dragster* Interview," *National Dragster,* Jan. 18, 1985, 7–8.

50. Rollain to Leslie Lovett, Sept. 21, 1984, Archives of the National Hot Rod Association, Glendora, Calif.; Dave Wallace, Jr., "Short but Sweet," *Drag Racing,* Mar. 1986, 63.

51. Jeff Morton, "*National Dragster* Interview: 'Diamond Dave,' " Oct. 11, 1985, 9, 11.

52. Snow qtd. in "The Snowman Cometh," *Drag Racing,* Sept. 1988, 30. The idea of using compressed air had also appealed to other racers, including Mickey Thompson and Art Malone.

14 *CHOICE*

1. Carrier quote in "Memo from Larry," *Super Stock,* July 1984, 6. While all sorts of implausible claims were made for performances by rockets, there is no doubt that some of them went faster than 300 miles per hour with elapsed times well down into the 4s. See Franklin Ratliff, "The Rocket Dragster: An Historical Perspective," *Drag Racing,* Feb. 1975, 25–27; and, on the first such machine, Ralph Guldahl, Jr., "X-1 Rocket Car," ibid., May–June 1968, 26–31ff.

2. The Beck quote in the epigraph is from Dave Overpeck, "Beck-Miner Team Whiz without Computer Whiz," *Indianapolis Star,* Sept. 2, 1984, 8D; the other quotes are from Thomas Pope, "The Crew Chief, Part II: Austin Coil," *Drag Racing Today,* June 16, 1989, 47; and Phil Burgess, "Head's Game," *National Dragster,* June 7, 1991, 14.

3. The phrase is from Eugene S. Ferguson's classic article "The Mind's Eye: Nonverbal Thought in Technology" (*Science,* Aug. 26, 1977, 827-36), which has informed my analysis throughout this book far more than one or two short citations can suggest.

4. Qtd. in Shav Glick, "Which Snake Will Show?" *Los Angeles Times,* Oct. 29, 1988, pt. 3, 13.

5. Ferguson, "Mind's Eye," 827.

6. For perspective on the use of *secret* to mean "technique," see Pamela O. Long, "Invention, Authorship, 'Intellectual Property,' and the Origin of Patents: Notes toward a Conceptual History," *Technology and Culture* 32 (Oct. 1991): 846–84, especially 860.

7. Mike Doherty, "Hydrazine: Magic Horsepower, but at What Cost?" *Drag Racing,* Jan. 1967, 41; see also "Dr. Dean Hill Discusses Hydrazine," *National Dragster,* Oct. 28, 1966 (offprint, AMDR).

8. "Belfatti Claims 'Rocket Fuel,'" *Drag World,* Apr. 9, 1965, 13.

9. Frank Maple to Jim Tice, qtd. in "Banned by One, Ignored by Another," *Tach* (AHRA), Jan. 1967, 20.

10. After it burned all the fuel, writes Dave Densmore, "it tended to burn everything else up, too" ("A Strange Tale about Nitro, Nitrous, Don Prudhomme and NHRA," *Super Stock,* Sept. 1989, 21). See also Norm Mayersohn, "No Laughing Matter," *Car Craft,* Apr. 1973, 44–45ff. Subsequently, several fuel racers were accused of concealing setups for injecting nitrous, including Harris. "I don't feel that nitrous is dangerous or destructive," Harris insisted ("Nitro plus Nitrous," *Drag Racing,* Sept. 1987, 110), but that claim was belied by his many engine explosions in the 1970s.

11. See Paul Stenquist, "Racing 1990: Top Fuel Technology," *Popular Mechanics,* Jan. 1990, 98–99; John Brasseaux, "Top Fuel Tech," *Drag Racing,* Nov. 1989, 28–30; Phil Elliott, "Top Fuel Tech," *Super Stock,* June 1987, 38.

12. Patrick Hale, "5189 Horsepower," *Drag Racing,* Apr. 1990, 56–59. With the assistance of Norman Drazy and Dale Armstrong, Hale designed the first personal computer (PC) software specifically for drag racers. Hale to the author, Jan. 24, 1986; "Quarter for Your PC" (Phoenix: Racing System Analysis, 1986); John Brasseaux, "Racing Data Software," *Drag Racing,* Oct. 1989, 60–63.

13. John Brasseaux, "Bird of Prey: The Eagle Is Hunting Elephants," *Drag Racing,* Jan. 1988, 82–91.

14. McGee qtd. in Don Gillespie, "Thunder Down Under," *Drag Racing Today,* Feb. 10, 1989, 14; see also Tom Dufor, "Four-Cam Fuelers," *Drag Racing,* May 1984, 66–69; Jim McFarland, "Nitro Engines of the Future," ibid., Apr. 1989, 24–31; Tim Marshall, "Quadraphonic Fuel Car," ibid., July 1989, 57–61; Gray Baskerville, "End of an Era," *Hot Rod,* Jan. 1989, 110–12. On the inception of American-style drag racing in Australia, see Peter Llewellyn, "Australia's First Drag Meet," *Hot Rod,* May 1960, 50–51; and Dave Cook, "The Kings of Drag Racing," *Australian Drag Racing News,* Mar. 1987, 36–57. The latter article is an exceptionally good survey of beginnings in the United States as well as Australia. Cook notes the presence of "recognisable 'modern' dragsters" at a Melbourne airstrip as early as 1956 (52), but another Australian authority warns that "there's a big gap between the events at Bondi Beach in 1930 [and] the events held here in Victoria in the late 50s" (Steve Munro to the author, Feb. 26, 1990).

15. Conniff, "In Florida," 16.

16. Norm Drazy and Pat Alexander, "Turn of the Screw," *Drag Racing,* Aug. 1988, 24.

17. Ibid., 27; see also Jeff Burk, "Superchargers," *Super Stock,* Feb. 1993, 13–19.

18. Qtd. in Dale Wilson, "Gone with the Wind," *Super Stock,* Feb. 1989, 84.

19. Dave Wallace, Jr., "Why the PSI Won't Die," *Super Stock,* Feb. 1990, 79; see also John Brasseaux, "Turning the Screws," *Drag Racing,* Sept. 1989, 46–53; and "Screwed Up: Is the PSI Down and Out of Drag Racing?" ibid., May 1990, 56–60.

20. Drazy reported penning this observation in a diary entry of Aug. 15, 1982; see Drazy and Alexander, "Turn of the Screw," 27. For an indication that Drazy realized that "domination" was out of reach, see "The PSI Supercharger: A Change for the Better" (Tempe, Ariz.: PSI, Inc., 1992).

21. Armstrong qtd. in Gray Baskerville, "Low Blow," *Drag Racing,* Aug. 1988, 37.

22. See *Drag News,* May 2, 1967. A decade later Keith Black tried to interest Prud-

homme in a system of telemetry, so he could monitor engine functions from the pits (Black interview, Feb. 6, 1989, DROHA, NMAH, 35), but Prudhomme was not interested.

23. Hib Halverson, "The Racepak Computer," *Drag Racing,* Oct. 1989, 24; see also Rick Voegelin, "Technology of Speed," *Popular Mechanics,* Mar. 1986, 91–92.

24. Jon Asher, "Who Needs 2 Speeds?" *Drag Racing,* Oct. 1988, 50–56; see also Cameron Benty, "Slider Clutch Science," *National Dragster,* Aug. 13, 1979, 18ff; John Bowen, "A Racer's Guide to Centrifugal Clutches," *Super Stock,* May 1986, 68–74; Dale Wilson, "The Man in Black," ibid., Mar. 1987, 70–72; John Brasseaux, "Locking Up," *Drag Racing,* June 1988, 31–39; Jeff Burk, "The Magnificent Seven," ibid., Oct. 1988, 44–49; Joe Sherk, "Clutch Performers," *Drag Racing Today,* July 28, 1989, 44–46. The first racer to go back to direct drive was Bill Mullins of Pelham, Alabama, a veteran whose skills included stretching a modest budget a long way; see Jeff Burk, "Direct Connection," *Drag Racing,* Jan. 1988, 22–25; see also "Behind the Scene with Bill Mullins," *National Dragster,* June 19, 1981, 5ff; and Dale Wilson, "Iron Man," *Super Stock,* Dec. 1985, 34–37ff.

25. Head qtd. in Mark Duvall, "Will Computers Ruin Racing?" *Super Stock,* Sept. 1986, 39.

26. Oswald qtd. in Jerry Bonkowski, "Driving by the Seat of His Pants," *Drag Racing Today,* Oct. 20, 1989, 10.

27. Garlits qtd. in Asher, "Who Needs 2 Speeds?" 51. Garlits added, protectively, "There's still going to be room for guys like me—people who know how to physically build engines. . . but if you're afraid of things like computers and sophisticated electronics, you're not going to make it as a crew chief."

28. Yates qtd. in "Gentlemen, Start Your Computers," *Panorama: A New Vision of Technology* 2 (Dec. 1987): 18.

29. John Brasseaux, "Blinded by Science," *Drag Racing,* Oct. 1989, 5.

30. Gibbs qtd. in Dave Wallace, Jr., "Will Computers Take the Driver Out of Drag Racing?" *Drag Racing,* Sept. 1986, 5.

31. "Waterman on Fuel," *Drag Racing,* Nov. 1986, 60; the other quotes are from Duvall, "Will Computers Ruin Racing?" 88.

32. Head qtd. in Burgess, "Head's Game," 62; Whiting qtd. in Jon Asher, "Ask the People Who Should Know!" *Drag Racing,* Oct. 1989, 30.

33. John Brasseaux, "Electronic Legalities," *Drag Racing,* Oct. 1989, 32–33; see also Brasseaux, "Cannon Blast," ibid., Apr. 1990, 78–79; and "For Better or for Worse: A Look at the New Air-Timer Systems," ibid., May 1990, 49–54.

34. Around 1984 it seemed that the funny car would become the sole archetype, but by 1990 most of the top funny car competitors were headed for the dragster ranks. For a sampling of an extensive commentary, see Jon Asher, "Top Fuel, 1985," 35–36; Dave Wallace, Jr., "Survival of a Species: Top Fuel Eliminator, 1986," *Drag Racing,* Mar. 1986, 2–3; Rich Scarcella, "Funny Car Division Is Not Dead Yet," *Reading Eagle* (Pa.), Sept. 17, 1989, C5; Asher, "Trouble in Funny Car Land," *Drag Racing,* Nov. 1989, 5; Jeff Burk, "One Fuel Class for NHRA?" *Super Stock,* Dec. 1989, 6ff; Gray Baskerville, "Defection," *Hot Rod,* Mar. 1990, 96–99; Dave Densmore, "Here Lies the Fuel Funny Car, 1964–1990," *Super Stock,* Feb. 1991, 18.

35. Burgess, "Head's Game," 62.

36. Wally Parks, "The Future of the Sport," in *Hot Rods,* ed. Wally Parks (Los Angeles: Trend, Inc., 1951), 156; Ralph Guldahl, Jr., "Bakersfield," *Popular Hot Rodding,* June 1964, 20.

37. Yates qtd. from *Journal of Motorsports News and Opinion* in *National Dragster,* Aug. 22, 1983, 4.

38. See B. L. Mellinger, "Japanese Interests Spur Drag Racing," *Drag World,* May 21, 1965, 10; Tony Nancy, "The International Drag Scene," *Car Craft,* Feb. 1967, 82; Dan Roulston, "Drag Racing around the World," ibid., July 1967, 20–24ff; Kjell Gustafson, "Drag Racing in Europe: How It Stands," *Drag Racing,* Apr. 1974, 48–52; Ron Clark,

"International Report" (Sweden), *Super Stock,* Jan. 1979, 10–11; Patrick Ekman, "A Look at Drag Racing in Finland," *National Dragster,* Jan. 28, 1983, 2ff; Andy Willsheer, "A Teutonic Tonic," *Drag Racing,* Dec. 1989, 12–14; Joe Sherk, "Foreign Invasion," *Drag Racing Today,* Dec. 1989, 29; George Phillips, "The Big Go, Far East," *Super Stock,* Jan. 1990, 81–83; Melvyn Record, "New Era Dawns in Land of the Rising Sun," *National Dragster,* Dec. 13, 1991, 11; and especially Karl Anders Alfeld, Jon Van Daal, Graeme Oliver, and Andy Willsheer, "Drag Racing around the World," *Drag Racing,* Mar. 1989, 20–28.

39. On Meyer's plans for the IHRA, see *International Hot Rod Association* (Waco and Bristol: IHRA, 1988); and Pat Ganahl, "Meyerplex," *Hot Rod,* Feb. 1988, 72–76ff.

40. Wally Parks to Tom Lemons, Sept. 12, 1979, AMDR. See also Pat Ganahl, "Re-Evolution: The Preservation, Restoration, and Re-running of the Classic Dragster—1950–1970," *Hot Rod,* Dec. 1983, 86–92.

41. See Tom Senter, "Don Garlits' Drag Racing Museum," *Popular Hot Rodding,* July 1981, 36–39ff; Dave Wallace, Jr., "Standing Thunder," *Drag Racing,* Sept. 1984, 86–89; "Drags Trip: The Museum of Drag Racing," *Nitro* (Paris), May 1989, 62–65.

42. Gray Baskerville, "Speed Is King," *S.C.T.A. Racing News* 4 (Apr. 1989): 11.

43. Gray Baskerville, "Hauling Al," *Hot Rod,* Apr. 1989, 33.

44. Ferguson, "The American-ness of American Technology," 5.

ESSAY ON SOURCES

An old friend once remarked that my research must have been impeded by the scarcity of "printed literature." He meant to say scholarly literature, and, of course, there is a scarcity of that. But the volume of popular periodical literature is overwhelming. Much of it is trivial, and, occasionally, it is deliberately slanted. No historical source of any sort can be trusted implicitly, however, and, with this particular kind of source, at least there are photos that invite "reading" apart from the text. Moreover, the commentary and analysis can be superb.

Articles in the periodicals I have cited are frequently cast in an interrogatory mode, and I have made heavy use of published interviews as well as interviews of my own and others conducted on my behalf by Don Jensen. In grounding much of my narrative in journalism and oral history, I realize that I am at risk concerning the issue of "objectivity."[1] So be it. I am inclined to celebrate the wealth of available sources and congratulate those who have sought to collect and preserve printed literature—Don Garlits and Dave Wallace, Jr., above all. Because much of this is decidedly fugitive, however, I have elected an unusual approach in the first part of this essay: not only to specify which serials I have searched but also to address the literature of drag racing as a historical topic in and of itself.

One begins with the flagship of the publishing empire founded by Robert E.

Petersen in 1947, *Hot Rod.* The first issue appeared in January 1948. Circulation topped 50,000 that year, and by the mid-1960s it was nearly 700,000, larger than either *Time* or *Newsweek.* When *Hot Rod* marked its twenty-fifth anniversary monthly sales were more than a million.[2] As with most of its many emulators, the topical emphasis veered hither and yon. For some two decades beginning in the early 1950s, however, drag racing was a staple: There were articles on pioneer strips such as Santa Ana, Tracy, Paradise Mesa, and Caddo Mills; on machines such as Art Chrisman's No. 25, featuring Rex Burnett's memorable renderings; and on questions of design written by key innovators such as Mickey Thompson. *Hot Rod* was particularly thorough in its coverage of major NHRA events, beginning with its first sanctioned race, at the Los Angeles County Fairgrounds in April 1953.

Wally Parks was the editor from 1950 until 1960, and among his successors was Terry Cook, who was also involved with several other periodicals mentioned on the following pages; there tended to be a musical chairs aspect to editorial slots with the slick monthlies and the tabloid weeklies as well. By 1990 *Hot Rod* paid little notice to drag racing, even though the publisher was a former strip manager and top-fuel driver, Harry Hibler. The attention it received seemed largely due to the enthusiasm of Gray Baskerville, a staffer who was perhaps hot rodding's quintessential buff.

Petersen founded two more automotive monthlies not long after *Hot Rod,* including *Car Craft* (*Honk!* for the first few issues), which hit the stands in May 1953. Several of *Hot Rod*'s principals were also involved with *Car Craft* at the outset, Wally Parks serving as editorial director. Cook became editor in the latter 1960s, pursuing his interest in both the technology and politics of drag racing for some five years.[3] But it was his predecessor, Alex Xydias, who published what in many ways was the best historical synthesis of hot rodding and drag racing: "The Hot Rod Story" (Sept.–Nov. 1966). The author was Dan Roulston, who also took a turn in the editor's chair of *Car Craft.* When *Hot Rod* went off in new directions *Car Craft* labeled itself "Drag Racing's Complete Magazine," and drag racing remained a major focus until the end of the 1970s.

The first specialty magazine outside the Petersen orbit was *Hop Up,* which first appeared in August 1951. *Hop Up* emanated from Enthusiasts' Publications in the Los Angeles suburb of Glendale. Unlike *Hot Rod,* it had to struggle to survive, and soon it came under the control of its former ad salesman, William S. Quinn.[4] The editorial staff included Dean Batchelor and Louis Kimzey, who would launch several other automotive magazines in the 1950s and 1960s. Along with Quinn, Kimzey began publishing *Rod and Custom* in May 1953, with drag racing being one of the subjects that interested editor Spencer Murray.

After Quinn and Kimzey sold *R&C* to Petersen in the summer of 1955 Murray stayed on, publishing a number of articles by automotive engineer Roger Huntington. Later Murray moved over to *Popular Hot Rodding,* founded in 1962 by Don Werner, a former Petersen staffer. Werner's firm was called Argus Publishing. In the early years *PHR* featured a series called "Pop Rod X-Ray," in which Dave Scott provided a technical analysis of various race cars which was relatively sophisticated, especially in comparison to material presented in *Rod and Custom.* While Petersen's automotive magazines were generally aimed at an under-twenty-one audience, in the 1960s *R&C* targeted a median age of fifteen, with technical material that was "light and easily comprehended." A later shift in editorial policy, however, resulted in publication of excellent material such as Mark Dees's 1973 series on "A Technical History of the Racing Flathead."

In the mid-1950s the monthlies had been joined by the first of a new genre, the newsprint tabloid, whose staple was up-to-date race results. *Drag News,* founded as a biweekly in March 1955 by Dean Brown, was purchased the next year by cam grinder Chet Herbert and became a weekly. The editorial staff included Dan Roulston. Herbert's sister Doris became editor and publisher in 1959, wearing both hats until the 1970s, when she began hiring others for the editorial job. By far the most noteworthy was Dave Wallace, Jr., who had been a regular contributor since 1964, when he was a high school sophomore and wrote the weekly account of events at San Fernando Raceway. Wallace was specially adept at tweaking the conscience of the NHRA, and between 1975 and 1977 he endowed his "Out of the Gate" column with a critical edge that was almost unprecedented in the world of drag racing journalism.[5]

In 1976 Herbert sold *Drag News* to Don Rackemann, who had been a racer, a strip manager, a publicist, a salesman, and above all a hustler. Rackemann owned Action Publishers, whose other serials included *Motorsports Weekly;* Action's general manager was Lou Baney.[6] Though Baney was certainly well connected throughout the world of drag racing, he and Rackemann were not able to sustain *Drag News,* and it folded in 1977. All told, it had lasted for twenty-two years, longer than all but one of the rivals that first began appearing in the late 1950s.

After Roulston left *Drag News* he founded *Drag Racer,* a serial with the enigmatic motto COVERAGE TO ALL—CONTROLLED BY NONE, which appeared in 1959–60. *Drag Racer* concentrated on the California scene—among the more valuable articles was "A History of S.C.T.A." (July 11, 1959)—but the biggest gap in the market pertained to coverage "back East." Even though *Drag News* reported on events nationwide, outside California it largely relied on volunteers, and coverage was spotty. This situation was partly alleviated by *Drag Times,* which began publication in April 1958 under the editorship of Wally Peterson (not to be confused with either Wally Parks or Robert Petersen). Headquartered in Chicago, *Drag Times* promised "the most complete midwest drag racing coverage," and, indeed, names such as Chris Karamesines appeared on its pages before they were ever seen in *Drag News. Drag Times* was the forerunner of many other regional tabloids, such as Jim Davis and Monk Reynolds's *Eastern Drag News,* founded in 1963. Under the ownership of Tod Mack and edited by Jack Redd, *EDN* evolved into a second paper called *Drag Times,* which lasted until 1977, the same year *Drag News* expired. Ironically, in the mid-1960s *Drag News* briefly put out a regional edition called *Drag West.*

Wallace has called the 1960s "the sport's golden age of print journalism."[7] More than a dozen new periodicals debuted, and there may even have been a dozen in business at the same time. Three new weeklies were launched in southern California alone. First was *Drag Sport Illustrated,* begun in March 1963 by Phil Bellomy and a photographer who had been one of *Drag News*'s mainstays, Jim Kelly. *Drag Sport* called itself "Drag Racing's Only Illustrated Newspaper." That was not true, but the paper was certainly more elegant than *Drag News.* Advertising revenues were always skimpy, though, and *DSI* expired in the summer of 1966. It was succeeded by Jerry Sutton's *Drag Digest and Drag Results.* Several veterans of *Drag Sport* eventually joined the staff, including Kelly as managing editor, but *Drag Digest* expired even sooner than *DSI,* after less than two years.

Yet drag racing journalism's ultimate success story also dates from the same epoch: The NHRA's *National Dragster,* which first appeared as a biweekly in February 1960, became a weekly in 1963, outstayed *Drag News,* and by 1990 was indisputably the preeminent serial publication in the field. During the seventeen years that *Dragster* and *Drag News* competed they almost never acknowledged that there was any other weekly paper, or, at any rate, another that was "independent, objective, and unbiased." *Dragster* was never independent—it reported NHRA news almost exclusively—but, eventually, that was nearly always where the big news was.[8]

Dragster had several direct predecessors. There was *Drag Link,* a newsletter described as "a service publication produced exclusively for NHRA-sanctioned drag strip sponsors and NHRA advisors." There was *Tie Rod,* a monthly begun in 1956. And there was also *Southeastern Drag News,* which called itself "an independent racing and hot rod publication," although it was limited to NHRA news and was edited by Ernie Schorb, one of NHRA's first two regional directors. Begun in April 1959, *SDN* kept going for about three years after *Dragster* started up. (And, interestingly, in the latter 1980s NHRA began publishing another regional southeastern paper, in the face of competition from an independent tabloid called *Quick Times Racing News,* edited in Vale, North Carolina, by Becky White.)

Over the years *Dragster* has had literally dozens of editors, including Dean Brown (founder of *Drag News*), Dan Roulston (a *Drag News* editor and founder of *Drag Racer*), Steve Evans (now drag racing's leading TV personality), even Wally Parks (who initially provided financing out of his own pocket).[9] In the 1980s Parks hired Neil Britt and John Raffa, both veterans of several monthlies, including *Car Craft.* Raffa soon moved on, but under Britt's guidance *Dragster* attained an unprecedented stability, adhering to production values hitherto quite foreign to the drag racing weeklies, with one possible exception.

The exception was *Drag World,* founded in March 1965 by Mike Doherty, with a working staff headed by Terry Cook, who had apprenticed as a regional correspondent for *Drag News. Drag World* published features that were richly informative on drag racing's politics and ethical dilemmas. Cook delighted in getting beyond mere rehashes of race results, and among much else his reportage has left us the only detailed account of the rise and fall of the United Drag Racers Association. But the Doherty-Cook heyday at *Drag World* lasted for only a couple of years; subsequently, *DW* became the American Hot Rod Association's answer to *National Dragster* (AHRA had formerly published a slick paper monthly called *Tach*), and a onetime *Drag World* columnist named Ben Brown founded *Drag Racer,* the second periodical so named and the second with a midwestern regional focus. Brown's associate was Jim Edmunds, who later became an NHRA publicist.

As AHRA's house organ, *Drag World* lost virtually all appeal as a historical source, filling its pages largely with unembellished race results and verbatim press releases. This was never so with *National Dragster,* which carried significant feature material right from the start. Yet efforts by the major associations, especially NHRA, always seemed to arouse antipathy. In 1967 Don Prieto, who had formerly been associated with *Drag Digest* and was a good friend of Terry Cook's, could write that the "house organs . . . and their puppet staffs offer little in the way of news even within their own organizations": "These papers are primarily a way of publicizing their own coming events and how great each is, and for all this you have to pay a quarter."[10]

Prieto was setting forth his views as managing editor of a slick bimonthly called *Drag Racing.* The first issue had appeared in January 1964, alternating with another magazine called *Modern Rod* (later *Drag Strip*), which had essentially the same focus. Both were published by Lou Kimzey, the would-be rival to Petersen a decade before, and the editorial staff included Mike Doherty. When Doherty left to launch *Drag World* John Durbin took over as editor of *Drag Racing,* and it soon became a monthly. Technical editors included Hayden Profitt, Keith Black, and Pete Robinson, and Scotty Fenn was a contributing editor. Ralph Guldahl, formerly *Drag News*'s star reporter (he covered the action at Lions), was editor during the critical years when funny cars invaded drag racing, but in 1968 Doherty returned. In 1973 Adrian Lopez bought *Drag Racing USA* (the name since 1970). Doherty stayed on as editor until the fall of 1974, when the editorial offices were moved to Alexandria, Virginia, where Lopez produced two other magazines, but this one did not survive the hard times precipitated by the first gasoline crisis in the mid-1970s.

For most of its eleven years *Drag Racing* had provided a rich source for historical research, though it had a close rival named *Super Stock & Drag Illustrated. Super Stock* had likewise been founded in 1964, by Monk Reynolds and Jim Davis. To have a "real writer" on staff they hired John Raffa, a graduate of the University of Maryland with a degree in American civilization. Raffa worked for Reynolds and Davis until 1966, then moved West to become managing editor (later editor and publisher) of Petersen's *Car Craft;* he returned to *Super Stock* for a year in 1972.[11] After that Jim Kelly was editor for a short time, Steve Collison for a long time (most of the 1980s), and Jeff Burk briefly with Dave Wallace, Jr., as his associate. *Super Stock* kept going into the 1990s with a new staff, as one of several magazines produced on the West Coast by the General Media Automotive Group. Unlike *Hot Rod* and the other early Petersen publications, it had only occasionally deviated from an exclusive focus on drag racing.

Petersen Publishing itself had taken one fling with a magazine devoted strictly to that topic. Wallace joined Burk at *Super Stock* after four years as editor of the second slick periodical named *Drag Racing.* Launched in 1984, it was an outgrowth of a series of annuals and a 160-page pictorial compiled by Wallace, *Petersen's History of Drag Racing* (1981), replete with interviews with people such as C.J. Hart and Wally Parks, and illustrations from Petersen Publishing's photo archives, the best in existence, with the exception of *National Dragster*'s.[12] By the 1980s Wallace rightly considered himself an old-timer, and *Drag Racing* presented him with an opportunity to address drag racing as history, a task he performed with aplomb. Another journalist suggested in 1986 that "Dave Wallace is the only reason this beautiful publication works," and that was essentially true. Wallace resigned in 1988, and *Drag Racing* folded in early 1991.

Paperback annuals, from which Petersen's *Drag Racing* had sprung, had a long tradition. Trend Books, a Petersen subsidiary, produced a series of *Hot Rod Annuals* in the 1950s under the editorial direction of Walter Woron, who was the initial editor of *Hot Rod* (he subsequently moved to *Motor Trend,* Petersen's second automotive magazine). At the outset these were full of valuable material, such as Wally Parks's "The History of the Hot Rod Sport," which appeared in the first annual (titled *Hot Rods*). Later NHRA put out a pictorial annual in conjunction with each Nationals, a *Nationals Yearbook* compiled by Barbara Parks, and a *Souvenir Yearbook* under the editorial direction of people such as Jim Edmunds and Neil Britt. *The Best of Drag Racing* was a 160-page NHRA production of 1975. Indispensable as a reference is the annual *NHRA Winston Drag Racing Media Guide* and, of course, NHRA's annual rule book.

Beginning in 1952, another set of annuals was produced irregularly by Fawcett Publications, with the direction of Griffith Borgeson. Under the editorship of John Lawlor and Spencer Murray, *Popular Hot Rodding* concentrated heavily on annuals from the beginning, sometimes publishing two different numbers in a single year—for example, *Drag Racing Handbook* and *Drag Racing Annual,* both in 1964. Bold Horizons Enterprises of Manasquan, New Jersey, produced a one-off pictorial yearbook in conjunction with the 1977 NHRA Nationals. Annual souvenir programs were published in hundreds of versions, one of the first appearing in conjunction with the joint NASCAR-NHRA event in 1960. The Kern County Racing Association published a particularly nice series of programs for its March championships.[13]

AHRA, the NHRA's major rival, expired in 1984 and with it *Drag World,* although a remnant survived in the American Drag Racing Association, with its *ADRA News.* There was also *United Racer,* whose lineage could be traced to the mid-1960s and a newsletter for members of the UDRA's Chicago chapter. Under the editorship of Ron Colson *United Racer* appeared haphazardly in the latter 1980s ("whenever there was enough advertising," suggested one observer), keeping alive the tradition of a tabloid headquartered in the Chicago area and with a midwestern regional focus.

Soon after founding the IHRA in 1970, Larry Carrier started his own biweekly tabloid, *Drag Review.* Sometimes it had a slick paper magazine format, but the content was rarely up to the level of *National Dragster*'s or even AHRA's *Drag World.* Carrier's confrontational editorials did add spice, however, and there were some informative interviews. After Carrier sold IHRA to Billy Meyer in

1988 *Drag Review* was renamed *Drag Racing Today* in an attempt to shed the house organ stigma. Editor Tim Berry ran some thoughtful features, and Meyer kept the journal when he in turn sold the IHRA (which resumed publication of *Drag Review* under the editorship of John Durbin). But *DRT* expired at the end of 1989, the run totaling twenty-three issues. Even so, that was a lot compared to *Drag Racing Illustrated,* which debuted at the beginning of 1989. There was well-conceived editorial material but virtually no advertising, and the publisher pulled the plug after just two issues.

Between *Drag Racing Illustrated* and *Hot Rod* (which now boasts an unbroken run in its fifth decade) dozens of other serials had come and gone, and I suspect that there are titles of which I am not even aware. Ed Sarkisian edited *Rod Journal* and wrote about drag racing for a number of other periodicals, including both *Rodding and Restyling* and *Cars,* and that alone makes them worthwhile. Other titles include *Rod Builder and Customizer, Speed Mechanics, Custom Rodder, Speed and Custom, Hotrod Parts Illustrated, Drag Scoop,* and *Quartermile Countdown.* Many drag strips put out their own publications, ranging from simple mimeographed newsletters, such as Gil Kohn's *Detroit Dragway News,* to Mel Reck's comparatively opulent *Strip Teaser* from Irwindale Raceway.

All told, the field of serial publications had thinned out markedly by the 1990s. With the expiration of Petersen's *Drag Racing, Super Stock* was the sole monthly aimed at the popular market. *Specialty and Custom,* founded in 1966 as "the business magazine for the specialty automotive market," was largely concerned with lobbying against restrictive legislation (as were all the popular magazines, to some extent). There were a few newcomers that appeared to be viable. In Clinton, Maryland, Bill Pratt began publishing *The Drag Racing List* in 1986, providing a handy statistical compendium while also reinforcing the impression that there is not much new under the sun in the way of drag racing periodicals: *The Drag Racing List* was anticipated by nearly fifteen years by *Drag Stat,* coming out of Newport Beach, California, with a computerized tabulation of past performances. Wendy Jeffries's *Bonneville Racing News,* begun in 1989, often ran items on drag racing history. This, the newest periodical noted here, in many ways carried on the traditions of the annual *Lakes Pictorial,* which preceded even *Hot Rod* and was likewise edited by a woman, Veda Orr.

Aside from Robert Petersen very few people have ever made much money from publishing, editing, or writing or taking pictures for specialty periodicals of the type I have noted. What we see here is another aspect of technological enthusiasm.[14]

Before moving on to monographic literature, there are two other facets of the periodical literature which merit brief mention. First, numerous articles have appeared in general interest magazines, though I have no idea of the full extent and can only be suggestive. The first with which I am familiar appeared even before Pearl Harbor: E. Lawrence, "Gow Jobs," *Colliers,* July 26, 1941. A piece titled "Hot Rods" appeared in *Life* for November 5, 1945, another titled "The Drag Racing Rage" on April 29, 1957, and yet another story featuring photos taken at Cordova, Illinois, appeared on October 5, 1963. There were articles in

Popular Mechanics in July 1953, February 1959, June 1965, and other occasions; Saturday Evening Post on May 12, 1962; Popular Science in December 1963; and Esquire in September 1970. There have been frequent features in Sports Illustrated, occasional ones in Time and Newsweek. I might also mention articles and essays of my own which I published while working on this book: "The Cars Won't Fly," Air and Space 1 (Aug.–Sept. 1986): 76–84; "Von Daimler und Benz zu Garlits und Beck: Woher kommen die allerschnellsten Rennwagen?" Kultur & Technik 10, no. 2 (1986): 114–23; "In Praise of Top Fuelers," American Heritage of Invention & Technology 1 (Spring 1986): 58–63; "The Machines of Nowhere," ibid., 8 (Spring 1992): 28–34; and "Strip, Salt, and Other Straightaway Dreams," in Possible Dreams: Technological Enthusiasm in Twentieth Century America, ed. John L. Wright (Dearborn: Henry Ford Museum, 1992), 98–109.

The other category is foreign periodicals. Australia has had pretty much the same range of serials as the United States, including a newspaper, Dragster Australia, and various magazines such as Australian Drag Racing News (which carried editor Dave Cook's excellent history of top-fuel racing in its March 1987 issue). The earliest Australian periodical was The Australian Hot Rodding Review, dating from the early 1960s, and there is also at least one pictorial dating from the 1960s, The Australian Drag Racing Pictorial. By the late 1980s there were also several European magazines that featured material on drag racing, including Internationales Dragster Rennen, published in Frankfurt, and Nitro: Magazine de l'Auto Extraordinaire, published in Paris.

The monographic literature can be covered much more succinctly. There is, to my knowledge, only one relevant academic thesis: James P. Viken, "The Sport of Drag Racing and the Search for Satisfaction, Meaning, and Self: Work and the Mastery of Accrued Skill in Suitable Challenge Situations" (Ph.D. diss., University of Minnesota, 1978). This is as turgid as the title suggests, although the mass of statistics derived from a set of questionnaires is not without utility in explicating "the etiological factors of high performance usage of automobiles and drag racing." A British academic, H. F. "Bert" Moorhouse at the University of Glasgow (like Viken, a sociologist), has written a book titled Driving Ambitions: A Social Analysis of the American Hot Rod Enthusiasm (Manchester: Manchester University Press, 1991) and has also published several articles on the subject, including: "The 'Work' Ethic and 'Leisure' Activity: The Case of the Hot Rod," in The Historical Meaning of Work, ed. Patrick Joyce (Cambridge: Cambridge University Press, 1986); "Organizing the Hot Rods: Sport and Specialist Magazines," British Journal of Sports History 3 (1986): 81–98; and "Racing for a Sign: Defining the 'Hot Rod,' 1945–1960," Journal of Popular Culture 20 (1986): 83–96. The only other significant academic address to this topic is much older: Gene Balsley, "The Hot Rod Culture," American Quarterly 2 (1950): 353–58.[15]

In Fallbrook, California, Don Montgomery has published four books: Hot Rods in the Forties (1987), Hot Rods as They Were (1989), Hot Rod Memories (1991), and Supercharged Gas Coupes (1992), which are most valuable for their photos but not insignificant in their historical analysis. Montgomery also wrote a series of articles on the Santa Ana drags, "Drag It Out," for *Street Rodder* in 1987. LeRoi "Tex" Smith and Tom Medley have projected a series titled *Hot Rod History,* the first volume of which appeared in 1990 under the imprint of Motorbooks International in Osceola, Wisconsin. The photos are generally not up to the quality of those in Montgomery's books, but the text is enlivened by punchy interviews. Melvyn Record, *Hot Rod Record Breakers* (Secaucus, N.J.: Chartwell Books, 1992), is splendidly illustrated.

The literature on Bonneville is extensive, and I will cite only a couple of recent, representative books: Peter J. R. Holthusen, *The Land Speed Record* (Newbury Park, Calif.: Haynes Publications, 1986); and George D. Lepp, *Bonneville Salt Flats* (Osceola, Wisc.: Motorbooks International, 1988). In a related genre, worth mentioning are Eugene Jaderquist, *How to Build Hot Rods* (New York: Fawcett Publications, 1957); Louis Hochman, *Hot Rod Handbook* (New York: Fawcett Publications, 1957); and John Christy, *Hot Rods: How to Build and Race Them* (New York: Bobbs-Merrill, 1960).

By far the best "how-to" book is Frank Hawley with Mark Smith, *Drag Racing: Drive to Win* (Osceola, Wisc.: Motorbooks International, 1989). This is partly autobiographical, but Hawley also addresses such diverse topics as sponsorship and the technology of timing. Hawley operates the Drag Racing School in Gainesville, Florida, which issues its own catalog and was also the subject of several articles by Randall McCarthy in *Super Stock* (Dec. 1985, Dec. 1986, and Jan. 1989).

The sole narrative history of drag racing from its inception, Wally Parks's *Drag Racing: Yesterday and Today* (New York: Trident Press, 1966), is dated but still useful because of the author's unique perspective. Lyle K. Engel's *The Complete Book of Fuel and Gas Dragsters* (New York: Four Winds Press, 1968) presents much of the same material that Mike Doherty published during his editorship of *Drag Racing.* Edward Radlauer's *Drag Racing: Quarter Mile Thunder* (London: Abelard-Schuman, 1966) is aimed at grade-schoolers, as are a number of other books such as Ross Olney's *Kings of the Drag Strip* (New York: G. P. Putnam's Sons, 1968) and *Great Dragging Wagons* (New York: G. P. Putnam's Sons, 1970); Charles Coombs, *Drag Racing* (New York: Scholastic Book Services, 1970); and Irwin Stambler's *Speed Kings* (Garden City, N.Y.: Doubleday, 1973) and *The Super Cars and the Men Who Race Them* (New York: G. P. Putnam's Sons, 1975). None of these is altogether useless.

There are several biographies worth reading. First in order of publication date is Mickey Thompson with Griffith Borgeson, *Challenger: Mickey Thompson's Own Story of His Life with Speed* (New York: Signet Key, 1964). Don Garlits and Brock Yates, *King of the Dragsters* (Radnor, Pa.: Chilton Book Co.) appeared in 1967. Garlits himself published updated versions in 1978 and 1990 entitled *"Big Daddy": The Autobiography of Don Garlits,* and with Darryl E. Hicks he authored *Close Calls* (Shreveport, La.: Huntington House, 1984). *Challenger* and the 1990 edition of *"Big Daddy"* are both richly evocative of the wellsprings of technological enthusiasm, but one should also see: Hal Higdon, *Six Seconds to Glory: Don Prudhomme's Greatest Drag Race* (New York: G. P. Putnam's Sons, 1975); Tom Madigan, *The Loner: The Story of a Drag Racer*

[Tony Nancy] (Englewood Cliffs, N.J.: Prentice-Hall, 1974); Frederick Katz, *Art Arfons: Fastest Man on Wheels* (New York: Routledge, 1965); and Ronnie Sox and Buddy Martin, *The Sox and Martin Book of Drag Racing* (Chicago: Henry Regnery Co., 1974). The only full-scale biography of an individual whose attainments were primarily as a manufacturer rather than a racer is Art Bagnall, *Roy Richter: Striving for Excellence,* published by the author (Los Alamitos, Calif., 1990), but in a related vein is Dan Iandola, *George Riley Racing Scrapbook,* also published by the author (Templeton, Calif., 1992).

Alex Xydias produced a documentary film entitled *The Hot Rod Story* in 1964, the earliest of its genre; this was subsequently transferred to videotape and is still available, as is *Gathering Speed: A Video History of Drag Racing,* produced by Diamond P Sports Video in 1985. Commercial videos are a realm unto their own, and I do not intend to go into these in detail.[16] Worth noting here, however, are several "videobiographies": *Garlits: The Legend* and *Garlits: The Legend Lives On* (Diamond P Sports Video, 1987 and 1988); *Close Calls: A Videobiography by Dean Papadeas* and *Swamp Rat* (Main Event Video, both 1987); *The Legend of Big Daddy Don Garlits* (O.E.F. Productions, 1987); *Shirley: The First Lady of Motorsports* (Diamond P Sports Video, 1987); *Don Prudhomme: Professional Racer* (Diamond P Sports Video, 1988); and *Thrill: The Eddie Hill Story by Dean Papadeas* (Main Event Video, 1988). There is every reason to assume that more of these will appear (Diamond P and Main Event already have literally dozens of commercial videotapes on the market), but it will take some doing to top Hollywood's 1983 venture into biography, *Heart like a Wheel.*

Actually, Hollywood has made several incursions into the field of hot rodding and drag racing, from *Rebel without a Cause* to *American Graffiti,* and in the 1950s there was a whole genre of "B" movies built around the hot rod subculture, such as *Hot Rod Girl.* Likewise, television shows such as "Dragnet," "Public Defender," and "The Life of Riley" had episodes that played the "rebel" image off against another more favorable image of hot rodders.

Then there is the realm of printed fiction. Here the leading light is Henry Gregor Felsen, whose titles include *Hot Rod* (New York: E. P. Dutton, 1950), *Street Rod* (New York: Random House, 1953), *Rag Top* (New York: Bantam Books, 1954), *Crash Club* (Random House, 1958), and *Road Rocket* (Bantam Books, 1963). There was also *Fever Heat* (New York: Dell Books, 1954), which was about stock car racing in the Midwest (and was made into a movie starring Nick Adams), but the others were about drag racing on the street and the horrible consequences that could ensue. "The lesson was usually to drive safely, although, as Felsen got older, he seems to have despaired of trying to teach kids to be careful."[17] Felsen had his followers, though the perils they addressed were different—for example, gamblers attempting to get control of organized drag racing, as in Ross Olney's *Drag Strip Danger* (Racine, Wisc.: Western Publishing Co., 1972). William C. Gault launched a series about racing with *Drag Strip,* first published by Dutton in 1959 and in its eleventh printing a decade later. Wrote A. J. Onia in *Old Cars,* "Take any W. C. Gault story, reduce it to seven or eight pages of script, change the characters, do it two or three times, and you had a hot rod comic book."[18] Titles included "Hot Rods and Racing Cars," "Hot Rod Cartoons," and "Drag 'n' Wheels."

Way beyond cartoonry artistically, but not an entirely distinct genre, is Tom Wolfe's "The Kandy-Colored Tangerine-Flake Streamline Baby," which originally appeared in *Esquire* and was reprinted in a 1967 Wolfe anthology published by Noonday Press. Wolfe gets at essences far better than the purely technical literature, of which there is plenty, to be sure, on the order of Andy Carl's *Nitromethane-based Racing Fuels and Detonation Additives, with Bibliography* (Fullerton, Calif.: ADC Performance Engineering, n.d.).

But I do not need to carry this essay further in that direction, nor is there any necessity to take it into other historiographical realms such as sports history and advertising history, no matter their close bearing (literature is cited in the notes which is not mentioned here). Perhaps I should add only three other printed sources: Steve Smith, *The Racer's Tax Guide;* and Pat Bentley, *Stalking the Motorsports Sponsor,* both published in Santa Ana, California, in 1978 by Steve Smith Autosports; and the *Sponsors Report Almanac,* published annually in Ann Arbor by Joyce Julius and Associates. This almanac "documents the amount of clear, in-focus exposure time and the number of mentions collected by sponsors" at various sporting events, including NHRA and IHRA drag races. Stalking sponsors, figuring tax angles, and tabulating "clear, in-focus exposure time" on television—with seven-figure values placed on that time—are exercises that epitomize the transformation of drag racing in the forty years since Pappy Hart first threw open the gates to "those tarmac torpedos" at Santa Ana in June 1950, and, for those fortunate enough to win, stood by to buy back trophies for $7.50 apiece.

NOTES

1. On oral history, cf. Edwin C. Bearss's characterization of this "challenging and provocative experience" in the National Park Service's *CRM Bulletin* 13, no. 2 (1990), and Peter Novick's observations about work being "often marked by uncritical overidentification with informants and the privileging of their perspective" (*That Noble Dream: The "Objectivity Question" and the American Historical Profession* [Cambridge: Cambridge University Press, 1988], 513).

2. Information on *Hot Rod* is derived largely from a 1965 report entitled "Petersen Publishing Company"; from Steve Kelley, "Our First 25 Years," *Hot Rod,* Jan. 1973; and from H. F. Moorhouse, "Racing for a Sign: Defining the 'Hot Rod,' 1945–1960," *Journal of Popular Culture* 20 (Fall 1986). *Hot Rod* always pursued an active political agenda, one not dissimilar in tone to that of another PPC mainstay, *Guns and Ammo,* tireless since 1958 in its campaign against "the dangers of overrestrictive gun laws."

3. On Cook, who had studied mechanical engineering at Lehigh University and was one of the two or three most incisive journalists ever involved in the field, see "On the Carpet: Terry Cook with Mike Doherty," *Drag World,* June 18, 1965. Thanks to Steve Cutcliffe of Lehigh for tracking down Cook's student record.

4. Separation of editorial and advertising was apparently not commonplace; moving on to *Hot Rod* after more than five years with *Car Craft,* Terry Cook noted in his farewell editorial how "blissful" it had been during the final third of his tenure to have had the two realms internally distinct from one another: "Believe me when I tell you that it is the exception to the rule today" ("Point of View," *Car Craft,* Mar. 1972, 6). In December 1986 the editor of a monthly defended the NHRA house organ's policy of running press

releases as news stories by saying that this was simply a means of supporting "the efforts of NHRA's many sponsors, who expect a little more for their financial involvement than a one-inch ad in the back of the paper." Whether rightly or not—much of the fire being fueled by Ralph Nader—by the 1990s the wider world of "automotive enthusiast journalism" was often perceived as ethically suspect. Cf. Joseph B. White, "Car Magazine Writers Sometimes Moonlight for Firms They Review," reprinted from the *Wall Street Journal* in the *Baltimore Sun* as "Car Journals' Cozy World," July 29, 1990; and the rebuttal by David E. Davis, Jr., in *Automobile,* Sept. 1990.

5. On the history of *Drag News,* see Wallace, "Out of the Gate," Apr. 5, 1975, 4; on NHRA, "Out of the Gate," Oct. 23, 1976, and Don Rackemann, "Open Forum," ibid.

6. "Baney & Rackemann Rejoin Forces," *Drag News,* Oct. 2, 1976, 4; see also "Out of the Gate," ibid., Jan. 17, 1976, 2.

7. Dave Wallace, Jr., "Celebrating 37 Years of Drag Racing Journalism," *Drag Racing,* May 1987, 4.

8. While I have read every issue of *Drag News,* I cannot claim to have seen every single *Dragster* (more than 1,500 of them), even though NHRA's vice president for publications, Neil Britt, has facilitated research by putting a substantial portion of the run on microfiche.

9. Parks details *Dragster*'s early vicissitudes in his Oct. 1989 interview for the NMAH Drag Racing Oral History Archive.

10. Prieto, "The Wave Maker," *Drag Racing,* Nov. 1967, 12.

11. See "Raffa Remembers Another Kind of School," *Super Stock,* Jan. 1990, 20–22; see also Raffa's "Staging Light," *National Dragster,* Apr. 15, 1985, 4.

12. On Petersen Publishing's photo archives, carefully kept by Jane Barrett for many years (but closed at the end of 1990), see Brian Hatano, "Starting Lines," *Drag Racing,* Feb. 1989, 5. Dave Wallace, Jr., has been organizing an archive of printed and manuscript material in California, the Wallace Family Archives, and it is now second only to Garlits's at his Museum of Drag Racing. See "Starting Lines," *Drag Racing,* Oct. 1988, 5.

13. So far as I know, Garlits's museum has the largest collection of programs and other ephemera such as press kits and rule books.

14. In 1960 the *Wall Street Journal* featured Petersen in a series called "Road to Riches," on "the new millionaires and how they made their fortunes" (July 22). Thirty-two years later Petersen donated $15 million to the Los Angeles County Museum toward funding an automotive museum named for himself (see *Old Cars News & Marketplace,* Sept. 15, 1992).

15. Balsley was a University of Chicago student of David Riesman's, whose 1950 book *The Lonely Crowd* (Chicago: University of Chicago Press) cast an admiring glance at hot rodders "working within the old American tradition of high-level tinkering."

16. Nor will I go into the realm of commercial phonograph records, even though dozens were produced, mostly in the 1960s, with titles such as "Hot Rods and Dragsters in Stereo" and "Hot Rod Heaven—Bakersfield Smokers."

17. Jim Morton, "Cars and Death: The World of Henry Gregor Felsen," *Pop Vord* 1 (1989): 18. Out of print for many years, Felsen's books were reissued in the late 1980s. Thanks to Joseph J. Corn of Stanford University for the reference to the Morton article.

18. A. J. Onia, "Hot Rodding in Two Dimensions," *Old Cars,* Dec. 13, 1990, 45.

INDEX

terized by Wally Parks, 102; characterizes successful racers, 55; chassis business, 100, 206; chassis designs, 53, 98; and Emery Cook, 83, 85, 143, 310; development of 426 Chrysler, 169; driving technique, 91, 195; emulates Mickey Thompson and Calvin Rice, 84; endorsements and sponsorships, 93, 100, 173; experiments with airfoils, 224, 227, 228; experiments with sidewinder, 73; experiments with (and injured by) two-speed transmission, 179–81; fires, 94, 98, 188; first dragster, 35, 84; frugal ways and financial success, 120, 165, 172; full envelope streamliner, 208–9, 216, 217, 222; and Darrell Gwynn, 237; helps define bounds of legitimacy, 161; installs supercharger, 91; installs zoom headers, 122; and International Timing Assn., 85, 86; and Ed Iskenderian, 52, 88, 90, 91, 93, 96; and Jocko Johnson, 206, 208–9; and Chris Karamesines, 96; and T. C. Lemons, 180, 182, 183, 194, 197; at Lions, 38, 112, 181, 182; and Tom McEwen, 181; and Art Malone, 94–95, 97; and Larry Minor, 286; and Shirley Muldowney, 270, 271–72, 277–79; Museum of Drag Racing, 328; at NHRA events, 100, 183, 195, 220; at NHRA/NASCAR Winter Nationals, 63, 97; and Herb Parks, 235; and Wally Parks, 54, 63, 78, 97, 100–101; plans and builds mid-engine dragster, 42, 182, 183, 189, 193, 202; plans three-wheeler, 201; and Setto Postoian, 85, 86, 96; professional ideal, 80, 85, 100–102; and Professional Racers Assn., 239–41; and Don Prudhomme, 195; and Pete Robinson, 199; runs first M&H slicks, 85; as star performer, 59; Swamp Rat I, 84; Swamp Rat XXX, xviii, 229, 230–32; Swamp Rat XXXI, 233; and Connie Swingle, 99–100, 101, 112, 182, 193, 194; as technological innovator, 79; as television commentator, 249, 293; and Richard Tharp, 183; transforms dragster into funny car, 143; tries super stocker, 136; views on computer use, 323; views on profitability of racing, 174; views on promoting tobacco use, 247; views on UDRA, 148; at World Series, 85, 88, 100
Garlits, Donna, 100, 102, 328
Garlits, Ed, 63, 84, 85, 86, 88, 89, 90, 94, 100
Garlits, Edward, 84
Garlits, Gay Lyn, 100

Garlits, Pat (Bieger), 84, 86, 181, 237, 280
Garrison, John, 113, 147, 296
Garza, Eloisa, 229
Gay, Don, 142
Gears, 19; differential, 3, 36, 123, 194, 198; direct drive, 49, 53, 85, 306, 322; two-speed transmissions, 24, 53, 121, 179–81, 237, 322
Gedjian, Blackie, 160
Gendian, Dave, 37
General Petroleum. *See* Socony-Mobil
Gibbs, Steve, 62, 151, 323
Gilbert, Rex, 96, 97
Gilmore, Woody, 110, 113, 168, 170, 205, 220; collaboration with Jocko Johnson, 208, 216; exponent of flexible chassis, 122, 303; and Lions boycott, 148; mid-engine dragsters, 192, 193, 194, 202; mid-engine funny car, 222, 223; predictions, 152, 206; price for chassis, 114; and zoom headers, 121–22
Giovannoni, Ray, 93
Gireth, Chuck, 91, 107, 110, 112, 162–63
Glendale Coupe and Roadster Club, 127
Glick, Shav, 64, 197
Glidden, Billy, 280
Glidden, Bob, 244, 280, 288
Glidden, Etta, 280
Glidden, Rusty, 280
Gold Agency. *See* Christ, Ben
Golden, Bill, 159, 160
Goleman, Gene, 187
Gonzales, Al, 82
Goodman, Ellen, xii
Goodyear Tire and Rubber Co., 121, 124, 179, 190, 197, 211, 313
Gorman, Bob, 312
Gotelli, Ted, 113, 285–86
Grabowski, Norm, 107
Graham, Athol, 211, 212
Graham, Marvin, 272
Granatelli, Andy, 191, 194, 262, 263
Grand American Series. *See* American Hot Rod Assn.
Green, Tom, 213
Greene, Bob, 214
Greer, Tommy, 110, 120, 162–66, 168, 173
Greth, Red, 39, 42, 95, 189
Ground effects. *See* Aerodynamics
Guillory, Lester, 206
Guldahl, Ralph, Jr., 195
Gwynn, Darrell, 230, 232, 234–37, 244, 249, 256–57, 279, 283, 287

Designed by Martha Farlow

Composed by The Composing Room of Michigan, Inc., in
Melior with display in Univers and Slipstream

Printed by The Maple Press Company on 60-lb. Glatfelter
Eggshell Offset and bound in ICG Arrestox

Library of Congress Cataloging-in-Publication Data

Post, Robert C.
   High performance : the culture and technology of drag racing, 1950–1990 / Robert C. Post.
      p.   cm. — (Johns Hopkins studies in the history of technology)
   Includes bibliographical references (p.   ) and index.
   ISBN 0-8018-4654-4
   1. Drag racing—United States—History. 2. Drag racing—Social aspects—United
States. I. Title. II. Series.
GV1029.P675   1994
796.7'2—dc20   93-4845